Solar Cell Device Physics

Solar Cell Device Physics
Second Edition

Stephen J. Fonash

AMSTERDAM • BOSTON • HEIDELBERG • LONDON
NEW YORK • OXFORD • PARIS • SAN DIEGO
SAN FRANCISCO • SINGAPORE • SYDNEY • TOKYO

Academic Press is an imprint of Elsevier

Academic Press is an imprint of Elsevier
30 Corporate Drive, Suite 400, Burlington, MA 01803, USA
The Boulevard, Langford Lane, Kidlington, Oxford, OX5 1GB, UK

Library of Congress Cataloging-in-Publication Data
Fonash, S. J.
  Solar cell device physics / Stephen J. Fonash. — 2nd ed.
      p. cm.
  Includes bibliographical references and index.
  ISBN 978-0-12-374774-7 (alk. paper)
  1. Solar cells. 2. Solid state physics. I. Title.
  TK2960.F66 2010
  621.31'244—dc22   2009045478

British Library Cataloguing-in-Publication Data
A catalogue record for this book is available from the British Library.

For information on all Academic Press publications,
visit our website: www.elsevierdirect.com

Printed in United States of America

10 11 12 13 14 15   9 8 7 6 5 4 3 2 1

To the memory of my parents, Margaret and Raymond,
who showed me the path of intellectual pursuits

To my wife Joyce for her continuing guidance and
support along the way

To my sons Steve and Dave, and their families,
for making the journey so enjoyable

Contents

Preface

As was the case with the first edition of *Solar Cell Device Physics*, this book is focused on the materials, structures, and device physics of photovoltaic devices. Since the first edition was published, much has happened in photovoltaics, such as the advent of excitonic cells and nanotechnology. Capturing the essence of these advances made writing both fun and a challenge. The net result is that *Solar Cell Device Physics* has been almost entirely rewritten. A unifying approach to all the developments is used throughout the new edition. For example, this unifying approach stresses that all solar cells, whether based on absorption that produces excitons or on absorption that directly produces free electron–hole pairs, share the common requirement of needing a structure that breaks symmetry for the free electrons and holes. The breaking of symmetry is ultimately what is required to enable a solar cell to produce electric power. The book takes the perspective that this breaking of symmetry can occur due to built-in electrostatic fields or due to built-in effective fields arising from spatial changes in the density of states distribution (changes in energy level positions, number, or both). The electrostatic-field approach is, of course, what is used in the classic silicon p–n junction solar cell. The effective-fields approach is, for example, what is exploited in the dye-sensitized solar cell.

This edition employs both analytical and numerical analyses of solar cell structures for understanding and exploring device physics. Many of the details of the analytical analyses are contained in the appendices, so that the development of ideas is not interrupted by the development of equations. The numerical analyses employ the computer code Analysis of Microelectronic and Photovoltaic Structures (AMPS), which came out of, and is heavily used by, the author's research group. AMPS is utilized in the introductory sections to augment the understanding of the origins of photovoltaic action. It is used in the chapters dedicated to different cell types to give a detailed examination of the full gamut of solar cell types, from inorganic p–n junctions to organic heterojunctions and dye-sensitized cells. The computer modeling provides the dark and light current voltage characteristics of cells but, more importantly, it is used to "pry open cells" to examine in detail the current components, the electric fields, and the recombination present during operation. The various examples discussed in the book are available on the AMPS Web site (www.ampsmodeling.org). The hope is that the reader will want to examine the numerical modeling cases in more detail and perhaps use them as a tool to further explore device physics.

It should be noted that some of the author's specific ways of doing things have crept into the book. For example, many texts use q for the magnitude of the charge on an electron, but here the symbol e is used throughout for this quantity. Also kT, the measure of random thermal energy, is in electron volts (0.026 eV at room temperature) everywhere. This means that terms that may be written elsewhere as $e^{qV/kT}$ appear here as $e^{V/kT}$ with V in volts and kT in electron volts. It also means that expressions like the Einstein relation between diffusivity D_p and mobility μ_p for holes, for example, appear in this book as $D_p = kT\mu_p$.

Photovoltaics will continue to develop rapidly as alternative energy sources continue to gain in importance. This book is not designed to be a full review of where we have been or of where that development is now, although each is briefly mentioned in the device chapters. The intent of the book is to give the reader the fundamentals needed to keep up with, and contribute to, the growth of this exciting field.

Acknowledgments

As with the first edition, this book has grown out of the graduate-level solar cell course that the author teaches at Penn State. It has profited considerably from the comments of the many students who have taken this course. All the students and post-docs who have worked in our research group have also contributed to varying degrees. Outstanding among these is Dr. Joseph Cuiffi who aided greatly in the numerical modeling used in this text.

The efforts of Lisa Daub, Darlene Fink and Kristen Robinson are also gratefully acknowledged. They provided outstanding assistance with figures and references. Dr. Travis Benanti, Dr. Wook Jun Nam, Amy Brunner, and Zac Gray contributed significantly in various ways, from proofreading to figure generation. The help of all these people, and others, made this book a possibility. The encouragement and understanding of my wife Joyce made it a reality.

List of Symbols

Element	Description (Units)
α	Absorption coefficient (nm^{-1}, cm^{-1})
β_1	Dimensionless quantity describing ratio of n-portion quasi-neutral region length to hole diffusion length
β_2	Dimensionless quantity describing ratio of n-portion quasi-neutral region length to the absorption length
β_3	Dimensionless quantity describing ratio of top-surface hole carrier recombination velocity to hole diffusion-recombination velocity in the n-portion
β_4	Dimensionless quantity describing ratio of the absorber thickness up to the beginning of the quasi-neutral region in the p-portion to absorption length
β_5	Dimensionless quantity describing ratio of p-portion quasi-neutral-region length to electron diffusion length
β_6	Dimensionless quantity describing ratio of the p-portion quasi-neutral-region length to absorption length
β_7	Dimensionless quantity describing ratio of back-surface electron carrier recombination velocity to the electron diffusion-recombination velocity
γ	Band-to-band recombination strength parameter ($cm^3 s^{-1}$)
Δ	Magnitude of the energy shift caused by an interface dipole (eV)
Δ	Thickness of dye monolayer in DSSC (nm)
Δ	Grain size in polycrystalline materials (nm)
Δ_C	Conduction-band offset between two materials at a heterojunction (eV)

Δ_V	Valence-band offset between two materials at a hetero-junction (eV)
$\Phi_0(\lambda)$	Photon flux per bandwidth as a function of wavelength ($m^{-2}s^{-1}$ per bandwidth in nm)
ϕ_B	Schottky barrier height of an M-S or M-I-S structure (eV)
ϕ_{BI}	Energy difference between E_C and E_F for an n-type material or the energy difference between E_F and E_V for a p-type material at the semiconductor surface in an M-I-S structure (eV)
Φ_C	Photon flux corrected for reflection and absorption before entering a material ($cm^{-2}s^{-1}$ per bandwidth in nm)
ϕ_W	Workfunction of a material (eV)
ϕ_{WM}	Workfunction of a metal (eV)
ϕ_{Wn}	Workfunction of an n-type semiconductor (eV)
ϕ_{Wp}	Workfunction of a p-type semiconductor (eV)
ε	Permittivity (F/cm)
η	Device power conversion efficiency
λ	Wavelength of a photon or phonon (nm)
μ_{Gi}	Mobility of charge carriers in localized gap states ($cm^2/V\text{-}s$)
μ_n	Electron mobility ($cm^2/V\text{-}s$)
μ_p	Hole mobility ($cm^2/V\text{-}s$)
ν	Frequency of electromagnetic radiation (Hertz)
ξ	Electric field strength (V/cm)
ξ_0	Electric field present at thermodynamic equilibrium (V/cm)
ξ'_n	Electron effective force field (V/cm)
ξ'_p	Hole effective force field (V/cm)
ρ	Charge density (C/cm^3)

σ_n Cross-section of a localized state for capturing an electron (cm^2)

σ_p Cross-section of a localized state for capturing a hole (cm^2)

τ_E Exciton lifetime (s)

τ_n Electron lifetime (dictated by τ_n^R, τ_n^L, or τ_n^A) for p-type material (s)

τ_n^A Electron Auger lifetime for p-type material (s)

τ_n^L Electron S-R-H recombination lifetime for p-type material (s)

τ_n^R Electron radiative recombination lifetime for p-type material (s)

τ_p Hole lifetime (dictated by τ_p^R, τ_p^L, or τ_p^A) for n-type material (s)

τ_p^A Hole Auger lifetime for n-type material (s)

τ_p^L Hole S-R-H recombination lifetime for n-type material (s)

τ_p^R Hole radiative recombination lifetime for n-type material (s)

χ Electron affinity (eV)

a Lattice constant (nm)

A_{abs} Absorbance

$A*$ Effective Richardson constant ($120\,A/cm^2/K^2$ for free electrons) ($A/cm^2/K^2$)

A_{1A}^A Rate constant for the Auger recombination shown in Figure 2.18a (cm^6/s)

A_{1B}^A Rate constant for the Auger recombination shown in Figure 2.18b (cm^6/s)

A_{1C}^A Rate constant for the Auger transition shown in Figure 2.18c (cm^6/s)

A_{1D}^A Rate constant for the Auger transition shown in Figure 2.18d (cm^6/s)

A_{1E}^{A}	Rate constant for the Auger transition shown in Figure 2.18e (cm^6/s)
A_{1F}^{A}	Rate constant for the Auger transition shown in Figure 2.18f (cm^6/s)
A_{2A}^{A}	Rate constant for the Auger generation corresponding to Figure 2.18a (s^{-1})
A_{2B}^{A}	Rate constant for the Auger generation corresponding to Figure 2.18b (s^{-1})
A_C	Solar cell area collecting photons in a concentrator cell (cm^2 or m^2)
A_C	Used in the density of states model $g_e^c(E) = A_c(E - E_c)^{1/2}$ ($cm^{-3}eV^{3/2}$)
A_S	Solar cell area generating current in a concentrator cell (cm^2 or m^2)
A_V	Used in the density of states model $g_e^v(E) = A_v(E_v - E)^{1/2}$ ($cm^{-3}eV^{3/2}$)
c	Speed of light (2.998×10^{17} nm/s)
d	Distance or position in a device (cm, nm)
D_E	Exciton diffusion coefficient (cm^2/s)
D_n	Electron diffusion coefficient or diffusivity (cm^2/s)
D_n^T	Electron thermal diffusion (Soret) coefficient (cm^2/K-s)
D_p	Hole diffusion coefficient or diffusivity (cm^2/s)
D_p^T	Hole thermal diffusion (Soret) coefficient (cm^2/K-s)
e	Charge on an electron (1.6×10^{-19} C)
E	Energy of an electron, photon, or phonon (eV)
E_C	Energy of the conduction-band edge, often called the LUMO for organic semiconductors (eV)
E_{Fn}	Spatially varying electron quasi-Fermi level (eV)
E_{Fp}	Spatially varying hole quasi-Fermi level (eV)
E_{gm}	Mobility band gap (eV)

E_G	Band gap (eV)
E_{pn}	Energy of a phonon (eV)
E_{pt}	Energy of a photon (eV)
E_0	Energy parameter in the model for the Franz-Keldysh effect defined by $E_0 = \frac{3}{2}(m^*)^{-1/3}(e\hbar\zeta)^{2/3} \times 6.25 \times 10^{18}$ with m^*, $\hbar$, and ζ expressed in MKS units (eV)
E_V	Energy of the valence-band edge, often called the HOMO for organic semiconductors (eV)
E_{VL}	Vacuum level energy (eV)
F_e	Total force experienced by an electron where $F_e = -e(\xi - (d\chi/dx) - kT_n(d\ln N_C/dx))$ [Computed using all terms in MKS units. Arises from the electric field and the electron effective field.] (Newtons)
F_h	Total force experienced by a hole where $F_h = e(\xi - (d(\chi + E)/dx) + kT_p(d\ln N_V/dx))$ [Computed using all terms in MKS units. Arises from the electric field and the hole effective field.] (Newtons)
g_A^A	Carrier thermal generation rate for Auger process of Figure 2.18a (cm^{-3}-s^{-1})
g_B^A	Carrier thermal generation rate for Auger process of Figure 2.18b (cm^{-3}-s^{-1})
$g(E)$	Density of states in energy per volume ($eV^{-1}cm^{-3}$)
$g_e^c(E)$	Conduction-band density of states per volume ($eV^{-1}cm^{-3}$)
$g_e^v(E)$	Valence-band density of states per volume ($eV^{-1}cm^{-3}$)
$g_{pn}(E)$	Phonon density of states ($eV^{-1}cm^{-3}$)
g_{th}^R	Number thermally generated electrons in the conduction band and holes in the valence band per time per volume due to band-to-band transitions (cm^{-3}-s^{-1})
$G(\lambda, x)$	Number of Processes 3–5 (see Fig. 2.11) absorption events occurring per time per volume of material per bandwidth (cm^{-3}-s^{-1}-nm^{-1})

G'	Exciton generation rate (cm^{-3}-s^{-1})
G_n''	Represents any electron generation rate (cm^{-3}-s^{-1})
G_p''	Represents any hole generation rate (cm^{-3}-s^{-1})
$G_{ph}^n(\lambda, x)$	Free electron generation rate per time per volume of material per bandwidth (cm^{-3}-s^{-1}-nm^{-1})
$G_{ph}^p(\lambda, x)$	Free hole generation rate per time per volume of material per bandwidth (cm^{-3}-s^{-1}-nm^{-1})
$G_{ph}(\lambda, x)$	Free carrier generation rate per time per volume of material per bandwidth. [Used when $G_{ph}^n(\lambda, x) = G_{ph}^p(\lambda, x)$.] ($cm^{-3}$-$s^{-1}$-$nm^{-1}$)
h	Planck's constant (4.14×10^{-15} eV-s)
$\hbar$	Planck's constant divided by 2π (1.32×10^{-15} eV-s)
$I(\lambda)$	Photon flux impinging on a device (cm^{-2}-s^{-1})
I	Electrical current produced by a device (A)
I	Exciton dissociation rate per area of interface (cm^{-2}-s^{-1})
$I(x)$	Intensity (photons per area per bandwidth) of light as it travels through a material (cm^{-2}-s^{-1}-nm^{-1})
I_0	Intensity of incident light (photons per area per bandwidth) (cm^{-2}-s^{-1}-nm^{-1})
J	Current density; terminal current density emerging from the device (A/cm^2)
J_0	Pre-exponential term in the multistep tunneling model $J_{MS} = -J_0 e^{BT} e^{AV}$ (A/cm^2)
J_{DK}	Dark current density (A/cm^2)
J_{FE}	Interface current density arising from field emission at a junction (A/cm^2)
J_I	Prefactor in the interface recombination current model $\{J_I(e^{V/n_I kT} - 1)\}$ (A/cm^2)

J_{IR}	Interface current density arising from trap-assisted interface recombination. [Also, specifically, current density lost to interface recombination at a heterojunction.] (A/cm^2)
J_{mp}	Current density at the maximum power point (A/cm^2)
J_{MS}	Current density arising from multistep tunneling at a junction (A/cm^2)
J_n	Conventional electron (conduction-band) current density (A/cm^2)
J_{OB}	Current density coming over an energy barrier at an interface (A/cm^2)
J_p	Conventional hole (valence-band) current density (A/cm^2)
J_{SB}	Current density lost to recombination at back contact under illumination (A/cm^2)
J_{SB}^D	Current density lost to recombination at a back contact in the dark (A/cm^2)
J_{sc}	Short-circuit current density (A/cm^2)
J_{SCR}	Prefactor in the space charge recombination current density model $\{J_{SCR}(e^{V/n_{SCR}kT} - 1)\}$ (A/cm^2)
J_{ST}	Current density lost to recombination at a top contact under illumination (A/cm^2)
J_{ST}^D	Current density lost to recombination at a top contact in the dark (A/cm^2)
k	Boltzmann's constant $(8.7 \times 10^{-5} eV/K)$
$\mathbf{k}$	Wave vector of a photon, phonon, or electron (nm^{-1})
$k_\parallel$	Component of a $\mathbf{k}$-vector that lies in the plane of a junction (nm^{-1})
L_{ABS}	Absorption length (defined in this text as distance needed for 85% of possible light absorption) $(\mu m, nm)$
L_C	Collection length for photogenerated charge carriers $(\mu m, nm)$

L_E^{Diff}	Exciton diffusion length (nm)
L_n	Electron diffusion length (μm, nm)
L_n^{Drift}	Electron drift length (nm)
L_p	Hole diffusion length (μm, nm)
L_p^{Drift}	Hole drift length (nm)
LUMO	Lowest unoccupied molecular orbital (energy level) (eV)
m^*	Effective mass of an electron (kg)
n	Conduction band free electron population per volume (cm^{-3})
n	Diode ideality (or n or quality) factor
n_0	Conduction-band free electron population per volume at thermodynamic equilibrium (cm^{-3})
n_i	Intrinsic carrier concentration (cm^{-3})
n_I	Diode ideality (or n or quality) factor for the interface recombination model $\{J_I(e^{V/n_I kT} - 1)\}$
n_1	Defined by $n_1 = N_C e^{-(E_c - E_T)/kT}$ where E_T is the location of gap states participating in S-R-H recombination (cm^{-3})
n_{p0}	Electron population in a p-type material at thermodynamic equilibrium (cm^{-3})
n_{SCR}	Diode ideality (or n or quality) factor for the space charge recombination model $\{J_{SCR}(e^{V/n_{SCR} kT} - 1)\}$
n_T	Number of acceptor states at some energy E occupied by an electron per volume (cm^{-3})
$\hat{n}_T$	Number of states at some energy E occupied by an electron per volume (cm^{-3})
N_A	Acceptor doping density (cm^{-3})
N_A^-	Number per volume of ionized acceptor dopant sites (cm^{-3})

N_C	Conduction band effective density of states (cm^{-3})
N_D	Donor doping density (cm^{-3})
N_D^+	Number per volume of ionized donor dopant sites (cm^{-3})
N_I	Density of trap sites at some energy E at an interface (cm^{-3})
N_T	Density of gap states at some energy E (cm^{-3})
N_{TA}	Density of acceptor gap states at some energy E (cm^{-3} or cm^{-3}-eV^{-1})
N_{TD}	Density of donor gap states at some energy E (cm^{-3} or cm^{-3}-eV^{-1})
N_V	Valence band effective density of states (cm^{-3})
p	Valence band free hole population per volume (cm^{-3})
p_0	Valence band free hole population per volume at thermodynamic equilibrium (cm^{-3})
p_D	Photogenerated dye molecule hole population in DSSC (cm^{-3})
p_{n0}	Valence-band free hole population per volume in an n-type material at thermodynamic equilibrium (cm^{-3})
p_1	Defined by $p_1 = N_v e^{-(E_T - E_v)/kT}$ where E_T is the location of gap states participating in S-R-H recombination (cm^{-3})
p_T	Number of donor states at some energy E unoccupied by an electron per volume (cm^{-3})
$\tilde{p}_T$	Number of states at some energy E unoccupied by an electron per volume (cm^{-3})
P_E	Number of excitons per volume (cm^{-3})
P_{IN}	The power per area impinging on a cell for a given photon spectrum $\Phi_0(\lambda)$; obtained from the integral of $\Phi_0(\lambda)$ across the entire photon spectrum (W/cm^2)
P_{OUT}	Power produced per area of a cell exposed to illumination (W/cm^2)

r_A^A	Auger recombination rate for path a of Figure 2.18 $(\text{cm}^{-3}\text{-s}^{-1})$
r_B^A	Auger recombination rate for path b of Figure 2.18 $(\text{cm}^{-3}\text{-s}^{-1})$
r_C^A	Auger transition rate for path c of Figure 2.18 $(\text{cm}^{-3}\text{-s}^{-1})$
r_D^A	Auger transition rate for path d of Figure 2.18 $(\text{cm}^{-3}\text{-s}^{-1})$
r_E^A	Auger transition rate for path e of Figure 2.18 $(\text{cm}^{-3}\text{-s}^{-1})$
r_F^A	Auger transition rate for path f of Figure 2.18 $(\text{cm}^{-3}\text{-s}^{-1})$
$R(\lambda)$	Reflected photon flux $(\text{cm}^{-2}\text{-s}^{-1})$
$\mathscr{R}^{AA}$	Net rate for Auger process a of Figure 2.18 $(\text{cm}^{-3}\text{-s}^{-1})$
$\mathscr{R}^{AB}$	Net rate for Auger process b of Figure 2.18 $(\text{cm}^{-3}\text{-s}^{-1})$
$\mathscr{R}^L$	Net S-R-H recombination rate $(\text{cm}^{-3}\text{-s}^{-1})$
$\mathscr{R}^R$	Net radiative recombination rate $(\text{cm}^{-3}\text{-s}^{-1})$
S_n	Electron contribution to the Seebeck coefficient, also called the thermoelectric power (eV/K)
S_n	Surface recombination speed for electrons (cm/s)
S_p	Hole contribution to the Seebeck coefficient, also called the thermoelectric power (eV/K)
S_p	Surface recombination speed for holes (cm/s)
T	Absolute temperature (K)
T	Transmitted photon flux $(\text{cm}^{-2}\text{-s}^{-1})$
T_n	Spatially varying electron effective temperature (K)
T_p	Spatially varying hole effective temperature (K)
v	Thermal velocity of electrons or holes (cm/s)
V	Voltage; terminal voltage (V)

V_{Bi}	Built-in potential (eV)
V_{mp}	Device voltage at the maximum power point (V)
V_n	Energy difference between the conduction band edge and the electron quasi-Fermi level at some point x (eV)
V_{oc}	Open-circuit voltage (V)
V_p	Difference between the hole quasi-Fermi level and the valence-band edge at some point x (eV)
V_{TEB}	Effective total electron barrier in the conduction band of a heterojunction (eV)
V_{THB}	Effective total hole barrier in the valence band of a heterojunction (eV)
W	Activation energy for charge carrier hopping between localized gap states (eV)
W	Width of the space-charge region (μm, nm)
x	Position in a device or layer (cm, nm)

List of Abbreviations

ALD	Atomic layer deposition
AM	Air mass
AR	Anti-reflection
a-Si:H	Hydrogenated amorphous silicon
AZO	Aluminum-doped zinc oxide
BCC	Body-centered cubic (lattice)
BHJ	Bulk heterojunction
CB	Conduction band
CM	Carrier multiplication
DSSC	Dye-sensitized solar cell
DSSSC	Dye-sensitized solid-state solar cell
EBL	Electron blocking layer
EPC	Electrochemical photovoltaic cell
EQE	External quantum efficiency (often expressed as a percentage)
ETL	Electron transport layer
FCC	Face-centered cubic (lattice)
FF	Fill factor $\equiv (J_{mp} V_{mp})/(J_{sc} V_{oc})$ (measures the rectangularity of the J-V characteristic, so ≤ 1)
HBL	Hole blocking layer
HJ	Heterojunction
HTL	Hole transport layer
IB	Intermediate band
IQE	Internal quantum efficiency (often expressed as a percentage)
ITO	Indium tin oxide
mc	Multicrystalline
MEG	Multiple exciton generation
M-I-S	Metal-insulator-semiconductor
MOCVD	Metal organic chemical vapor deposition
M-S	Metal-semiconductor

nc	Nanocrystalline–polycrystalline material composed of crystal grains each $<100\,nm$
P3HT	Poly(3-hexylthiophene)
PCBM	Phenyl C_{61} butyric acid methyl ester
PEDOT-PSS	Poly(3,4-ethylenedioxythiophene)-poly(styrene-sulfonate)
PHJ	Planar heterojunction
poly-Si	Polycrystalline silicon
QD	Quantum dot
RT	Room temperature
SAM	Self-assembled monolayer
SB	Schottky barrier (Barrier depleting majority-carriers in a semiconductor caused by a metal contact)
SC	Simple cubic (lattice)
SH	Simple hexagonal (lattice)
S-I-S	Semiconductor-intermediate layer-semiconductor
S-R-H	Shockley-Read-Hall recombination
TCO	Transparent conducting oxide
TE	Thermodynamic equilibrium
VB	Valence band
μc	Microcrystalline–polycrystalline material composed of grains $<1000\,\mu m$ to $100\,nm$

Chapter One

Introduction

1.1 PHOTOVOLTAIC ENERGY CONVERSION

Photovoltaic energy conversion is the direct production of electrical energy in the form of current and voltage from electromagnetic (i.e., light, including infrared, visible, and ultraviolet) energy. The basic four steps needed for photovoltaic energy conversion are:

1. a light absorption process which causes a transition in a material (the absorber) from a ground state to an excited state,

2. the conversion of the excited state into (at least) a free negative- and a free positive-charge carrier pair, and

3. a discriminating transport mechanism, which causes the resulting free negative-charge carriers to move in one direction (to a contact that we will call the cathode) and the resulting free positive-charge carriers to move in another direction (to a contact that we will call the anode).

The energetic, photogenerated negative-charge carriers arriving at the cathode result in electrons which travel through an external path (an electric circuit). While traveling this path, they lose their energy doing something useful at an electrical "load," and finally they return to the anode of the

DOI: 10.1016/B978-0-12-374774-7.00001-7

cell. At the anode, every one of the returning electrons completes the fourth step of photovoltaic energy conversion, which is closing the circle by

4. combining with an arriving positive-charge carrier, thereby returning the absorber to the ground state.

In some materials, the excited state may be a photogenerated free electron–free hole pair. In such a situation, step 1 and step 2 coalesce. In some materials, the excited state may be an exciton, in which case steps 1 and 2 are distinct.

A study of the various man-made photovoltaic devices that carry out these four steps is the subject of this text. Our main interest is photovoltaic devices that can efficiently convert the energy in sunlight into usable electrical energy. Such devices are termed solar cells or solar photovoltaic devices. Photovoltaic devices can be designed to be effective for electromagnetic spectra other than sunlight. For example, devices can be designed to convert radiated heat (infrared light) into usable electrical energy. These are termed thermal photovoltaic devices. There are also devices which directly convert light into chemical energy. In these, the photogenerated excited state is used to drive chemical reactions rather than to drive electrons through an electric circuit. One example is the class of devices used for photolysis. While our emphasis is on solar cells for producing electrical energy, photolysis is briefly discussed later in the book.

1.2 SOLAR CELLS AND SOLAR ENERGY CONVERSION

The energy supply for a solar cell is photons coming from the sun. This input is distributed, in ways that depend on variables like latitude, time of day, and atmospheric conditions, over different wavelengths. The various distributions that are possible are called solar spectra. The product of this light energy input, in the case of a solar cell, is usable electrical energy in the form of current and voltage. Some common "standard" energy supplies from the sun, which are available at or on the earth, are plotted against wavelength (λ) in W/m^2/nm spectra in Figure 1.1A. An alternative photons/m^2-s/nm spectrum is seen in Figure 1.1B. The spectra in Figure 1.1A give the power impinging per area (m^2) in a band of wavelengths 1 nm wide (the bandwidth $\Delta\lambda$) centered on each wavelength λ. In this figure, the AM0 spectrum is based on ASTM standard E 490

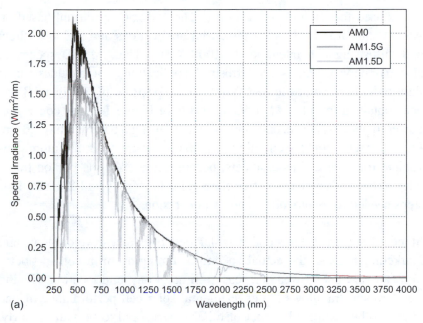

(a)

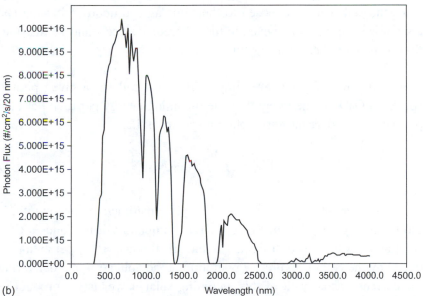

(b)

FIGURE 1.1 Solar energy spectra. (a): Data expressed in watts per m² per 1 nm bandwidth for AMO (from Ref. 1, with permission) and for AM1.5G, and AM1.5D spectra (from Ref. 2, with permission). (b): The AM1.5G data expressed in terms of impinging photons per second per cm² per 20 nm bandwidth.

and is used for satellite applications.[1] The AM1.5G spectrum, based on ASTM standard G173, is for terrestrial applications and includes direct and diffuse light. It integrates to 1000 W/m². The AM1.5D spectrum, also based on G173, is for terrestrial applications but includes direct light only. It integrates to 888 W/m².[2] The spectrum in Figure 1.1B has been obtained from the AM1.5G spectrum of Figure 1.1A by converting power to photons per second per cm² and by using a bandwidth of 20 nm. Photon spectra $\Phi_0(\lambda)$, exemplified by that in Figure 1.1B, are more convenient for solar cell assessments, because optimally one photon translates into one free electron–free hole pair via steps 1 and 2 of the four steps needed for photovoltaic energy conversion.

Standard spectra are needed in solar cell research, development, and marketing because the actual spectrum impinging on a cell in operation can vary due to weather, season, time of day, and location. Having standard spectra allows the experimental solar cell performance of one device to be compared to that of other devices and to be judged fairly, since the cells can be exposed to the same agreed-upon spectrum. The comparisons can be done even in the laboratory since standard distributions can be duplicated using solar simulators.

The total power P_{IN} per area impinging on a cell for a given photon spectrum $\Phi_0(\lambda)$ is the integral of the incoming energy per time per area per bandwidth over the entire photon spectrum; i.e.,

$$P_{IN} = \int_\lambda \frac{hc}{\lambda} \Phi_0(\lambda) d\lambda \qquad (1.1)$$

where an example $\Phi_0(\lambda)$, expressed as photons/time/area/bandwidth, is plotted in Figure 1.1B. In Equation 1.1 the quantity h is Planck's constant and c is the speed of light. The electrical power P_{OUT} per area produced by the cell of Figure 1.2 operating at the voltage V and delivering the current I as a result of this incoming solar power is the product of the current I times V divided by the cell area.

Introducing the current density J defined as I divided by the cell area allows P_{OUT} to be written as

$$P_{OUT} = JV \qquad (1.2)$$

A plot of the possible J-V operating points (called the "light" J-V characteristics) of the cell of Figure 1.2 is seen in Figure 1.3. The points labeled J_{sc} and V_{oc} represent, respectively, the extreme cases of no voltage produced between the anode and cathode (i.e., the illuminated solar cell is short-circuited) and of no current flowing between the anode and cathode (i.e., the illuminated solar cell is open-circuited).

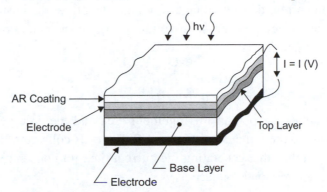

FIGURE 1.2 Cross-section of a typical solar cell. The area of photon impingement and the area of current production are the same. The anti-reflection (AR) coating has the function of reducing reflection losses. The collecting electrodes (cathode and anode) are shown with the top electrode being transparent.

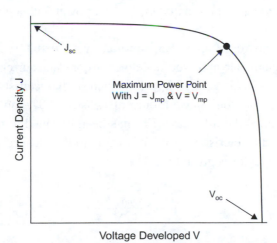

FIGURE 1.3 The current density-voltage (J-V) characteristic of the photovoltaic structure of Figure 1.2 under illumination. The short-circuit current density J_{sc} and open-circuit voltage V_{oc} are shown. The maximum power point (largest J-V product) is also shown. Device efficiency η is defined as $\eta = (J_{mp} V_{mp})/P_{IN}$ where P_{IN} is the incoming power per area.

At any of the operating points seen in Figure 1.3, P_{OUT} is given by the JV product.

The quantity P_{OUT} has its best value at the maximum power point labeled by the current density J_{mp} and the voltage V_{mp} on the light J-V characteristic in Figure 1.3. This operating point gives the maximum obtainable current density-voltage product. Therefore, the best thermo-dynamic efficiency η of the photovoltaic energy conversion process for the cell of Figure 1.2 is:

$$\eta = \frac{(J_{mp}V_{mp})}{P_{IN}} \tag{1.3}$$

which assumes the photon impingement area and the area generating the current are the same, as in Figure 1.2. For cells collecting light over a larger area than that generating the current (i.e., for concentrator solar cells), this expression is replaced by

$$\eta = \frac{A_S}{A_C} \frac{(J_{mp}V_{mp})}{P_{IN}} \tag{1.4}$$

where A_S is the solar cell area generating current and A_C is the area collecting the photons. The advantage of a concentrator configuration lies in its being able to harvest more incoming solar power with a given cell size.

As can be seen from Figure 1.3, the ideally shaped J-V characteristic would be rectangular and would deliver a constant current density J_{sc} until the open-circuit voltage V_{oc}. For such a characteristic, the maximum power point would have a current density of J_{sc} and a voltage of V_{oc}. A term called the fill factor (FF) has been invented to measure how close a given characteristic is to conforming to the ideal rectangular J-V shape. The fill factor is given by

$$FF = \frac{J_{mp}V_{mp}}{J_{sc}V_{oc}} \tag{1.5}$$

By definition, $FF \leq 1$.

The J-V characteristics seen in Figure 1.3 merit one additional comment. What has been plotted is that part of a cell's J-V characteristics for which power is being produced. Put simply, in this first quadrant plot, conventional current is emerging from the anode of the cell. Conventional current enters the anode of the archetypical power-consuming devices, the resistor and the diode. In this book, the power-producing quadrant of a solar cell will henceforth be switched to the fourth quadrant, to be consistent with resistor and diode plots, which will be in the first and third quadrants.

1.3 SOLAR CELL APPLICATIONS

Solar photovoltaic energy conversion is used today for both space and terrestrial energy generation. The success of solar cells in space applications is well known (e.g., communications satellites, manned and unmanned space exploration). On earth, solar cells have a myriad of applications varying from supplementing the grid to powering emergency call boxes. However, the need for much more extensive use of solar cells in terrestrial applications is becoming clearer with the growing understanding of the true cost of fossil fuels and with the widespread demand for renewable and environmentally acceptable terrestrial energy resources. As long as 120 years ago, visionaries looking through the soot and smoke of the early industrializing world saw the need for a renewable and environmentally acceptable energy source. Writing in 1891, Appleyard foresaw "the blessed vision of the Sun, no longer pouring his energies unrequited into space, but, by means of photo-electric cells and thermo-piles, these powers gathered into electrical storehouses to the total extinction of steam engines, and the utter repression of smoke."[3] It is interesting to note Appleyard's specific mention of what he calls photo-electric cells. This energy conversion approach was known even then due to Becquerel's discovery of photovoltaic action in 1839.[4]

To increase the use of terrestrial solar photovoltaics, more efforts are needed to enhance cell energy-conversion efficiency η, to increase module (a grouping of cells) lifetimes, to reduce manufacturing costs, to reduce installation costs, and to reduce the environmental impact of manufacturing and deploying solar cells. The last three may be combined into "true costs." Looked at it this way, increasing the use of terrestrial solar photovoltaics depends on increasing a "figure of merit" defined by

$$\frac{\text{energy conversion efficiency}}{\text{true costs}} \times \text{lifetime}$$

Developing the knowledge base needed to further increase this figure of merit, and thereby bringing Appleyard's vision to fruition, are the objectives of this book.

REFERENCES

1.   ASTM Standard E490.
2.   ASTM Standard G173-03.
3.   R. Appleyard, *Telegraphic J. Electr. Rev.* 28, 124 (1891).
4.   E. Becquerel, *Compt. Rend.* 9, 561 (1839).

Chapter Two

Material Properties and Device Physics Basic to Photovoltaics

2.1 INTRODUCTION

In order to conceive new photovoltaic energy-conversion schemes, improve existing configurations, develop and improve cell materials, and understand the origins of the technical and economic problems of solar cells, the basics behind photovoltaic device operation must always be kept in the forefront. With that in mind, an overview of the material properties and physical principles underlying photovoltaic energy conversion is

DOI: 10.1016/B978-0-12-374774-7.00002-9

presented in this chapter. The mathematical models for phenomena that are fundamental to solar cell operation, such as recombination, drift, and diffusion, are discussed rather than just presented. This is done with the firm conviction that awareness of the assumptions behind the various models better enables one to judge their appropriateness and to make adjustments as necessary, when analyzing and developing new solar cell structures. This is particularly the case for solar cells today, which can involve combinations of a variety of features or phenomena, such as nano-scale morphology, amorphous materials, organic materials, plasmonics, quantum confinement, and exciton-producing absorption.

2.2 MATERIAL PROPERTIES

Both solid and liquid materials are used in solar cells. Homojunction, heterojunction, metal-semiconductor, and some dye-sensitized solar cells use all-solid structures, whereas liquid-semiconductor and many dye-sensitized cells use solid–liquid structures. These materials can be inorganic or organic. The solids can be crystalline, polycrystalline, or amorphous. The liquids are usually electrolytes. The solids can be metals, semiconductors, insulators, and solid electrolytes.

2.2.1 Structure of solids

The solids used in photovoltaics can be broadly classified as crystalline, polycrystalline, or amorphous. Crystalline refers to single-crystal materials; polycrystalline refers to materials with crystallites (crystals or equivalently grains) separated by disordered regions (grain boundaries); and amorphous refers to materials that completely lack long-range order.

2.2.1.1 CRYSTALLINE AND POLYCRYSTALLINE SOLIDS

The distinguishing feature of crystalline and polycrystalline solids is the presence of long-range order, represented by a mathematical construct termed the lattice, and a basic building block (the unit cell), which, when repeated, defines the structure of the lattice. The atoms or molecules of the crystal have their positions fixed with respect to the points of the lattice. Amazingly, there are only 14 crystal lattices possible in a three-dimensional universe.[1] Four common ones are shown in Figure 2.1. Different planes in a crystal can have different numbers of atoms residing

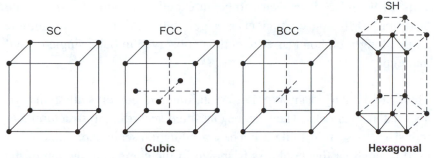

FIGURE 2.1 Some important unit cells characterizing crystalline solids. The simple cubic (SC) unit cell has lattice points only at the cube corners, the face-centered cubic (FCC) unit cell has additional lattice points in the center of each cube face, whereas the body-centered cubic (BCC) unit cell has lattice points at the cube corners and an additional lattice point at the center of the cube. The simple hexagonal (SH) unit cell has lattice points at the corners defining each hexagonal face and at the center of these two faces.

on them, as may be deduced by comparing, for example, the simple cubic and the body-centered cubic lattices in Figure 2.1. In solids that are compounds (e.g., the semiconductors CdS or CdTe), different planes can even be composed of different atomic species. Miller indices are a convenient convention for labeling different planes.[1]

Polycrystalline solids differ from single-crystal solids in that they are composed of many single-crystal regions. These single-crystal regions (grains) exhibit long-range order. The various grains comprising a polycrystalline solid may or may not have their lattices randomly oriented with respect to one another. If there is correlation in the orientations of the grains, the material is referred to as being an oriented polycrystalline solid. The transition regions in a polycrystalline solid that exist between the various single crystals are what we termed grain boundaries. These regions of structural and bonding defects can extend for perhaps a fraction of a nanometer or more and may even contain voids. Grain boundaries can have a significant influence on physical properties. For example, they can getter dopants or other impurities, store charge in localized states arising from bonding defects, and, through the stored charge, give rise to electrostatic potential energy barriers that impede transport.[2] The grain boundaries of polycrystalline materials can be broadly classified as either open or closed. An open boundary is easily accessible to gas molecules; a closed boundary is not. However, even a closed boundary is expected to be an excellent conduit for solid-state diffusion. Diffusion coefficients are generally an order of magnitude larger along such boundaries than those observed in bulk, single-crystal material.[2]

There are actually many types of crystalline and polycrystalline materials used in solar cells. They can be classified according to structural feature sizes, as seen in Table 2.1. The classification scheme of Table 2.1 is used throughout this book. As is shown in the comments section of the table, there are large differences in the terminology currently in use.

Table 2.1 Some Material Structure Types

Material type	Size of single-crystal region	Comments
Nanoparticle	Particle size <100 nm. May be single crystal, polycrystalline, or amorphous.	Shapes include spherical and columnar-like. Quantum size effects possible for particles <10 nm. Semiconductor-type called quantum dots when size effects are present.
Nanocrystalline (nc) material	Made up of single-crystal grains each <100 nm.	Polycrystalline. In the case of Si, also called microcrystalline silicon (μc-Si). In the case of Si, nc-Si has small grains of crystalline silicon within an amorphous Si phase. Quantum size effects possible for grains <10 nm.
Microcrystalline (μc) material	Single-crystal grains <1000 μm to 100 nm.	Polycrystalline. Often simply called polycrystalline material.
Multicrystalline (mc) or semi-crystalline material	Single-crystal grains >1 mm.	Polycrystalline. Grains can be many cm or larger in size.
Single-crystal material	No grains nor grain boundaries.	Whole material is one crystal.

2.2.1.2 AMORPHOUS SOLIDS

Amorphous solids[†] are disordered materials that contain large numbers of structural and bonding defects. They possess no long-range structural order, which means there is no such thing as a unit cell and a lattice. Amorphous solids are composed of atoms or molecules that display only short-range order, at best. There is no necessity for uniqueness in the amorphous phase. For example, there are a myriad of amorphous silicon-hydrogen (a-Si:H) materials that vary according to Si defect density, hydrogen content, and hydrogen-bonding details.

Solids can also exist in a form that contains regions of crystalline and amorphous phases, as seen in Figure 2.2. The example used in the figure is a polymeric solid that has amorphous domains containing disordered polymer chains joined to crystalline regions where the chains form an ordered array. The existence of such mixed-phase solids, containing amorphous and microcrystalline regions, often depends on the material fabrication procedure.

2.2.2 Phonon spectra of solids

Because of the interactions among its atoms, a solid has vibrational modes. The quantum of vibrational energy is termed the phonon. At a given temperature T, atoms of a solid are oscillating about their equilibrium sites; therefore, there are phonons present in the solid. In thermodynamic equilibrium, the distribution of phonons among allowed modes of vibration (phonon energy levels, E_{pn}) is dictated by Bose-Einstein statistics.

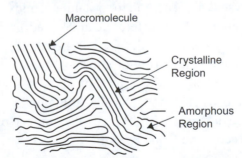

Macromolecule

Crystalline Region

Amorphous Region

FIGURE 2.2 Organic solid containing crystalline and amorphous regions. Some of the polymer molecules constituting this solid are found in both regions.

[†]Glasses are a subset of amorphous materials which possess a glass transition temperature. Above this temperature glasses can flow.

Phonons can be involved in heat transfer, carrier generation (thermal or in conjunction with light absorption), carrier scattering, and carrier recombination processes. They behave like particles. For example, when an electron in a solid interacts with a vibrational mode, the event is best viewed as an interaction between two types of particles, electrons and phonons. Phonons have a dispersion relationship $E_{pn} = E_{pn}(\mathbf{k})$, which relates the phonon energy E_{pn} to the wave vector $\mathbf{k}[|(\mathbf{k})| = (2\pi/\lambda)]$ of the vibrational mode. This is analogous to the dispersion relationship for light, which relates photon energy E_{pt} to the wave vector $\mathbf{k}$ of the light. For free space, this dispersion relationship for light has the extremely simple form $E_{pt} = \hbar c |(\mathbf{k})|$ where $\hbar$ is Planck's constant divided by 2π. In the case of phonons, the function $E_{pn} = E_{pn}(\mathbf{k})$ is more complicated and gives what is termed the phonon spectrum or phonon energy bands in a solid. In the case of both phonons and photons, $\hbar\mathbf{k}$ has the interpretation of particle (phonon or photon) momentum.[1]

2.2.2.1 SINGLE-CRYSTAL, MULTICRYSTALLINE, AND MICROCRYSTALLINE SOLIDS

In single-crystal, multicrystalline, and microcrystalline materials, both total energy and total momentum are conserved in phonon–electron interactions.[1] For example, in a "collision" between an electron and phonon in a crystal (or in a crystal region in the case of polycrystalline material), the change in the $\mathbf{k}$-vector of the phonon and the electron must conserve total (electron plus phonon) momentum as well as energy. This constraint is termed the $\mathbf{k}$-selection rule. In addition, in single-crystal, multicrystalline, and microcrystalline solids, only specified values of $\mathbf{k}$ are allowed and therefore can be used in the $E_{pn} = E_{pn}(\mathbf{k})$ relation.[1] Since only certain modes (certain $\mathbf{k}$-vectors) are permitted in a given crystal, there is a density of allowed phonon states in $\mathbf{k}$-space which translates into a density $g_{pn}(E)$ of allowed phonon states per energy per volume, as shown graphically in Figure 2.3. The density of states $g_{pn}(E)$ is such that $g_{pn}(E)\,dE$ gives the number of phonon states per volume between some E and E + dE.

The $\mathbf{k}$-space just mentioned for plotting $E_{pn} = E_{pn}(\mathbf{k})$ in crystalline solids is also called reciprocal space. This space has a lattice, too, and its distances have the dimensions of reciprocal length. Directions in reciprocal space correspond to directions in the real crystal. The reciprocal-space lattice can be viewed as the Fourier transform of the real-space

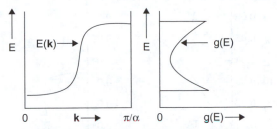

FIGURE 2.3 Relationship between the **k** values permitted to be used (not shown) in a dispersion function and the resulting density of states g(E). This dispersion relation in **k**-space and its resulting density of states in energy may be that of phonons or electrons in a crystalline solid. It turns out that the density of states in energy g(E) is the more basic concept than the allowed states in **k**-space, since it applies to noncrystalline materials, too; i.e., its validity does not depend on the existence of a **k**-space.

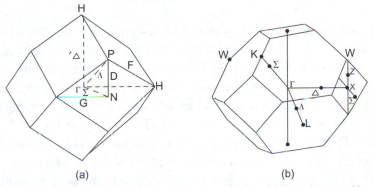

FIGURE 2.4 Brillouin zones for: (a) the BCC crystal lattice; (b) the FCC crystal lattice.

lattice of a crystal. Because the real-space lattice is so structured, its reciprocal-space lattice is equally structured. Just as all the information on the structure of a crystalline solid is contained in the unit cell of the real-space lattice, all the information on the dispersion relation $E_{pn} = E_{pn}(\mathbf{k})$ is contained in the unit cell of the reciprocal lattice. The unit cell in reciprocal space is termed the first Brillouin zone or simply the Brillouin zone.[1] The rest of reciprocal space repeats this $E_{pn} = E_{pn}(\mathbf{k})$ information. Figure 2.4 shows the Brillouin zones corresponding to the FCC and BCC real-space lattices seen in Figure 2.2. The notation is standard in Figure 2.4 and denotes symmetry points and axes.[1]

A part of the phonon spectra $E_{pn} = E_{pn}(\mathbf{k})$ for two materials of interest to solar cell applications, crystalline silicon and gallium arsenide, is presented as an example in Figure 2.5. Here the function $E_{pn} = E_{pn}(\mathbf{k})$ is depicted for

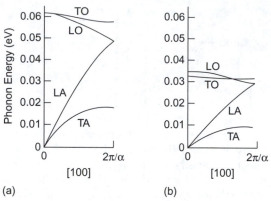

FIGURE 2.5 Phonon bands in two crystalline solids: (a) silicon, and (b) gallium arsenide. Phonon bands are depicted for **k** values lying along the Γ to X directions in the Brillouin zone of an FCC lattice. (From Ref. 3, with permission.)

k values lying along the Γ to X direction (and equivalent directions) of the Brillouin zone for FCC crystals. This Brillouin zone is utilized for both materials, since both have the FCC direct (real-space) lattice. The data in Figure 2.5 specifically give the phonon energies or bands (the functions are multivalued) found in these crystalline materials for **k**-vectors varying from Γ ($|\mathbf{k}| = 0$) to X ($|\mathbf{k}| = 2\pi/a$) in their Brillouin zone. In Figure 2.5, the notation O refers to optical branches (in polar materials these modes can be strongly involved in optical properties); the notation A refers to acoustic branches (so called because frequencies audible to the human ear are on these branches at about the origin in Fig. 2.5). The notations T and L refer to the transverse and longitudinal modes, respectively. The largest values of $|\mathbf{k}|$ in the Brillouin zone will depend on the lattice constant a of the semiconductor; however, using reasonable values of a, it is seen that $|\mathbf{k}|_{max}$ in Figure 2.5 is of the order of 10^8 cm^{-1}. If we were to extend the ordinate and superimpose the plot for all photon energies $E_{pt} < 3\,eV$ (which easily covers the solar spectrum of Fig. 1.1) versus k onto Figure 2.5, we would see that such a plot would fall essentially on the ordinate. We can take a very important point from this: the momentum of the photons constituting the majority of the solar spectrum (see Fig. 1.1) is very small compared to the momentum of phonons. From Figure 2.5 it may also be inferred that phonon energies in solids are of the order of 10^{-2} eV to perhaps 10^{-1} eV. Photon energies, at least those in the near infrared, visible, and near ultraviolet range, where the spectra of Figure 1.1 are at their richest, are of the order of 1 eV.

2.2.2.2 NANOPARTICLES AND NANOCRYSTALLINE SOLIDS

As particle or grain size becomes smaller, the surface-to-volume ratio obviously increases and surface-stress effects on bulk and surface phonon modes become more important. However, when a nanoparticle's or nanocrystalline grain's characteristic dimension approaches some multiple (<5–50) of the lattice constant, very fundamental change is also possible; i.e., the vibrational modes can start to change, due to their being limited in spatial extent,[4,5] and the constraint of a phonon energy corresponding to a well defined **k**-vector can disappear, also because of spatial limitations[6]—these changes are the result of what is termed phonon confinement. The removal of the constraint of a precise $\hbar\mathbf{k}$ for a phonon of a given energy makes sense if viewed in terms of Heisenberg's Δx $\Delta k = \theta(2\pi)^{\ddagger}$; i.e., as Δx becomes smaller due to confinement, the phonon momentum becomes ill-defined. The implication for phonon-electron "collisions" is that a phonon can now supply its energy and a range of momenta. As a particle becomes even smaller, the vibrational modes can become discrete in energy, as would be seen for a molecule. The impact of phonon confinement at a given nanoparticle or nanocrystalline grain size will depend on the degree to which isolation is accomplished. For nanocrystalline materials, this means that the proximity of other nanocrystal grains will act to reduce the influence of phonon confinement.

2.2.2.3 AMORPHOUS SOLIDS

In amorphous solids, a vibrational mode may extend over only a few nanometers. It therefore again follows from $\Delta x \, \Delta k = \theta(2\pi)$ that phonons in disordered materials are not characterized by a well-defined wave vector **k** and there is no phonon **k**-selection rule. The quantity **k** is no longer a "good quantum number." In amorphous solids there is no Brillouin zone in reciprocal space because there is no unit cell in real space, since there is no crystal lattice. Also, in these materials, it becomes difficult to distinguish between acoustic and optical phonons. However, in amorphous solids, phonons play the same critical roles in electron transport, heat conduction, etc., as they do in crystalline solids.

It follows that the concept of density of **k** states in reciprocal space is not valid for amorphous materials. However, the concept of a density

---

[‡]The notation $\theta(2\pi)$ is being used to signify of the order of 2π.

of phonon states in energy $g_{pn}(E)$ is still valid. In fact, the $g_{pn}(E)$ of an amorphous solid will conform to that of the corresponding crystalline material to a degree depending on the importance of second nearest-neighbor, third nearest-neighbor, etc., forces.

2.2.3 Electron energy levels in solids

A very helpful approach that is usually valid for many solids is the Born-Oppenheimer or adiabatic principle. This principle asserts that, if one is solving the Schrödinger equation for the collection of the cores (nucleus plus core electrons) and valence electrons that make up a crystalline, polycrystalline, nanocrystalline, or amorphous solid, then one can separate the core motion (the vibration field we just discussed) from that of the valence electron motion.[7] In this picture, a single electron "sees" an effective potential resulting from (1) the cores in their average positions and from (2) all the other valence electrons. Solving this single-electron problem gives rise to what are termed single-electron energy levels. However, as we note in this section, it is not always possible to separate the Schrödinger equation for a solid into one problem dealing with phonons and into another dealing with electrons treated as single particles immersed in an effective potential. For example, the multi-particle core-electron interactions in which the cores polarize to shield an electron's charge give rise to multi-particle solutions to the overall Schrödinger equation which are called polarons. Multi-particle electron–electron interactions can give rise to solutions termed excitons. Polarons and excitons are examples of multi-particle states.

2.2.3.1 SINGLE-CRYSTAL, MULTICRYSTALLINE, AND MICROCRYSTALLINE SOLIDS

(a) Single-electron states

Because the unit cell of the direct lattice of a single-crystal, multicrystalline, and microcrystalline material completely specifies the structure, it completely determines the environment of an electron in a crystalline solid. Essentially, the Schrödinger equation for single-electron states in a crystal need be solved for only one unit cell, subject to boundary conditions that represent the periodicity of the structure. The dispersion relation $E = E(\mathbf{k})$ that comes out of this solution specifies the energy E available to an electron in a single-particle state with wave vector $\mathbf{k}$. This wave vector $\mathbf{k}$, when multiplied by $\hbar$, may be viewed as the momentum of the electron, just as a phonon of wave vector **k** could be viewed as having the momentum $\hbar\mathbf{k}$ in a crystalline material. As in the case of the phonon dispersion relation for a crystal, the periodicity of the direct lattice ensures that the electron dispersion relation $E = E(\mathbf{k})$ is periodic in reciprocal space. Hence, all the $E = E(\mathbf{k})$ information is completely contained in the first Brillouin zone appropriate to the crystal. That information just repeats throughout the rest of **k**-space. As is also the case for phonons, only certain **k**-vectors in **k**- (reciprocal) space are permitted to electrons in a crystal. Consequently, there is a density of allowed electron states in **k**-space. Through the dispersion relationship, this can be transformed into a density of states in energy per energy per volume.[1] We designate this density of allowed electron single-particle states per energy per volume as the quantity $g_e(E)$. This is another example of the density of states in energy concept seen in Figure 2.3.

The $E = E(\mathbf{k})$ relationships between a single-electron allowed energy level E and the wave vector **k** for silicon and gallium arsenide are seen in Figure 2.6 for specified directions in the Brillouin zone. Both materials

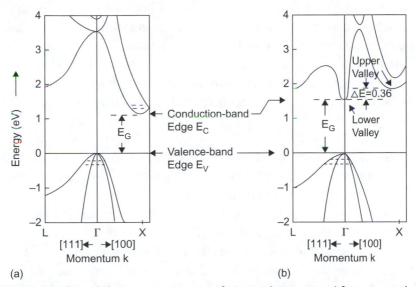

(a)   (b)

FIGURE 2.6 Allowed electron energies versus **k**-vectors (wave vectors) for two crystalline, inorganic semiconductors: (a) silicon; (b) gallium arsenide. Silicon is indirect gap (the maximum in the valence band E_V and the minimum in the conduction band E_C have different **k**-values); gallium arsenide is direct gap (the maximum in the valence band E_V and the minimum in the conduction band E_C have the same **k**-value). The valence band edges are aligned here in energy for convenience only. (After Ref. 3, with permission.)

have the same Brillouin zone, as noted in conjunction with the phonon plots of Figure 2.5. From the electron energy dispersion relationships for these two example materials, it can be seen that both have a lower band of allowed electron energy levels (actually two bands), then a gap of forbidden energies of width E_G, and finally a band (or, more precisely, overlapping bands) of allowed energy levels. This type of band structure—with allowed levels in a band, an energy gap with no allowed states, and then allowed levels in a band—typifies an ideal crystalline semiconductor, with the stipulation that the valence electrons completely fill that lower band at $T = 0\,K$. In other words, this lower band is the home of all the valence electrons; consequently, it is called the valence band (VB). The highest allowed energy in this lower grouping of bands (valence band) is termed the valence band edge E_V. The lowest allowed energy in the upper grouping of bands (conduction band, CB) is termed the conduction band edge E_C. The energy levels in these bands are single-particle states that are delocalized; i.e., they extend throughout the crystal. Formally, bands like those seen in Figure 2.6 (which shows the locations in energy of E_C and E_V) are representative of a region in the crystal that is large compared to a unit cell but small compared to the characteristic length over which material composition (such as in an alloy semiconductor) or some electrostatic potential (such as in an electrostatic field barrier region) may vary. As an imposed electron potential energy or the material composition changes with position in a solid, energies like E_C and E_V will move up or down in energy with respect to a fixed reference.

As we noted, in a semiconductor the VB is filled with valence electrons at absolute zero, and therefore the CB is empty of valence electrons at absolute zero. When electrons leave the states of the valence band due to picking up thermal energy or due to photon absorption, the ensemble of remaining electrons acts like positive particles (holes). The number of these valence band holes equals the number of vacated states. When electrons enter the conduction band due to picking up thermal energy or due to photon absorption, they act like electrons and their number equals the number of electrons in that band.[1] The holes in the valence band are "free" carriers in that they are in delocalized states and can move in an electric field (drift) or diffuse. The electrons in the conduction band are "free" carriers in that they, too, are in delocalized states and can move

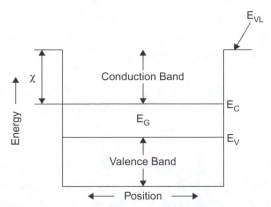

FIGURE 2.7 Schematic showing the energy bands available in a semiconductor as a function of position. Reference energy here is the vacuum energy E_{VL}.

in an electric field (drift) or diffuse. If essentially all the electrons per volume in the conduction band n came from the holes per volume p in the valence band, then the semiconductor is said to be intrinsic and n = p. If an intrinsic material is in thermodynamic equilibrium (TE), n = p = n_i where n_i, the intrinsic carrier population in TE, is dictated by the energy gap and the temperature.[3]

Frequently, the detailed E = E(**k**) information relating E and **k** shown in the examples of Figure 2.6 is not needed. In many applications it is necessary only to know the location in energy of the conduction band edge E_C, of the valence band edge E_V, and of the local vacuum level E_{VL} (energy needed to escape the material at x) as a function of position. Such a diagram is given, as a function of position x in the crystal and with respect to an arbitrarily chosen reference energy, in Figure 2.7. This diagram also introduces a new quantity, the electron affinity χ. This is the energy required to promote an electron from the bottom of the conduction band to the vacuum level at some position x. The energy required to go from the top of the valence band to the vacuum level at some position x is the hole affinity $E_G + \chi$, where E_G is the forbidden energy gap.

An ideal insulator differs from the ideal semiconductor of Figure 2.7 only in that $E_G \geq 2.5$ eV, where the lower bound of 2.5 eV is somewhat arbitrary. The real point is that, if a material is an insulator at room temperature, then its E_G is too wide to give any significant intrinsic carrier population at $300\,K$. From the perspective of Figure 2.7 a metal arises either when the gap E_G is zero or when the valence electrons only partially fill the lowest band at $T = 0$.[1]

Figure 2.7 represents the electron energy band scheme for a uniform composition, ideal crystal. Such a material only has the delocalized (extended) states of the valence and conduction bands. However, real materials have surfaces, and surfaces introduce localized states (with atomic-like wave functions physically localized at the surfaces), which may be populated by carriers.[3] In addition, the bulk of a real crystal is not perfect, since in the bulk there will be impurities and defects that will also introduce localized, atomic-like states at the positions of the impurities and defects.[3] Further, if a solid is nano-, multi- or microcrystalline, the structural and bonding defects that are the grain boundaries cause localized states to exist in the grain boundary regions. A localized state is represented in energy band diagrams like Figure 2.7 by introducing a short horizontal line drawn at its position in energy and space. Localized states may have energies that lie in the range of the bands or of the energy gap. The former case is uninteresting from a technological point of view because the levels are degenerate with those of delocalized band states. Consequently, carriers in such localized states immediately transfer to the delocalized states of the band and are free to move about. The latter case of states that lie in the energy gap is very important technologically because the carriers in these states cannot move—unless they are demoted or promoted to delocalized states. These states in the energy gap are often referred to as localized gap states or simply gap states. Drift and diffusion are not possible for carriers in gap states unless the gap state density is extremely high. Hopping transport, discussed shortly, can be possible for high densities. Gap states can be purposefully present to dope a material. Gap states can act as conduits between the conduction and valence bands (i.e., support generation and recombination) and they can store charge, thereby affecting the electric field in a device. If they principally do the latter but are not dopants, they are referred to as traps. As we discuss later, whether a gap state is principally a recombination center or a trap depends on its energy position and its capture cross-sections for electrons and holes.

Localized gap states found in crystalline, polycrystalline, amorphous, nanocrystalline, and nanoparticle solids may be broadly classified as acceptor-like, donor-like, or amphoteric in nature. Acceptor and donor-like states are single-electron states with the following definitions[3]:

1. *Acceptor states* Neutral when unoccupied by an electron and, therefore, negative when occupied by an electron (ionized).

2. *Donor states* Neutral when occupied by an electron and, therefore, positive when unoccupied by an electron (ionized).

Amphoteric gap states can be occupied by none, one, or two valence electrons. Their charge state depends on their occupancy; e.g., one electron could be the neutral state, no electrons the positively charged state, and two electrons the negatively charged situation.

The charge per volume resulting from acceptor-like states at energy E in the gap depends on whether they have captured an electron and is given by $-en_T$ where n_T is the number of acceptor states per volume at the energy E that have been successful in capturing an electron. Correspondingly, the charge per volume resulting from donor-like states at energy E in the gap depends on whether they have lost an electron and is given by ep_T where p_T is the number of donor states per volume at energy E that have been successful in losing an electron. Amphoteric states, as noted, can give rise to positive or negative space charge, depending on their occupancy.

As mentioned, gap states can serve as sources (dopants) of carriers for the bands.[3] They are purposefully introduced for this role (doping), when desired, or can be inadvertently present and have this function. Due to trapping and doping, the electrons in the conduction band per volume n and the holes in the valence band per volume p may not equal each other as they do in an intrinsic (i.e., defect-free) material. When this occurs, the material is referred to as extrinsic. Interestingly, it is possible to have so many gap states present in a material that they form a band within the energy gap. This is the high density of gap states situation mentioned earlier. These states, depending on the distance between the physical sources of the states, can be delocalized or localized in nature. Such a band within the energy gap is termed an intermediate band (IB).

(b) Excitons

Single-crystal, multicrystalline, and microcrystalline solids can also harbor the solutions to the Schrödinger equation that are known as polarons

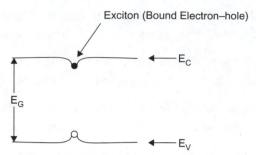

FIGURE 2.8 A representation of an exciton ground state superimposed on the single-electron levels of a band diagram.

and excitons. The former, an electron-phonon interaction as noted earlier, increases the inertia of electrons, thereby giving them a larger effective mass. Polarons are generally only of any significance in crystals with some degree of ionicity.[1] Excitons are of much more interest, since they are a multi-particle phenomenon that can be involved in the light absorption process in solar cell materials. Specifically, excitons are multi-electron solutions that may be viewed as an electron bound to a hole via Coulombic attraction.[1] Figure 2.8 shows an exciton ground state superimposed on the single-electron levels of a band diagram. The binding energy is the energy needed to have the exciton in its ground state dissociate into a free electron in the conduction band and a free hole in the valence band. Since the binding energy is dictated by the Coulombic attraction, materials that polarize more have lower binding energies; i.e., the binding energy correlates inversely with the dielectric constant. Excitons can be created by photon absorption and they can be mobile in a solid. When they move, energy, but not net charge, moves. Since they are not charged, they can only move by diffusion. The spatial extent of an exciton can be over several lattice constants (a Wannier exciton) or essentially localized at one atom or molecule (a Frenkel exciton). If excitons are produced by light absorption in a solar cell (step 1 in Chapter 1) and are to be utilized, then some process must also be present (step 2 in Chapter 1) to convert the exciton into at least one free negative charge carrier–one free positive charge carrier pair. (We return to this and specifically to the "at least one" part in Section 2.2.6.)

2.2.3.2 NANOPARTICLES AND NANOCRYSTALLINE SOLIDS

The single-electron energy levels can evolve in the case of nanoparticles and nanocrystals from the band picture of single-crystal, multicrystalline,

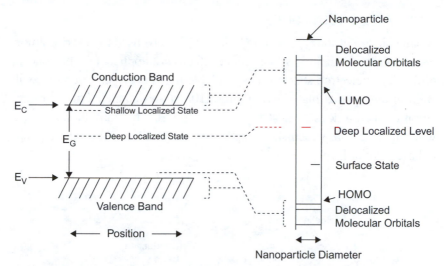

FIGURE 2.9 Schematic single-electron energy level diagram showing, with increasing confinement, the evolution from bands to molecular orbitals. (After Ref. 8.)

and microcrystalline solids to a molecular orbital picture as the size parameter is decreased,[8] as shown in Figure 2.9. The extent of the evolution depends on the confinement. Analogous to phonon confinement, reducing the size of nano-scale structures can cause changes in the allowed states and a relaxation of the electron **k**-selection rules. These effects, in the case of electrons, are usually called quantum confinement effects. The spacing in energy between the occupied molecular orbitals and the unoccupied molecular orbitals (the energy gap) increases with confinement, as would be anticipated from the particle in a box problem. As may be noted from Figure 2.9, the highest occupied molecular orbital is called HOMO while the lowest unoccupied molecular orbital is called the LUMO. We will use this terminology from time to time in place of the corresponding terminology from solid-state physics and electrical engineering; i.e., valence band edge and conduction band edge. We also note that the energies and binding energies of excitons change in these materials with confinement.[8]

The impact of the quantum confinement effects depicted in Figure 2.9 for a given nanoparticle or nanocrystalline grain size will depend on the degree to which isolation is accomplished. For nanocrystalline materials, this means that the proximity of other nanocrystal grains will act to reduce the influence of quantum confinement. In the case of molecules, the quintessential nanoparticles, interaction among the orbitals may allow transport from molecule to molecule.

2.2.3.3 AMORPHOUS SOLIDS

A band picture of the type seen in Figure 2.7 can be used to describe amorphous solids, but diagrams like Figure 2.6 are not applicable. The concept of density of electron states (states available per energy per volume) is still valid in amorphous materials. Figure 2.10a shows a schematic representation of such a density of states function $g_e(E)$. This may be contrasted with Figure 2.10b, which shows $g_e(E)$ for a crystalline solid that happens to have a donor level present.

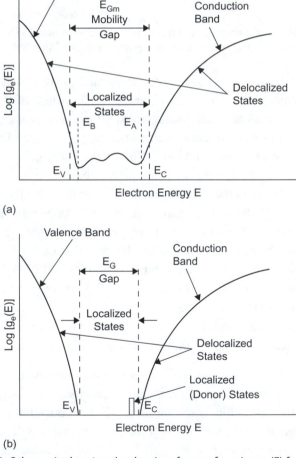

(a)

(b)

FIGURE 2.10 Schematic showing the density of states function $g_e(E)$ for (a) an amorphous solid and (b) a crystalline solid.

From Figure 2.10 it is clear that amorphous materials can contain large numbers of localized gap states. These are essentially of two types: intrinsic and extrinsic. The intrinsic localized states are defined as arising from the distribution of bond angles and interatomic distances statistically occurring in an amorphous solid. Extrinsic localized states are defined as arising from defects (broken chemical bonds) and impurities. Intrinsic localized states are expected to be the principal contribution to the localized states occurring near the band edges (see Fig. 2.10a); defect and impurity states are believed to be the source of the large density of localized states typically seen over the remainder of the gap in many amorphous materials. As long as the density of these defect states is large, doping will have little effect on dislodging the Fermi level position in the gap and recombination and trapping will be severe problems. In the case of amorphous silicon, hydrogen is used to suppress the number of these states by forming chemical bonds to the defects. So much hydrogen is used that the material is actually a silicon-hydrogen alloy denoted as a-Si:H.

Since the localized state density is so large in many amorphous materials, transport via these gap states is possible; i.e., an electron can move from one localized site to another. We discuss this in detail later. It suffices to note at this point that transport via these localized states would give rise to small mobilities compared to transport involving the delocalized states. Hence, Figure 2.10a shows a mobility gap, which denotes the switch from small to larger mobilities, rather than a true gap. This mobility gap is expected to have sharp boundaries E_C and E_V, which serve as the demarcation between localized and delocalized states.

2.2.3.4 ORGANIC SOLIDS

Organic solids are usually classified as either small-molecule or polymeric solids. The molecular orbital picture (Fig. 2.9) is generally accepted as applicable for organic materials of either type. Both types of organics have been used in solar cells. In solids, the molecules interact to varying degrees, as noted earlier, and the result can be that, in some cases, the orbital levels form a band. While the density of delocalized states $g_e(E)$ of single-crystal and amorphous materials is parabolic as seen in Figure 2.10, at least near the band edges, the density of states $g_e(E)$ for organic materials is expected to have levels within a Gaussian-like envelope.[9]

2.2.4 Optical phenomena in solids

2.2.4.1 ABSORPTION PROCESSES

In this section we examine the photon interactions with solids that give rise to absorption. Figure 2.11 is a sketch showing the various processes that absorb electromagnetic radiation in solids and their range of influence. Process 1 in the figure is free-carrier absorption; it arises from photon-induced, electron (or hole) transitions within a band from one single-particle state to another. These intraband transitions, depicted diagrammatically in Figures 2.12a and b, are important in metals and semiconductors whenever there is a significant density of carriers in a band. Intraband transitions of the type seen in Figure 2.12a would be forbidden in crystalline solids (due to **k** conservation), were it not for the scattering processes (due to defects, impurities, phonons), which allow for the conservation of momentum. Process 2 of Figure 2.11 is phonon absorption; i.e., light is absorbed by exciting phonon modes in a material. Electrons are not involved in this process. Because of the low energies possessed by phonons (see Fig. 2.5), this absorption process occurs in the infrared region of a light spectrum.

Process 3 of Figure 2.11 includes all the photon-induced transitions between states in the gap as well as between gap states and a band. Example transitions are shown in Figures 2.12d and e. Process 4 is exciton-producing absorption. Absorption in organic materials, such as

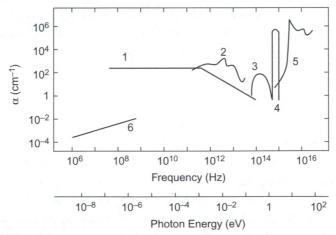

FIGURE 2.11 A schematic portrayal of the range of optical absorption processes in solids. The processes and the absorption coefficient α are discussed in the text.

the small-molecule dyes of dye-sensitized solar cells and the polymer absorbers of organic solar cells, is generally an exciton process. Excitons can also play the key role in absorption in nanoparticles. Process 5 is the band-to-band transitions, which are seen in Figure 2.12c. Process 6 is a loss observed in some amorphous materials; it probably arises due to electrons that are hopping (see Section 2.3.1.3) from localized site to localized site.[10]

The absorption coefficient $\alpha(\lambda)$ or equivalently $\alpha(\nu)$, which is the ordinate in Figure 2.11, is critical to solar cell performance. The Process 3, 4, and 5 contributions to α are the mechanisms we will be interested in because they can lead to free electrons and free holes. While Process 4 or Process 5 or both will be present in an absorber material, Process 3 may or may not be present at all. We can define a function $G(\lambda, x)$, which determines the number of Process 3, Process 4, or Process 5 events taking place at x per volume per time per bandwidth. This bandwidth is around the wavelength λ of the light impinging on the absorbing material.

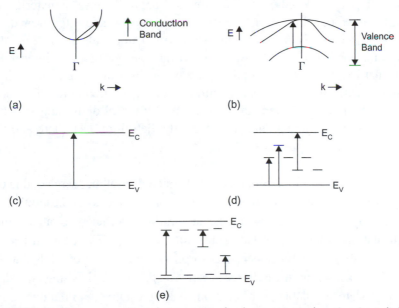

FIGURE 2.12 Electron transitions between single-electron states that give rise to light absorption: (a, b) free-carrier (intraband) transitions give rise to Process 1 of Figure 2.11; (c) band-to-band transitions give rise to Process 5; (d, e) band-localized–state transitions and localized-state–localized-state transitions are the source of Process 3. The type of Process 3 seen in (d) can give rise to free carriers.

It can be shown that $G(\lambda, x) \propto \alpha(\lambda, x)\xi^2(\lambda, x)$, where $\xi(\lambda, x)$ is the (optical frequency) electric field of light of wavelength λ at point x.[11]

If reflection and interference wave effects are not important in a solar cell structure, then monochromatic light entering into the cell's absorber at $x = 0$ with intensity $I_0(\lambda)$ (photons per area per time per bandwidth) is expected to have the intensity $I(\lambda, x)$ at some point x in the material given by[10]

$$I(\lambda, x) = I_0(\lambda)e^{-\alpha(\lambda)x} \qquad (2.1)$$

Equation 2.1 is known as the Beer-Lambert law. This expression, when valid, allows $G(\lambda, x)$ to be written as

$$G(\lambda, x) = \alpha(\lambda)I_0(\lambda)e^{-\alpha(\lambda)x} \qquad (2.2)$$

since the spatial derivative of Eq. 2.1 gives the photons lost. Comparison with the more general expression above shows that the Beer-Lambert law is based on $\xi^2(\lambda, x)$ falling off exponentially into a material, as would be expected if internal reflection, scattering, and interference are being neglected. Such expectations are often not realistic in thin film structures. When Eq. 2.1 is valid, $\alpha(\lambda)$ has the very simple interpretation that $1/\alpha(\lambda)$ is the absorption length for light of wavelength λ in a material with $\alpha(\lambda)$. A general comment is also merited here on absorption coefficient data: there are two slightly different definitions of $\alpha(\lambda)$ in use for experimental data. These are the definition used here and the one discussed in Appendix A.

In single-crystal, multicrystalline, and microcrystalline semiconductors and insulators, the onset of the band-to-band Process 5 absorption is marked by a sharp increase in the absorption coefficient α at the fundamental absorption edge that occurs for $h\nu = E_G$. The detailed behavior of α above the fundamental absorption edge in these solids can be explained using Figure 2.6. We recall that this figure depicts single-particle electron states available, as a function of $\mathbf{k}$, in a direct-gap (gallium arsenide) and in an indirect-gap (silicon) crystalline semiconductor. In direct-gap material, an electron can be excited from the top of the valence band to the bottom of the conduction band with essentially no change in $\mathbf{k}$-vector required. Since a photon has a $\mathbf{k} \approx 0$ on the scale of Figure 2.6

to contribute to the momentum, there is no difficulty in **k**-vector conservation in electron-photon interactions that result in an electron's being excited across a direct gap. This is not the case for an indirect-gap material, as Figure 2.6 also shows. In this case, an electron excited across an indirect gap must change its **k**-vector. The photon simply cannot supply the momentum; hence phonon involvement is required. As seen in Figure 2.5, phonons can supply the necessary momentum but conservation of total energy and total momentum for the three particles (electron, photon, and phonon) must now apply. The necessity of a three-body interaction in indirect-gap materials tends to lessen the magnitude and crispness of $\alpha = \alpha(\lambda)$ at and above the fundamental absorption edge. This issue of conserving the total **k**-vector during band-to-band transitions can disappear for small-enough crystalline nanoparticles and nanocrystalline materials for two reasons: the constraint of a well-defined **k**-vector can disappear for phonons and electrons with confinement, as we have discussed. Because of this, nanocrystalline silicon, for example, exhibits stronger absorption than its single-crystal, multicrystalline, and microcrystalline counterparts.[6]

The distinction among Processes 3, 4, and 5 of Figure 2.11 generally becomes difficult to discern from $\alpha(\lambda)$ for amorphous solids. This is because, as seen in Figure 2.10, amorphous solids can contain large densities of localized gap states at the delocalized state band edge. As a result, the absorption edge in amorphous materials is often broad. Overall, due to the relaxation of the **k**-vector conservation rules, absorption at and above the fundamental absorption edge can be stronger for amorphous materials than for corresponding crystalline material.

One more comment about absorption: a strong electric field ξ, can shift the fundamental absorption edge (Process 5) in solids. This phenomenon is known as the Franz-Keldysh effect, and it is depicted in Figure 2.13. As seen, the delocalized valence-band wave functions leak into the forbidden gap in the presence of a strong electric field. The same is true for the wave functions of the delocalized states in the conduction band. Therefore there is a probability P that a photon can promote an electron at the valence band edge at point x to the conduction band edge at point x'. The electric field dependence of this probability goes as[12]

$$P \sim \exp - \left(\frac{E'}{E_0} \right) \tag{2.3}$$

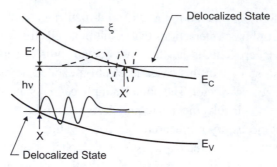

FIGURE 2.13 Franz-Keldysh effect allows band-to-band absorption for $h\nu < E_G$. As shown, the process involves tunneling and photon absorption in the presence of an electric field ξ.

where in this case E_0 stands for

$$E_0 = \frac{3}{2}(m^*)^{-1/3}\ (e\hbar\xi)^{2/3} \tag{2.4}$$

Here the energy E' is defined at x in Figure 2.13, m^* is the effective mass, and ξ is the electric field. As seen in Figure 2.13, an electron in the valence band can make the transition into such a tail from the conduction band with the absorption of a photon $h\nu < E_G$. This gives rise to the shift in the fundamental absorption edge. In amorphous materials one would expect such tails could extend further into the gap due to overlapping with localized gap states. In addition, Figure 2.13 does not consider any possible exciton role.

2.2.4.2 INTERFERENCE, REFLECTION, AND SCATTERING PROCESSES

While we have been discussing photons, the "particles" of light, and their interaction with materials, the wave properties of light are also very important in solar cell structures. Illumination of wavelength λ can be reflected as it enters a cell and it can undergo additional reflection at each interface. Interference patterns can be set up within the cell. Scattering can occur at small structures or features. An example of contrasting $G(\lambda, x)$ behavior that can result from the varying importance of these wave phenomena is presented in Figure 2.14 for two values of λ. These two values of λ are impinging on an organic heterojunction solar cell.[11] As is shown, exponentially varying $G(\lambda, x)$ behavior with position in the absorber, as predicted from the Beer-Lambert law, is followed reasonably well for $\lambda = 300\,\text{nm}$. This is due

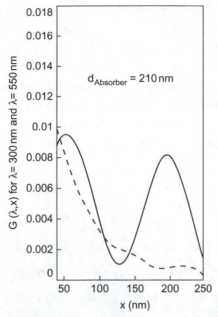

FIGURE 2.14 The function G(λ, x), which counts the number of Process 3, Process 4, or Process 5 (or some combination of these) events taking place at x per volume per time for a bandwidth around the wavelength λ as a function of position in the 210 nm thick absorber of an indium tin oxide (ITO)/Poly(3,4-ethylenedioxythiophene)-poly(styrene-sulfonate) (PEDOT-PSS)/ organic blend absorber/C_{60}/Al heterojunction solar cell for λ = 300 (-------) and 550 nm. The absorber is seen to run from 40 nm to 250 nm. (Adapted from Ref. 11, with permission.)

to very strong absorption at λ = 300 nm and consequently to the fact that there is little opportunity for this wavelength of light to penetrate to interfaces to set up reflection and interference effects. However, the weaker absorption at λ = 550 nm permits very significant reflection and interference effects, as seen in the figure. The Beer-Lambert law fails at this wavelength for this structure.

Photonic structures can be used to exploit the wave (physical optics) behavior of light to shape the $\xi^2(\lambda, x)$ distribution within a solar cell absorber.[13] This can be very advantageous since, as we have noted, G(λ, x) $\propto$ $\alpha(\lambda)\xi^2(\lambda, x)$. These structures utilize the periodicity of two-dimensional or three-dimensional "photonic crystals", composed of repeating regions of materials with differing dielectric constants, to control light.[14] The length-scale of the periodicity of these photonic crystal structures must be a significant fraction of the wavelength of the light to be manipulated.[15]

Given the wavelengths of the photon-rich part of the solar spectrum (Fig. 1.1), these structures must have features at the nano-scale (i.e., ~100 nm) or somewhat larger. Figures 2.15a and b show an example of using such a photonic crystal structure to affect reflectance. This particular example shows the impact of an array of CdS nanocolumns of different heights on reflectance of light impinging on the array. These experimental light-trapping data were obtained prior to the addition of CdTe to fabricate a CdS-CdTe heterojunction solar cell with the CdS nanocolumns. The reflectance changes somewhat with the addition of CdTe.[16]

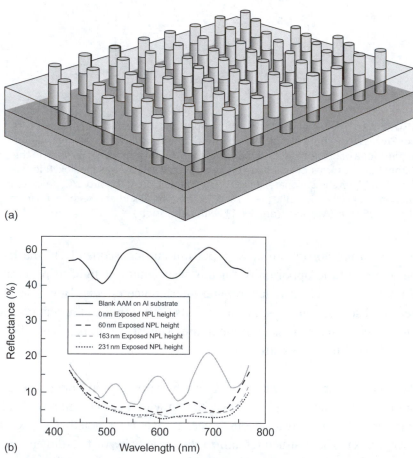

(a)

(b)

FIGURE 2.15 (a) An array of CdS nano-pillars (NPLs) grown in an anodized aluminum (AAM) template. Center-to-center spacing is about 150 nm. (b) Experimentally measured reflectance from such an array, with the NPLs protruding from the AAM template at heights of 0, 60, 163, and 231 nm. (From Ref. 16, with permission.)

Texturing at contacts of a cell with features on the order of a micron can also lead to light trapping and reduced reflection from the cell. This light trapping is easily explained using geometrical optics (beam tracing) and the recognition that there will be longer path lengths through the absorber for light entering front texturing or reflected at the back texturing. The longer path lengths lead to more absorption. The approach becomes problematic for thin-film solar cells, since the absorber thickness can be of the order of, or smaller than, the texturing features required.[15,17] The photonic crystal approach is totally different, since it relies on physical optics phenomena (reflection, interference, and diffraction) and employs much smaller structures.[15]

Plasmonics is another phenomenon that offers interesting potential for solar cell applications.[17] The term refers to the cooperative electron oscillations (quantized into plasmons) that can be set off by light impinging on nano-scale metal features (e.g., nanoparticles, etched structures) and the re-radiation of light (scattering) from these features.[1,17] In the case of solar cell structures, it also appears that there can be a coupling of this re-radiated light into the waveguide modes that can exist in a cell structure and the feedback of the light into the metal features.[17] Plasmonic effects can give rise to very strong near fields in the immediate vicinity of the metal features, resulting in greatly enhanced absorption in these volumes.[18] Consequently, plasmon scattering can be designed to create an advantageous $\xi^2(\lambda, x)$ distribution and thereby to enhance absorption and control its distribution. A listing of calculated plasmon frequencies, expressed as wavelengths, for selected metal nanoparticles (in a medium of index of refraction of 1.33) is found in Table 2.2.[19] Light at or near the wavelengths listed in Table 2.2 can couple into the plasmon scattering mechanism. These wavelengths listed in the table can shift due to medium, substrate, and waveguide effects.[17] Both randomly arranged metal nanoparticles and periodic metal nano-scale features ("plasmon crystals") have been used to modify the $\xi^2(\lambda)$ distribution in solar cells by exploiting the plasmon effect. In the case of metal nanoparticles, several nanoparticle positioning schemes have been explored, including placing the metal nanoparticles at and near the material interface in heterojunctions[20] and placing the particles in a two-dimensional lattice on the top (light-entering) surface of a cell.[21] Figures 2.16a and b show the results of a simulation study for an a-Si:H p–i–n cell with Au nanoparticles positioned on the top surface, as shown in

Table 2.2 Some Plasmon Wavelengths

Metal	Localized surface plasmon (resonance) wavelength (nm)
Pd	~250
Ag	~390
Ba	~400
Eu	~380 (with a shoulder to ~500)
Ca	~500
Y	~430 (broad peak)
Au	~525
Pt	~230
Cu	~210 (with a shoulder to ~580)
Cs	>700

From Ref 19.

Figure 2.16a. In such an a-Si:H cell, the red and near infrared wavelength photons ($\lambda > 650$ nm) have very long absorption lengths (exceeding 1μm) and normally cannot be efficiently absorbed by the 500-nm a-Si:H absorber used in the simulation. However, absorption is enhanced in Figure 2.16b across a range of light wavelengths (except for those near or at $\lambda = 500$ nm) when 70–80 nm radius Au nanoparticles with a separation (pitch) of 650 nm are present. The disadvantageous behavior at $\lambda \sim 500$ nm is due to reflection at the top surface for these wavelengths.[21] Similar simulations have found that the use of a two-dimensional metal "plasmon crystal" with nano-scale features at the back of a cell (see Fig. 2.17) can avoid loss of long-wavelength absorption and actually enhance absorption in this range, as seen in Figure 2.17.[21] The deleterious short-wavelength (values around 500 nm) loss seen for the nanoparticle case in Figure 2.16 does not appear here since these short wavelength photons are absorbed before they reach the back of the cell.

2.2.5 Carrier recombination and trapping

Carriers, free electrons in a semiconductor conduction band (CB) and the free holes in a semiconductor valence band (VB), are what are

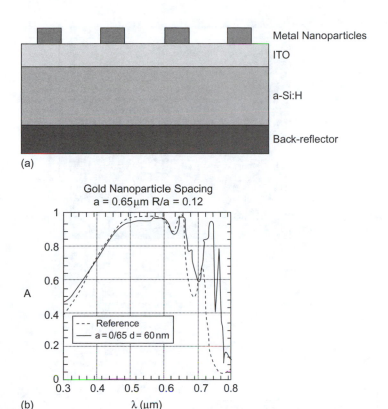

(a)

(b)

FIGURE 2.16 (a) Diagram of a nanoparticle-coated a-Si:H solar cell. (b) Absorption simulation results for this cell with nanoparticles present (solid line) and for the corresponding cell with no nanoparticles present (dashed line). The a-Si:H absorber layer thickness was taken to be 0.5 μm. The low absorption at short wavelenghs is due to front surface reflection. (From Ref. 21, with permission.)

needed to make a solar cell work. Once electrons are excited to the conduction levels and the corresponding holes are created in the valence levels, or once excitons dissociate, producing electrons in the conduction levels and holes in the valence levels, the aim is to get them to do work via Step 3 of Chapter 1. However, excited electrons can return to their ground state or drop in energy into some gap state without doing anything useful. This can happen in one of several ways: an electron can give up its energy through a radiative mechanism (radiative recombination), which involves emission of a photon (this may or may not also involve phonons); an electron can give up energy by emission of phonons, a process which involves gap states too (Shockley-Read-Hall or gap-state–assisted recombination); or an electron can give up its energy through an Auger mechanism (Auger recombination), which involves

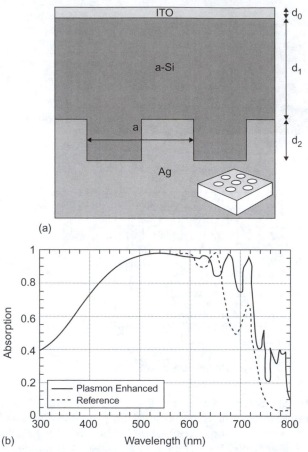

FIGURE 2.17 (a) A Ag "plasmon crystal" at the back of an a-Si:H solar cell. (b) Absorption simulation results for this cell with plasmon crystal present (solid line) and for the corresponding cell with no plasmon crystal present (dashed line). The a-Si:H absorption layer thickness is taken to be $d_1 = 0.5\,\mu m$. The plasmon crystal has a lattice pitch a ~750 nm and an etch depth d_2 of 200 nm. The radius of the holes is taken to be 225 nm. The low absorption seen at short wavelengths is due to front surface reflection. (From Ref. 21, with permission.)

the transfer of its energy to another electron or hole. Multi-particle states also have similar mechanisms. For example, excitons can lose energy by radiation or by an Auger process (Auger recombination). These three general recombination mechanisms all have corresponding opposite processes; these are thermal generation mechanisms. Recombination and thermal generation are always present in a material or device. Photogeneration can also be present under illumination. In thermodynamic equilibrium (TE), there is no applied or developed voltage, illumination, nor temperature gradient present. The principle of detailed balance applies in TE and it requires that each thermal generation process and its corresponding recombination process statistically balance one another. When additional (excess) carriers are created in a material, as occurs under light or carrier injection (due to biasing), the material is no longer in the state of TE and there is net recombination, where net recombination is defined as recombination minus thermal generation.

Free electron–hole recombination has the overall result of annihilating a free electron in the conduction band and a free hole in the valence band. In steady-state operation, which is what we are interested in, it is the simultaneous wiping out of a free electron–hole pair. Recombination is driven by the carrier populations in the bands. A unimolecular recombination process is driven by just n or p. A bimolecular process is driven by the np product. While recombination is the annihilation of a free electron and of a free hole, free carriers are also subject to trapping. The term trapping, strictly speaking, is used in two ways: (1) to mean capture by a gap state as part of a series of processes in localized state assisted recombination or (2) to mean the act of an electron or hole getting stuck at a localized state. When used in the latter sense, trapping is the stringing together of one or more energy-loss processes with the net result that an electron from the conduction band or a hole from the valence band finds itself in a deadend in a gap state. Recombination, when it is localized state assisted, can involve gap states existing in the bulk, at surfaces, and at grain boundaries of materials. Trapping involves these gap states too. Carriers in gap states, whether there due to trapping or passing through in a recombination process, give rise to charge according to the state occupation and to whether the states are acceptor-like, donor-like, or amphoteric.

2.2.5.1 RECOMBINATION MODELING

(a) Radiative recombination

As discussed in detail in Appendix B, net radiative recombination for free electrons and holes may be expressed as

$$\mathscr{R}^R = \left(\frac{g_{th}^R}{n_i^2}\right)(pn - n_i^2) \tag{2.5}$$

where n_i^2, the square of intrinsic number density, is always equal to n_0p_0 (the subscript zero refers to thermodynamic equilibrium (TE) values), whether a material is intrinsic or extrinsic.[3] The dimensions of $\mathscr{R}^R$ are number (of free holes or equivalently of free electrons) annihilated per volume per time. Obviously, $\mathscr{R}^R$ is zero in TE.

If a material is p-type and if, even under illumination, p is still essentially p_0, then Eq. 2.5 can be written as

$$\mathscr{R}^R = \left(\frac{g_{th}^R}{n_i^2}\right)(np_0 - n_0p_0) \qquad (2.6)$$

Defining $[p_0 g_{th}^R/n_i^2]^{-1}$ as τ_n^R allows Eq. 2.6 to be finally written as

$$\mathscr{R}^R = \frac{n - n_0}{\tau_n^R} \qquad (2.7)$$

where τ_n^R is the called the electron radiative recombination lifetime or simply the electron lifetime, if this mechanism dominates over other recombination paths. Similarly, if a material is n-type and if, even under illumination, n is still essentially n_0, then it follows that Eq. 2.5 can be written as

$$\mathscr{R}^R = \frac{p - p_0}{\tau_p^R} \qquad (2.8)$$

where $\tau_p^R = [n_0 g_{th}^R/n_i^2]^{-1}$ and is called the hole radiative recombination lifetime or simply the hole lifetime, if this is the dominant recombination mechanism. When these linearized forms of Eq. 2.5 (Eqs. 2.7 and 2.8) can be utilized, the bimolecular process of radiative recombination becomes controlled by one carrier; i.e., it becomes unimolecular.

(b) Shockley-Read-Hall recombination

As we noted, a second process by which free electrons and holes can relax to lower energies and recombine involves phonon emission. As may be seen from Figure 2.5, phonons have energies ≤ 0.1 eV and consequently band-to-band recombination strictly by simple phonon emission would require essentially simultaneous multiple-phonon involvement, which is believed to be unlikely. However, the situation changes if we bring in the localized states present in the material. Charge carriers can give up their energy in collisions with the physical entity that gives rise to this localized level, and become trapped by it. The energy may be released in this collision (trapping) event as phonons or photons or both. The trapping of an electron, for example, and then the trapping of a hole completes the recombination. This what we have called gap-state or equivalently localized-state assisted recombination. We use the symbol $\mathscr{R}^L$ for the net recombination rate from such processes. The superscript emphasizes the role played by localized gap states. This mechanism is termed Shockley-Read-Hall (S-R-H) recombination.[3]

As shown in Appendix C, the quantity $\mathscr{R}^L$, the net recombination through gap states at energy E, can be expressed in steady state as

$$\mathscr{R}^L = \frac{v\sigma_n\sigma_p N_T \left(np - n_i^2\right)}{\sigma_p(p + p_1) + \sigma_n(n + n_1)} \tag{2.9}$$

where σ_n is the capture cross-section (attractiveness) of these states for electrons and σ_p is their capture cross-section (attractiveness) for holes. If a material is p-type and if, even under illumination, p is still essentially p_0, then Eq. 2.9 can be written as

$$\mathscr{R}^L = \frac{v\sigma_n\sigma_p N_T(np_0 - n_0p_0)}{\sigma_p(p_0 + p_1) + \sigma_n(n + n_1)} \tag{2.10}$$

or as

$$\mathscr{R}^L = \frac{v\sigma_n\sigma_p N_T p_0(n - n_0)}{\sigma_p(p_0 + p_1) + \sigma_n(n + n_1)} \tag{2.11}$$

Defining $\tau_n^L = [v\sigma_n\sigma_p N_T p_0/\sigma_p(p_0 + p_1) + \sigma_n(n + n_1)]^{-1}$ and assuming it does not depend strongly on n finally allows Eq. 2.9 to be linearized to

$$\mathscr{R}^L = \frac{n - n_0}{\tau_n^L} \tag{2.12}$$

where τ_n^L is the called the electron S-R-H recombination lifetime or simply the electron lifetime, if this is the dominant recombination mechanism. Similarly, if a material is n-type and if, even under illumination, n is still essentially n_0, then it follows that Eq. 2.9 can be linearized to

$$\mathscr{R}^L = \frac{p - p_0}{\tau_p^L} \qquad (2.13)$$

where $\tau_p^L = [v\sigma_n\sigma_p N_T n_0 / \sigma_p(p + p_1) + \sigma_n(n_0 + n_1)]^{-1}$, assumed not to depend strongly on p, is the hole S-R-H recombination lifetime or simply the hole lifetime, if this is the dominant recombination mechanism. The linearization that takes Eq. 2.9 to 2.12 or 2.13, when valid, causes the mechanism to move from being bimolecular to being unimolecular; i.e., to being driven by one carrier.

It must be stressed that all of the above equations for S-R-H recombination have been written for one localized state group at some energy E in the energy gap. If a gap state distribution is present, there must be corresponding statements for the other gap states at other energies. Therefore, the total S-R-H recombination must be the sum of all these contributions. This requires that the above equations all become sums over the distribution.

(c) Auger recombination

The distinguishing feature of an Auger mechanism is that electrons and holes or excitons take energy from, or give energy to, other electrons and holes or excitons. There are many types of Auger processes. Some of the single-particle paths, from among many possible, are seen in Figure 2.18.[22] Process (a) of the figure shows an electron-hole Auger recombination process in which an electron falls from the conduction band to the valence band by giving its energy to another electron in the conduction band. We expect the Auger recombination rate r_A^A for this path to obey $r_A^A = A_{1A}^A n^2 p$. The corresponding generation process has an energetic electron in the conduction band relaxing to the band edge and generating an electron-hole pair with the released energy. It may be modeled as $g_A^A = A_{2A}^A n$. This corresponding Auger generation process is often called impact ionization.[3] Process (b) of the figure shows the equivalent Auger recombination process for a case in which a hole takes away the energy. It follows that we would expect $r_B^A = A_{1B}^A np^2$ for this path. The corresponding generation process has an energetic hole in the valence band relaxing to the band edge and generating an electron-hole pair with the released energy. It may be modeled as $g_B^A = A_{2B}^A p$. This too can be termed impact ionization. Processes (c)–(f) are not, of themselves, recombination processes, since an electron-hole pair does not disappear. As can be seen, (c)–(e) get carriers into or out of localized states; i.e., they cause trapping or de-trapping. Process (f) allows a trapped carrier to drop down lower in energy. Parts (c), (d), and (f) of the figure show localized state analogues to process (a) with $r_C^A = A_{1C}^A n^2 \tilde{p}_T$, $r_D^A = A_{1D}^A \tilde{n}_T np$ and $r_F^A = A_{1F}^A n \tilde{n}_T \tilde{p}_T$. Obviously there are free hole equivalents to these paths. Process (e), which results in the de-trapping of two carriers, can be modeled by $r_E^A = A_{1E}^A \tilde{n}_T^2 p$. As explained also in Appendix C, $\tilde{p}_T$ counts the single electron states per volume at energy E that are missing an electron and $\tilde{n}_T$ counts the single electron states per volume at energy E that are occupied by an electron. The quantity $\tilde{p}_T$ contributes to the p_T in the expression ep_T for gap-state positive space charge density only if the empty states are donor-like. The quantity $\tilde{n}_T$ contributes to the n_T in the expression $-en_T$ for gap-state negative space charge density only if the occupied states are acceptor-like.

As we see from our discussion of Figure 2.18, there are two free-carrier Auger recombination processes. We can develop models for them by first using the principle of detailed balance to relate A_{1A}^A with A_{2A}^A for process (a) and to relate A_{1B}^A with A_{2B}^A for process (b). These relationships, with some straightforward algebra, then allow the net recombination for process (a), $\mathscr{R}^{AA}$, and the net recombination for process (b), $\mathscr{R}^{AB}$, to be written as

$$\mathscr{R}^{AA} = A_{2A}^A \left[\frac{n^2 p}{n_0 p_0} - n \right] \tag{2.14a}$$

and

$$\mathscr{R}^{AB} = A_{2A}^A \left[\frac{p^2 n}{n_0 p_0} - p \right] \tag{2.14b}$$

The strong dependence seen in Eq. 2.14 on n or p means Auger recombination can become dominant in solar cells at high carrier concentrations.

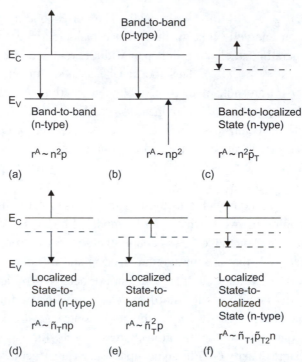

FIGURE 2.18 Some possible Auger transition processes in solids. The dependence of the different energy loss paths on the various number densities involved is indicated. Here $\tilde{p}_T$ gives the number of empty localized states at energy E per volume and $\tilde{n}_T$ gives the number of occupied localized states at energy E per volume. (Adapted from Ref. 22.)

Often it is hoped that the above expressions for Auger recombination can be adequately represented by a linearized model and that the processes of Eq. 2.14 can be written as

$$\mathscr{R}^A = \frac{n - n_0}{\tau_n^A} \tag{2.15a}$$

for p-type material and as

$$\mathscr{R}^A = \frac{p - p_0}{\tau_p^A} \tag{2.15b}$$

for n-type material. When this is done, τ_n^A and τ_p^A are the electron and hole Auger lifetimes.

Table 2.3 Differences between $G(\lambda, x)$ and $G_{ph}(\lambda, x)$

Phenomenon	Comments
Intermediate band absorption	Can allow Process 3 of Figure 2.11 to contribute to $G_{ph}(\lambda, x)$
Exciton dissociation	Must occur for exciton contribution to $G(\lambda, x)$ to appear in $G_{ph}(\lambda, x)$
Auger multiplication	For example, energetic excitons can undergo Auger process, producing a second exciton, and thereby the possibility of carrier multiplication.

2.2.6 Photocarrier generation

The function $G(\lambda, x)$ was introduced in Section 2.2.4.1 to count the number of Process 3, Process 4, and Process 5 absorption events at some point x in an absorber due to light of wavelength λ. $G(\lambda, x)$ is not necessarily the photocarrier generation function $G_{ph}(\lambda, x)$ per volume per time because of the three reasons listed in Table 2.3. First, some part of the Process 3 contribution to $G(\lambda, x)$ may be found in $G_{ph}(\lambda, x)$ but only if there are gap states of sufficient number and of sufficiently advantageous energy position to assist in carrier promotion from the VB to the CB by photons. Gap states need to be present in large numbers to ensure a significant contribution from this path and, if so, may form an intermediate band (IB). If there are enough gap states present, however, there may be a down side: the possibility of enhanced recombination. With a properly energetically positioned IB, VB-IB and subsequent IB-CB transitions can possibly lead to a net positive effect of enhanced carrier generation, due to the involvement of sub-band gap photons.[23] Turning to Processes 4 and 5, we note that the Process 5 contribution to $G(\lambda, x)$ shows up directly in $G_{ph}(\lambda, x)$. In the case of Process 4, the excitons produced must dissociate into free carriers to contribute to $G_{ph}(\lambda, x)$. This can be accomplished at heterojunction interfaces and, in principle, at regions of high electric fields. However, the latter possibility is reported by some to be inconsequential.[24] Where and to what degree exciton dissociation happens determine the Process 4 component of $G_{ph}(\lambda, x)$. For example, when exciton dissociation occurs at an interface, electrons appear on one side of the interface and holes on the other. In that case $G_{ph}(\lambda, x)$ becomes two expressions: $G_{ph}^{n}(\lambda, x)$ and $G_{ph}^{p}(\lambda, x)$. This will be addressed further in subsequent chapters.

The Process 4 contribution can, in principle, also be amplified by Auger exciton multiplication in a manner analogous to the single-carrier Auger generation mechanism corresponding to process (a) of Figure 2.18. That is, an energetic exciton can relax in energy to a lower state by creating a second exciton.[25] This process can occur in semiconductor nanoparticles (quantum dots) and is usually termed carrier multiplication.[25] Strictly speaking, the carrier population has not been multiplied until the excitons dissociate.

2.3 TRANSPORT
2.3.1 Transport processes in bulk solids

Thinking of semiconductors as being composed of a bulk and interface (e.g., contact and grain boundary) portion allows us to address transport systematically in these regions. We undertake the modeling of bulk-region conduction band and valence band transport in this book using the concepts of drift and diffusion. The assumptions behind, and the development of, this drift-diffusion formalism are presented in detail in Appendix D. The key transport equations that come out of the drift-diffusion approach of Appendix D are summarized here for convenience. These equations are rigorously valid for single-crystal solids. In cases of multicrystalline and microcrystalline inorganic solids, the drift-diffusion model is strictly valid within crystals so long as the scattering length is less than the characteristic dimension of the crystal. In polycrystalline materials, intragrain drift-diffusion transport may be in series with injection, recombination, or tunneling processes at grain boundaries. For nanocrystalline materials, a drift-diffusion model can suffice by using an effective mobility, which depends on grain size. The drift-diffusion approach also works well for amorphous inorganic and amorphous and crystalline organic materials. In the case of amorphous materials, there can also be transport via the gap states in parallel with conduction band and valence band transport. For organic materials, transport between molecules controlled by tunneling mechanisms can dominate in some cases. For nanoparticles, transport is expected to be interface dominated and also to depend on the matrix in which the particles are imbedded. In the latter two situations, carriers may have to percolate through a solid as they search for optimum interface tunneling paths.

2.3.1.1 BULK REGION CONDUCTION BAND TRANSPORT

As established in Appendix D, the general expression for n(x), the free electron population per volume in the conduction band at some point in a material system, which is valid even when out of TE, is given by

$$n = \int_{E_c}^{\infty} \left[\frac{1 + \exp(E - E_{Fn})}{kT_n} \right]^{-1} A_c (E - E_c)^{1/2} \, dE \qquad (2.16)$$

In this expression, E_{Fn} is the spatially varying electron quasi-Fermi level, T_n is the spatially varying electron temperature, and $A_c(E - E_c)^{1/2}$ is a model for the conduction band density of states $g_c(E)$. The quantity A_c may vary with position. This parabolic model for the conduction band density of states is rigorously valid for crystalline and polycrystalline materials and works well for amorphous inorganics, too. A Gaussian distribution is probably more appropriate for organics. Equation 2.16 may be simplified to

$$n = N_C \exp \left[\frac{-(E_c - E_{Fn})}{kT_n} \right] = N_C \exp \left[\frac{-V_n}{kT_n} \right] \qquad (2.17)$$

if the electron quasi-Fermi level E_{Fn} lies at least several kT_n below E_c. In Eq. 2.17, N_C is the conduction band effective density of states and the quantity V_n locates the conduction band edge with respect to the electron quasi-Fermi level at point x (see Appendix D). When the criterion for the use of Eq. 2.17 is fulfilled, Boltzmann statistics, rather than Fermi-Dirac, is being used to determine the conduction band population.

It follows from Appendix D that the conventional current density J_n being carried by these n electrons per volume at point x can be written as

$$J_n = e\mu_n n \frac{dE_{Fn}}{dx} - e n \mu_n S_n \frac{dT_n}{dx} \qquad (2.18)$$

where the material parameters μ_n and S_n are the electron mobility and Seebeck coefficient, respectively. In Eq. 2.18 both drift and diffusion are built into the quasi-Fermi level gradient term. As also shown in Appendix D, Eq. 2.18 can be rewritten as

$$J_n = e\mu_n n \left(\xi - \frac{d\chi}{dx} - kT_n \frac{d \ln N_C}{dx} \right) + eD_n \frac{dn}{dx} + eD_n^T \frac{dT_n}{dx} \qquad (2.19a)$$

or alternatively as

$$J_n = e\mu_n n\xi + e\mu_n n\xi'_n + eD_n \frac{dn}{dx} + eD_n^T \frac{dT_n}{dx} \qquad (2.19b)$$

Equations 2.19a and b explicitly use the parabolic density of states model but have equivalent forms for other density of states models. Equation 2.19b clearly shows the drift and diffusion components. The electron diffusion coefficient D_n and the electron thermal diffusion coefficient (or Soret coefficient) D_n^T have been introduced into Eqs. 2.19a and b, where, as defined in Appendix D,

$$D_n = kT_n\mu_n \qquad (2.20)$$

and

$$D_n^T = \left[\frac{\mu_n n(V_n + S_n T_n)}{T_n}\right] \qquad (2.21)$$

It is important to note from Eqs. 2.19a and b that the ξ'_n, electron effective force field, and ξ, the electrostatic field, play the same role in electron transport, where the electron effective force field is seen to be defined by

$$\xi'_n = -\left(\frac{d\chi}{dx} + kT_n \frac{d\ln N_C}{dx}\right) \qquad (2.22)$$

This quantity ξ'_n arises from changes in electron affinity χ and from changes in conduction band effective density of states N_C with position. The quantity ξ arises, of course, from electric charge density. This realization of the identical roles played by ξ'_n and ξ in causing electron drift currents is very important to the understanding of photovoltaic action.[26] From the perspective of Eq. 2.19, the electron conventional current density J_n can be viewed as resulting from thermal diffusion, diffusion, and drift, with the latter arising from a total force on an electron given by

$$F_e = -e\left(\xi - \frac{d\chi}{dx} - kT_n \frac{d\ln N_C}{dx}\right) \qquad (2.23)$$

The Eq. 2.19 formulations are seen to reduce to the usual[3]

$$J_n = e\mu_n n\xi + eD_n \frac{dn}{dx} \qquad (2.24)$$

if the material does not have a varying electron affinity χ or varying conduction band effective density of states N_C and if there is no electron temperature gradient. As discussed in Chapters 4–7, in solar cell structures there often are abruptly stepped or graded material regions and abruptly stepped or graded heterojunctions. In those cases, the models for J_n given by Eqs. 2.19a and b are required to capture the device physics. As noted earlier, although valid for single-crystal or amorphous regions, the equations of this section can also work well for inorganic and organic multicrystalline and microcrystalline solids by accounting for the interface transport or by using effective mobilities. We will use Eq. 2.19 in our device analyses in Chapters 4–7 but assume the electron temperature can be characterized with a system temperature T having negligible variation with position.

2.3.1.2 BULK REGION VALENCE BAND TRANSPORT

The general expression for p(x), the free hole population per volume in the valence band at some point in a material system, which is valid even when out of TE, is given by

$$p = \int_{-\infty_c}^{E_v} \left[1 + \exp \frac{-(E - E_{Fp})}{kT_p} \right]^{-1} A_V (E_V - E)^{1/2}\, dE \qquad (2.25)$$

In this expression, E_{Fp} is the spatially varying hole quasi-Fermi level, T_p is the spatially varying hole temperature, and $A_V(E_V - E)^{1/2}$ is a model for the valence band density of states and its prefactor may vary with position. This parabolic model for the valence band density of states is rigorously valid for crystalline and polycrystalline materials and works well for amorphous inorganics, too. A Gaussian distribution can be utilized instead for organics. Eq. 2.25 may be simplified to

$$p = N_V \exp \left[\frac{-(E_{Fp} - E_V)}{kT_p} \right] = N_V \exp \left[\frac{-V_p}{kT_p} \right] \qquad (2.26)$$

if the hole quasi-Fermi level E_{Fp} lies at least several kT_p above E_V. In Eq. 2.26, N_V is the valence band effective density of states and the quantity V_p locates the hole quasi-Fermi level with respect to E_V (see Appendix D). When the criterion for the use of Eq. 2.26 is fulfilled, the implication is that Boltzmann statistics are valid for determining the valence band population.

The results from Appendix D for the conventional current density J_p for holes are analogous to those obtained for the free electrons of the conduction band. To be specific, in the presence of a spatially varying hole quasi-Fermi level E_{Fp} and spatially varying hole temperature T_p, the hole conventional current density J_p can be written as

$$J_p = e\mu_p p \frac{dE_{Fp}}{dx} - e p \mu_p S_p \frac{dT_p}{dx} \qquad (2.27)$$

where the material parameters μ_p and S_p are the hole mobility and Seebeck coefficient, respectively. Equation 2.27 includes both hole drift and diffusion. To show this explicitly, Eq. 2.27 is rewritten in Appendix D as

$$J_p = e\mu_p p \left(\xi - \frac{d(\chi + E_G)}{dx} + kT_p \frac{d\ln N_V}{dx} \right) - eD_p \frac{dp}{dx} - eD_p^T \frac{dT_p}{dx}$$

$$(2.28a)$$

or alternatively as

$$J_p = e\mu_p p \xi + e\mu_p p \xi'_p - eD_p \frac{dp}{dx} - eD_p^T \frac{dT_p}{dx} \qquad (2.28b)$$

Equations 2.28a and b use the parabolic density of states model but have equivalent forms for other density of states models. The hole diffusion coefficient D_p and the hole thermal diffusion coefficient (or Soret coefficient) D_p^T have been introduced into Eqs. 2.28a and b where, as defined in Appendix D,

$$D_p = kT_p \mu_p \qquad (2.29)$$

and

$$D_p^T = \left[\frac{\mu_p p(S_p T_p - V_p)}{T_p} \right] \qquad (2.30)$$

In Eqs. 2.28a and b the quantity ξ is the electrostatic field but ξ'_p is seen to be the hole effective force field defined by

$$\xi'_p = -\left(\frac{d(\chi + E_G)}{dx} \right) - kT_p \frac{d\ln N_V}{dx} \qquad (2.31)$$

The hole effective force field ξ'_p arises from changes in hole affinity $\chi + E_G$ and from changes in valence band effective density of states N_V, as discussed in detail in Appendix D. It is important to note from Eqs. 2.28a and b that the ξ'_p, hole effective force field, and ξ, the electrostatic force field, play the same role in hole transport.[26] From the perspective of Eq. 2.28, the hole conventional current density J_p can be viewed as resulting from thermal diffusion, diffusion, and drift, with the latter arising from a total hole force given by

$$F_h = e\left(\xi - \frac{d(\chi + E_G)}{dx} + kT_p\frac{d\ln N_V}{dx}\right) \qquad (2.32)$$

Interestingly, it can be seen from a comparison of Eqs. 2.22 and 2.31 that the electron and hole effective force fields can be very different. It can also be seen from Eqs. 2.19b and 2.28b that the electrostatic field causes the electron and hole conventional currents to flow in the same direction, which means it causes electrons and holes to move in opposite directions. Comparison of these same two equations further shows that properly designed changes in electron affinity and band density of states and hole affinity and band density of states can also cause electrons and holes to move in opposite directions. Put succinctly, both the electric field and effective fields can be used to break symmetry in a solar cell structure. Both are equally able to give rise to photovoltaic action.[26] This is a very important point. If one is willing to accept that electrostatic and effective fields need not both be present in a cell and to generalize these ideas to all sorts of solar cell structures, from p–n homojunctions to dye-sensitized solar cells, then a very general picture of photovoltaic action emerges: electrostatic fields, effective fields, or both may be used to break symmetry and to give rise to photovoltaic action. This perspective is the path we follow in this book.

The expressions of Eq. 2.24 are seen to reduce to the usual[3]

$$J_p = e\mu_p p\xi - eD_p\frac{dp}{dx} \qquad (2.33)$$

if the material does not have a varying hole affinity $\chi + E_G$ or varying valence band effective density of states N_V and if there is no hole temperature gradient. As we will see, there often are graded material regions and graded heterojunctions in solar cell structures, necessitating the use of Eq. 2.28 to capture all the device physics taking place. Although valid for single-crystal or amorphous regions, the hole transport equations of this section can also work well for inorganic and organic multicrystalline and microcrystalline solids by accounting for the interface transport or by using effective mobilities. We will use Eq. 2.28 in our device analyses in Chapters 4–7 but henceforth assume there is one temperature T for electrons and holes and its variation with position is negligible.

2.3.1.3 AMORPHOUS MATERIALS

Transport in amorphous solids is complex because current can be carried not only by electrons and holes in the conduction and valence bands, respectively, but also by carriers in gap states; i.e., the gap states may be of a high enough density that they too can support transport in some amorphous materials. To acknowledge this, the current density J must be written, in general, as

$$J = J_n + J_p + \sum_i J_{Gi} \tag{2.34}$$

Here J_n is the conduction-band contribution (from the delocalized states above the mobility gap in Fig. 2.10), J_p is the valence-band contribution (from delocalized states below the mobility gap in Fig. 2.10), and J_{Gi} is the current density from the ith group of gap states. In the case of J_{Gi}, the motion of electrons in gap states may often be best described in terms of their hopping from one site to another. Hopping involves tunneling, but, since these sites may not all be at the same energy, hopping is a phonon-assisted tunneling process. Every time the localized electron moves, it may emit or absorb a phonon. This makes hopping a thermally activated process and causes the mobility μ_G to be of the form

$$\mu_{Gi} = \mu_{Gi0} e^{-W(E)/kT} \tag{2.35}$$

where the quantity W is the activation energy.[27,28] Since the mobilities of the delocalized states are expected to be orders of magnitude larger than those of localized states (the mobility gap), and since, among the localized states, mobilities can vary by orders of magnitude due to the $e^{-W(E)/kT}$

factor, the conductivity of amorphous materials can be tremendously affected by shifts in populations. For example, if light appreciably shifts the carrier population into localized states with large hopping mobilities or into band states, then large changes in the conductivity will result, and, in fact because of this, many amorphous semiconductors, both organic and inorganic, are strong photoconductors.

2.3.2 Transport processes at interfaces

There are a number of processes unique to interfaces that can allow carriers to cross the boundary between two materials or between two grains. These processes are in series with the bulk transport mechanisms just discussed. In this section we focus on these interface transport mechanisms. A metal-semiconductor structure in forward bias is used in Figure 2.19 to illustrate them. From the figure it is seen that mechanisms a to e involve the semiconductor majority carriers, while mechanism f involves both majority and minority carriers. Mechanism g involves only minority carriers. Although here we discuss these mechanisms in the context of a metal-semiconductor contact, the general features of these various interface transport mechanisms will also be found at other interfaces.

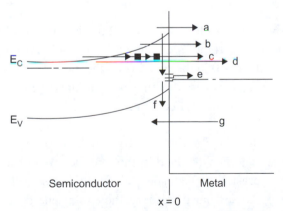

FIGURE 2.19 Interface transport mechanisms illustrated using a forward-biased, metal-semiconductor junction. Path a is thermionic emission, path b is thermally enhanced field emission, path c is multistep tunneling, path d is field emission, path e involves trapping and subsequent emission, path f is interface recombination, and path g is minority-carrier injection.

2.3.2.1 THERMIONIC EMISSION

Thermionic emission is a classical process (no tunneling) by which a carrier transfers from an allowed state in one material to an allowed state in another material with ideally no change in total energy. The general model, valid for semiconductor-semiconductor or metal-semiconductor interfaces, for the net current density J_{OB} coming over the barrier at an interface, may be written for electrons, for example, as.[29]

$$J_{OB} = -A^* \, T^2 e^{-\phi_B/kT} [e^{E_{Fn}(0^-)/kT} - e^{E_{Fn}(0^+)/kT}] \qquad (2.36)$$

Here ϕ_B is the barrier height, $E_{Fn}(0^-)$ is the shift in the quasi-Fermi level position at $x = 0^-$, and $E_{Fn}(0^+)$ is the shift in the quasi-Fermi level position at $x = 0^+$ with current flowing. These are measured positively up from the Fermi level TE position at the interface. This expression once again shows the current-driving role of quasi-Fermi levels first seen in Eqs. 2.18 and 2.27.

In the case of our example interface, the metal-semiconductor structure of Figure 2.19, this mathematical model for path a may be expressed as

$$J_{OB} = -A^* \, T^2 e^{-\phi_B/kT} [e^{V/kT} - 1] \qquad (2.37)$$

The minus sign arises in Eqs. 2.36 and 2.37 due to our taking net electron emission to the right as a negative conventional current. In this expression, $E_{Fn}(0^+) = 0$ in the metal (i.e., the electron quasi-Fermi level is at the Fermi level) because the high population of carriers in the metal is not disturbed by the flow of current.

In these expressions, A^* is the effective Richardson constant[30]; it is a function of the materials involved in the interface. For example, in crystalline and polycrystalline materials, electrons crossing the interface in conservative processes must preserve their total energy as well as the component of their **k**-vector that lies in the plane of the junction. The number of electrons in a materials system that are able to do this will depend on the $E = E(\mathbf{k})$ functions of the materials involved, and this is reflected in the value of A*. There can be materials systems where electrons must interact with phonons to adjust their **k**-vectors so they can fit into the E = E(**k**) of their new host. In so doing, their energy is changed. For amorphous materials, this **k** conservation is relaxed, since **k** is not a "good quantum number." The effective A* must incorporate these variations from materials system to materials system.[29,30]

2.3.2.2 THERMALLY ENHANCED FIELD EMISSION

Process b of Figure 2.19 is thermally enhanced field emission or thermionic field emission. Thermally enhanced field emission is a direct tunneling process by which an electron is transferred from an allowed state in one material to an allowed state in another. It differs from field emission, which is also a direct tunneling process, in that field emission (path d) is not thermally assisted.[30] Except for the higher semiconductor doping levels and lower temperatures, thermionic field emission is not expected to dominate over the parallel path of thermionic emission at semiconductor-metal interfaces.[30]

2.3.2.3 MULTISTEP TUNNELING

Process c of Figure 2.19 is referred to as multistep tunneling. Such processes are what is termed indirect tunneling and may or may not be conservative, although the particular example shown is. Multistep tunneling may be thought of as the interface analogue of the bulk transport process of hopping discussed in Section 2.3.1.3. Since this path does not involve direct tunneling but rather tunneling from one defect to another in the barrier region, it can involve phonons and occur for a range of barrier thicknesses and doping levels. This transport process has been proposed to have a forward J-V characteristic of the form

$$J_{MS} = -J_0 e^{BT} e^{AV} \qquad (2.38)$$

where A and B are constants.[31] Here V is the band-bending change in the semiconductor barrier region for the forward bias direction. The quantity $J_0 e^{BT}$ multiplying the voltage term in Eq. 2.38 is seen to give, when plotted versus T, linear dependence on T. This is quite different from the temperature-activated forms of the voltage independent terms in Eqs. 2.36 and 2.37.

2.3.2.4 FIELD EMISSION

Process d of Figure 2.19 is pure field emission; i.e., direct tunneling through the semiconductor barrier by majority carriers from the bottom of the band. The field emission current density J_{FE} has a very strong dependence on any changes that take place in the semiconductor band bending, since this modifies the barrier shape and therefore the tunneling probability. Since path d necessitates direct tunneling through the semiconductor barrier at its widest, it is appreciable only for very high doping levels.[30]

2.3.2.5 TRAPPING AND SUBSEQUENT EMISSION

Path e requires that an electron in the conduction band at $x = 0$ in our example interface of Figure 2.19 be trapped by a localized state at or near the interface and then be subsequently emitted into the metal. In general, such processes depend on the population of carriers in the initial states, on the population of carriers in the intermediary states, and on the population of carriers in the final states, as well as on capture cross-sections.

2.3.2.6 INTERFACE RECOMBINATION

Path f of Figure 2.19 is trap-assisted interface recombination. The current density J_{IR} flowing in the conduction band at $x = 0^-$ due to this process is obtained by recasting Eq. 2.9 into the form

$$ J_{IR} = \frac{ev\sigma_n\sigma_p N_I \left[n(0^-)p(0^-) - n_i^2 \right]}{\sigma_p \left[p(0^-) + p_1 \right] + \sigma_n \left[(n(0^-) + n_1 \right]} \tag{2.39} $$

In this expression N_I is the density, per area of interface, of the trap states at energy E. If there is a distribution of gap states at an interface, then this expression must be summed over this distribution.

2.3.2.7 MINORITY CARRIER INJECTION

Process g seen in the forward-biased metal-semiconductor interface of Figure 2.19 is minority carrier injection. In the case of this figure, holes are generated in the valence band (by electrons moving to the right into the metal) at the metallurgical junction $x = 0$. These holes supply path f and any remaining holes move to the left into the bulk of the semiconductor by drift and diffusion. In many interfaces similar to our example, path g is able to provide as many holes as are required by path f and by the bulk. In such a situation, the quasi-Fermi level for holes $E_{Fp}(0^-)$ must equal $E_{Fp}(0^+)$. If path g were not able to supply the holes required by path f and the bulk, then $E_{Fp}(0^-)$ would lie above $E_{Fp}(0^+)$ in energy.

2.3.2.8 THE SURFACE RECOMBINATION SPEED MODEL

Often all these various transport possibilities at an interface are modeled by using the following two expressions,[3] which are linear in the carrier population:

$$J_n(x) = \pm e\, S_n\, [n(x) - n_0(x)] \qquad (2.40a)$$

and

$$J_p(x) = \pm e\, S_p\, [p(x) - p_0(x)] \qquad (2.40b)$$

The $n_0(x)$ and $p_0(x)$ in these expressions are the TE values of the free carrier populations at point x which is just to the right (interface to the left) or left (interface to the right) of the interface. The parameters S_n and S_p characterize this interface and are called surface recombination speeds. The sign choices seen in these equations depend on whether the interface is to the left or right of x as seen in Figure 5.20.

Equation 2.37 can be rigorously put into this from by rewriting it as

$$J_{OB} = -A^*\, T^2 e^{-\phi_B/kT}\, \frac{N_C}{N_C}\, (e^{V/kT} - 1) = -\frac{A^*T^2}{N_C}(n - n_0) \qquad (2.40c)$$

which can be used to show that a surface recombination speed of $\sim 10^7$ cm/s adequately represents thermionic emission. Equation 2.39 can be put into the form of Eq. 2.40 as well by using the linearization assumptions discussed in Section 2.2.5.1.

2.3.2.9 GENERAL COMMENT

Figure 2.20 shows a very general solar cell structure in which electrons collected from the right contact (cathode) traverse the external circuit and do

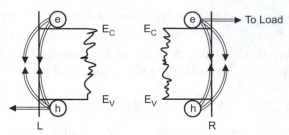

FIGURE 2.20 Contacts to a very general solar cell structure. Electrons that survive the loss mechanisms at R, all of which are due to holes at contact R, go off to the external circuit to do work at the load. Holes are collected at interface L in numbers that match electrons arriving from the load. The electrons at contact L are the source of loss there. In the surface recombination picture of Eq. 2.40, S_n characterizes the three electron paths and S_p the two hole paths for the right contact. An S_n used in Eq. 2.40 characterizes the two electron paths and an S_p the three hole paths for the left contact.

work at the load when the device is in the power quadrant. The number of electrons available for work in the external circuit is equal to the total number of arriving electrons at this cathode minus the arriving electrons lost to holes, as seen in the figure. Since some of the electrons recombine with holes at the contact surface or are thermionically emitted and then annihilated by holes—or suffer some other loss mechanism at the contact—we always need to calculate the total J_n and J_p at the contact to get the net J that flows through the external circuit. As seen from Figure 2.20, the algebraic sum of $J_n + J_p$ evaluated at R subtracts out the electrons lost to holes and gives the number of electrons entering the external circuit. The analogous situation happens for holes at contact L (the anode). The number available to meet the electrons arriving from the load is equal to the arriving hole number minus those lost to electrons at L. Of course, the solar cell must arrange its carrier populations, electric field distribution and current flows in steady state to make sure all of this happens.

2.3.3 Continuity concept

The continuity concept is a very powerful part of our device physics toolbox. It is accounting applied to electrons and holes. For example, considering free electrons first, the free electron conventional current density $J_n(x_1)$ at some point x_1 and the free electron conventional current density at some point x_2 (where $x_2 > x_1$) must be equal in steady state unless free electrons are (1) gained by photogeneration or (2) lost to net recombination in the region x_1 to x_2. Expressed mathematically, this means

$$\frac{J_n(x_1)}{e} = \frac{J_n(x_2)}{e} + \int\limits_{x_1}^{x_2}\int\limits_\lambda G_{ph}(\lambda,x)\,dx\,d\lambda - \int\limits_{x_1}^{x_2}(\mathscr{R}^R + \mathscr{R}^L + \mathscr{R}^A)\,dx$$

(2.41)

where the necessity of converting conventional electron current density J_n into electron particle density $-J_n/e$ has been taken into consideration. The definitions of Section 2.2.5 and Section 2.2.6 have been used in Eq. 2.41. This equation is the steady-state integral form of the continuity concept for free electrons. If we wanted the general time-dependent version, we would have to add a term accounting for the possible build-up of electrons in the region x_1 to x_2 and terms accounting free electrons arising from trapping and de-trapping traffic into and out of the conduction band. Fortunately, we are generally interested in steady-state behavior in solar cells. Equation 2.41 can also be expressed in a steady-state differential form:

$$\frac{\partial J_n/\partial x}{e} = -\int\limits_\lambda G_{ph}(\lambda,x)\,d\lambda + \mathscr{R}^R + \mathscr{R}^L + \mathscr{R}^A$$

(2.42)

The continuity concept applied to free holes in the valence band for the same region x_1 to x_2 gives

$$\frac{J_p(x_2)}{e} = \frac{J_p(x_1)}{e} + \int\limits_{x_1}^{x_2}\int\limits_\lambda G_{ph}(\lambda,x)\,dx\,d\lambda - \int\limits_{x_1}^{x_2}(\mathscr{R}^R + \mathscr{R}^L + \mathscr{R}^A)\,dx$$

(2.43)

The definitions of Section 2.2.5 and Section 2.2.6 have again been used. This equation is the steady-state integral form of the continuity concept for holes. If we wanted the general time-dependent version, we would have to add a term accounting for the build-up of holes in the region x_1 to x_2 and a term accounting for free holes arising from trapping and detrapping. Equation 2.43 can also be expressed in differential form:

$$\frac{\partial J_p/\partial x}{e} = \int\limits_\lambda G_{ph}(\lambda,x)\,d\lambda - (\mathscr{R}^R + \mathscr{R}^L + \mathscr{R}^A)$$

(2.44)

2.3.4 Electrostatics

A solar cell has a lot of things going on inside: generation, recombination, currents flowing, and perhaps trapped charge. As demonstrated by Eqs. 2.19 and 2.28, effective fields and the electric field in the cell are involved in the current flow. The electric field is in turn modified by the currents flowing and the charge in the delocalized states, traps and recombination centers. All of this activity is included in Poisson's equation, which may be written as

$$\frac{\partial(\varepsilon\xi)}{\partial x} = e\left[p - n + \sum p_T - \sum n_T + N_D^+ - N_A^-\right] \qquad (2.45)$$

where ε is the permittivity, which may be a function of position. The right-hand side of this equation is called the space charge or the charge density. The terms in the large parentheses in Eq. 2.45 account for the charge per volume due to free carriers and for charge per volume in localized states (dopants, recombination centers, and traps). As seen, the two summations used in the expression are needed to account for all the positive charge and all the negative charge in defect gap (recombination centers and traps) states at different energies in the gap. As also seen, dopants are treated separately since they are purposefully introduced. One donor and one acceptor dopant level have been used for definitiveness.

2.4 THE MATHEMATICAL SYSTEM

Collecting all our previous work, we can now write down the full mathematical system that describes all the physics taking place in a solar cell in steady state. This mathematical system is made up of, unfortunately, a large set of coupled, non-linear equations. This set includes Eqs. 2.19 and 2.28, the equations of Section 2.2.5, the continuity equations for electrons and holes, and Poisson's equation. The solution to this set for n, p, J_n, J_p, etc., must be consistent with the boundary conditions imposed by the interface transport discussed in Section 2.3.2. We have to obtain this solution in order to establish the cell's current density-voltage (J-V) characteristic (see Fig. 1.3), which is required for cell evaluation, design, and optimization. Since we have been staying with one dimension, this total conventional current density J of the J-V characteristic is a constant and is obtained from

$$J = J_n + J_p \qquad (2.46)$$

where the components J_n and J_p are evaluated at the same plane (any plane) in the cell. The terminal voltage V produced by the cell at some operating point while delivering the current density J is given by the integral over the structure of the difference between the electric field distribution present at the selected operating point $\xi(x)$ and the electric field distribution present in TE which we term $\xi_0(x)$; i.e.,

$$V = \int_{\text{structure}} [\xi(x) - \xi_0(x)]dx \qquad (2.47)$$

It must be stressed that this expression is valid regardless of whether the cell has an exciton-producing absorber or free electron-hole–producing absorber or is any cell structure from a dye sensitized solar cell to a conventional p–n junction device. Put succinctly, it is valid for all types of cells. Equation 2.47 is of such general validity because it calculates the relative shift of the contact Fermi levels present at the selected operating point. This shift is the voltage V at this operating point. The sign convention assumed in writing Eq. 2.47 is that the cathode is the left-hand contact.

As we will see, establishing the current densities J_n and J_p—and therefore the J-V characteristic—for a solar cell can be done analytically in some situations, with the help of assumptions. We will explore this approach in Chapters 4–7. However, turning a computer loose to tackle the full mathematical system, and not making any of those assumptions, is very powerful and insightful and we will do that too in Chapters 4–7. In the computer analysis, we will make extensive use of numerical solutions to the full mathematical system. In this numerical modeling, the specific versions of the equations of the mathematical system that will be utilized are the following:

$$J_n = e\mu_n n\left(\xi - \frac{d\chi}{dx} - kT\frac{d\ln N_C}{dx}\right) + eD_n\frac{dn}{dx} \qquad (2.48a)$$

$$J_p = e\mu_p p\left(\xi - \frac{d(\chi + E_G)}{dx} + kT\frac{d\ln N_V}{dx}\right) - eD_p\frac{dp}{dx} \qquad (2.48b)$$

$$\frac{\partial J_n/\partial x}{e} = -\int_\lambda G_{ph}(\lambda,x)\,d\lambda + \mathcal{R} \qquad (2.48c)$$

$$\frac{\partial J_p/\partial x}{e} = \int_\lambda G_{ph}(\lambda, x)\, d\lambda - \mathscr{R} \tag{2.48d}$$

$$\frac{\partial(\varepsilon\xi)}{\partial x} = e\left(p - n + \Sigma p_T - \Sigma n_T + N_D^+ - N_A^-\right) \tag{2.48e}$$

In these equations, $\mathscr{R}$ can be any of the unimolecular or bimolecular recombination mechanisms of Section 2.2.5. In our numerical modeling we employ the full non-linearized formulations discussed in Section 2.2.5 for whichever $\mathscr{R}$ is assumed. For example, we will often assume $\mathscr{R}$ is controlled by S-R-H recombination and use

$$\mathscr{R}^L = \frac{v\sigma_n\sigma_p N_T(np - n_i^2)}{\sigma_p(p + p_1) + \sigma_n(n + n_1)} \tag{2.48f}$$

summed over some gap state distribution. For Eq. 2.48e the trapped charges n_T and p_T at energy E as well as the ionized dopant concentrations N_A^- and N_D^+ are computed using Fermi-Dirac statistics, as discussed in Appendix C. As seen in that Appendix, these quantities become functions of n and p in steady state. Keeping that in mind, we see that set 2.48 has 5 equations and 5 unknowns.

The quantity $G_{ph}(\lambda, x)$ in this set is the photogeneration function of Section 2.2.6. In the case of free electron-hole pair–caused absorption, it is modeled in the numerical simulations by using $\alpha(\lambda)I_0(\lambda)e^{-\alpha(\lambda)x}$ (see Eq. 2.2); i.e., the absorption mechanism is taken to follow the Beer-Lambert law. In the case of exciton-caused absorption, $G_{ph}(\lambda, x)$ is modeled by a delta-function-like distributions at the region of exciton dissociation. This point is discussed further in Chapter 5.

In the numerical results presented in this text, the computer solves the equations subject to the boundary conditions

$$J_n(L) = eS_n[n(L) - n_0(L)] \tag{2.48g}$$

and

$$J_p(L) = -eS_p\,[p(L) - p_0(L)] \qquad (2.48h)$$

for the left (L) contact to the cell structure and subject to the boundary conditions

$$J_n(R) = -eS_n\,[n(R) - n_0(R)] \qquad (2.48i)$$

and

$$J_p(R) = eS_p\,[p(R) - p_0(R)] \qquad (2.48j)$$

for the right (R) contact. These are taken from Eq. 2.40 and assume a coordinate system with x increasing from left to right. As we establish in Chapter 4, these boundary conditions can be made inconsequential by various layers we can place adjacent to contacts. By simultaneously solving this whole set of equations numerically, the computer then assembles the J-V characteristics for no illumination (dark J-V) and illumination (light J-V) situations in our discussions of Chapter 4–7. When the results are tied to the device coordinate system, the conventional current density is taken as positive when it flows in the positive x-direction of the device coordinates. Except where noted, voltage is taken as positive if the Fermi level of the right contact is above that of the left contact. In discussions tied to plots of J versus V, J is taken as negative in the power quadrant. The resulting numerical solutions give us the J-V behavior but also allow us to peer inside a cell as it is operating and thereby allow us to explore the roles of diffusion, electrostatic fields, effective fields, drift, recombination, trapping, interfaces, etc.

2.5 ORIGINS OF PHOTOVOLTAIC ACTION

As discussed earlier, Eqs. 2.19 and 2.28 give an understanding of what can make J occur in a solar cell. These equations show that the answer lies in (1) the presence of an electric field, (2) the presence of effective force fields (changes with position in affinities and band effective densities of states), (3) diffusion, and (4), if there is temperature gradient, thermal diffusion. If we neglect temperature gradients, which can be set up under the presence of light, there are three possible causes of photovoltaic action. As discussed in Chapter 3, there really are only two important causes: the two terms in these equations that make one direction different than the other; i.e., the terms that tend to push electrons one way and holes the other, thereby separating carriers. These symmetry breaking terms are the electric field and effective force field terms.

REFERENCES

1. C. Kittel, Solid State Physics, eighth ed., John Wiley & Sons, New York, 2005, pp. 3–22, 89–102, 164–218, 420–435.

2. T. Kamins, Polycrystalline Silicon for Integrated Circuit Applications, Kluwer Academic Publishers, Boston, MA, 1988, pp. 92–96, 178–179.

3. S. Sze, K.K. Ng, Physics of Semiconductor Devices, third ed., John Wiley & Sons, Hoboken, NJ, 2007, pp. 7–56, 246–258.

4. M. Rajalakshmi, A.K. Arora, B.S. Bendre, S. Mahamuni, Optical phonon confinement in zinc oxide nanoparticles, J. Appl. Phys. 87 (2000) 2445.

5. H. Richter, Z.P. Wang, L. Ley, The one phonon Raman spectrum in microcrystalline silicon, Solid State Commun. 39 (5) (August 1981) 625–629.

6. A. Kaan Kalkan, S.J. Fonash, Control of Enhanced Absorption in Poly-Si, Proceedings of the Spring Materials Research Society Meeting, Amorphous and Microcrystalline Silicon Technology, Materials Research Society, vol. 467:415, 1997.

7. M. Born, R. Oppenheimer, Zur Quantentheorie der Molekeln., Ann. Phys. 84 (1927) 457–484.

8. A.D. Yoffe, Low-dimensional systems: quantum size effects and electronic properties of semiconductor microcrystallites (zero-dimensional systems) and some quasi-two-dimensional systems, Adv. Phys. 51 (2002) 799. G.

9. J.J.M. Halls, J. Cornill, D.A. dos Santos, R. Silbey, D.-H. Hwang, A.B. Holmes, J.L. Brebas, R.H. Friend, Charge- and energy-transfer processes at polymer/polymer interfaces: A joint experimental and theoretical study, Phys. Rev. B 60 (1999) 5721.

10. E.A. Davis, in: P.G. LeComber, J. Mort (Eds.), Electronic and Structural Properties of Amorphous Semiconductors, Academic Press, New York, 1973.

11. N.-K. Persson, in: O. Inganas, S.-S. Sun, N.S. Sariciftci (Eds.), Organic Photovoltaics, CRC Press, Boca Raton, FL, 2005, pp. 114–129.

12. J.I. Pankove, Optical Processes in Semiconductors, Prentice-Hall, Englewood Cliffs, NJ, 1971, p. 29.

13. D. Zhou, R. Biswas, Photonic crystal enhanced light trapping in solar cells, J. Appl. Phys. 103 (2008) 093102.

14. E. Yablonovich, Inhibited spontaneous emission in solid-state physics and electronics, Phys. Rev. Lett., 58, 2059 (1987); S. John, Strong localization of photons in certain disordered dielectric superlattices, Phys. Rev. Lett., 58, 2486 (1987).

15. J.D. Joannopoulos, P.R. Villeneuve, S. Fan, Photonic crystals: putting a new twist on light, Nature 386 (1997) 143.

16. Z. Fan, H. Razavi, J. Do, A. Moriwaki, O. Ergen, Y.L. Chueh, P.W. Leu, J.C. Ho, T. Takahashi, L.A. Reichertz, S. Neale, K. Yu, M. Wu, J.A. Ager, A. Javey, Three-dimensional nanopillar-array photovoltaics on low-cost and flexible substrates, Nat. Mater. 8 (2009) 648–653.

17. H.R. Stuart, D.G. Hall, Absorption enhancement in silicon-on-insulator waveguides using metal island films, App. Phys. Lett., 69, 2327 (1996); H.R. Stuart, D.G. Hall, Enhanced dipole-dipole interaction between elementary radiators near a surface, Phys. Rev. Lett., 80, 5663 (1998).

18. A. Kaan Kalkan, S.J. Fonash, Laser-activated surface-enhanced Raman scattering substrates capable of single molecule detection, Appl. Phys. Lett. 89, 233103 (2006).

19. J.A. Creighton, D.G. Eadon, Ultra-visible absorption spectra of the colloidal elements, J. Chem. Soc. 87 (1991) Faraday Trans. (24, 3881–3891).

20. S. Forrest, J. Xue, Strategies for solar energy power conversion using thin film organic photovoltaic cells, Conference Record of the Thirty-first IEEE Photovoltaic Specialists Conference, Orlando, FL, January 2005.

21. R. Biswas, D. Zhou, B. Curtin, N. Chakravarty, V. Dalal, Surface plasmon enhancement of optical absorption of thin film a-Si:H Solar cells with metallic nanoparticles, Thirty-fourth Photovoltaic Specialists Conf., Philadelphia, PA, June 2009.

22. P.T. Landsberg, Non-radiative transitions in semiconductors, Phys. Status Solidi. 41, 457 (1970).

23. A. Luque, M. Marti, Increasing the efficiency of ideal solar cells by photon induced transitions at intermediate levels, Phys. Rev. Lett. 78, 5014 (1997).

24. B.A. Gregg, Excitonic solar cells, J. Phys. Chem. B 107, 4688 (2003).

25. R.J. Elingson, M.C. Beard, J.C. Johnson, P. Yu, O.I. Micic, A.J. Nozik, A. Shabaev, A.L. Efros, Highly efficient multiple exciton generation in colloidal PbSe and PbS quantum dots, Nano. Lett. 865 (2005); A. Zunger, A. Franceschetti, J.-W. Luo, V. Popescu, Understanding the physics of carrier-multiplication and intermediate band solar cells based on nanostructures—whats going on? Thirty-fourth IEEE Photovoltaic Specialist Conference, Philadelphia, PA, June, 2009.

26. S.J. Fonash, S. Ashok, An additional source of photovoltage in photoconductive materials, Appl. Phys. Lett., 35, 535 (1979); S.J. Fonash, Photovoltaic devices, CRC Critical Reviews Solid State Mater, 9, 107 (1980).

27. N.F. Mott, E.A. Davis, Electronic Processes in Non-Crystalline Materials, Oxford University Press, London and New York, 1971.

28. L. Friedman, Transport Properties of Organic Semiconductors, Phys. Rev. A 133, 1668 (1964).

29. S.J. Fonash, The role of the interfacial layer in metal–semiconductor solar cells, J. Appl. Phys. 46, 1286 (1975); S.J. Fonash, General formulation of the current-voltage characteristic of a p–n heterojunction solar cell, J. Appl. Phys. 51, 2115 (1980).

30. E.H. Rhoderick, R.H. Williams, Metal-Semiconductor Contacts, second ed., Oxford University Press, New York, 1988.

31. A.R. Riben, D.L. Feucht, Electrical transport in nGe–pGaAs heterojunctions, Int. J. Electron. 20, 583 (1966); A.R. Riben, D.L. Feucht, nGe–pGaAs heterojunctions, Solid-State Electron 9, 1055 (1966).

Chapter Three

Structures, Materials, and Scale

3.1 INTRODUCTION

The key material in a solar cell is the absorber. These materials are capable of absorption-caused excited states produced by photons with energies in the photon-rich range of the solar spectrum (Fig. 1.1). The resulting excited states must be mobile; i.e., free electron-hole pairs, which can be separated, or excitons, which can be disassociated into free electrons and free holes and separated. Absorber materials can be organic or inorganic semiconductors, dye molecules, or quantum dots, which are the man-made inorganic particle equivalent of dye molecules. In some configurations, the absorption and separation are both accomplished in the same

DOI: 10.1016/B978-0-12-374774-7.00003-0

material. In these cases there is a region in the absorber with a built-in electric field designed to break symmetry thereby forcing electrons in one direction and holes in the other. We will term this region a junction. In other configurations a second material is used with the absorber to set up the symmetry-breaking region. In this case the symmetry-breaking may be accomplished with an electric field, effective fields, or both. We will also refer to such regions as junctions.

Besides the absorber and any junction-forming materials, there is a supporting cast of other material components in a solar cell structure. They may include materials that block one carrier while supporting transport of the other to help make one direction of carrier motion different from the other. These materials obviously augment the symmetry-breaking region. Ideally, they are either hole transport–electron blocking layer (HT-EBL) or electron transport–hole blocking layer (ET-HBL) materials. Another important component is the antireflection material. Antireflection materials serve as an optical impedance matching medium; i.e., they are used to couple light efficiently into the solar cell structure. Plasmonic (metals) or photonic (insulators, semiconductors, metals) materials may also be employed as optical components for controlling scattering, reflection, interference, and diffraction thereby dictating the optical electric field strength and thus the absorption distribution in the absorber. Conducting materials provide the ohmic (ideally no voltage drop) contacts of the cell electrodes and the grids needed for carrying the current to the outside world. These materials must produce minimal electrical and optical loss. Contact materials may be metals or transparent conducting oxides (TCOs). The latter type of contact material provides low electrical resistance yet allows the transmission of light into a cell. Finally, protection of the overall device, while not hindering optical coupling, is the function of encapsulating materials.

While the materials system for a solar cell can be involved, the core of the cell operation is the absorber and the symmetry-breaking region, the "charge separation engine." Of course, for power to be drawn from a solar cell, contacts are needed to allow current to leave. The core plus contacts is the simplest solar cell structure. Everything else is added to enhance performance.

We just discussed that the creation of the symmetry-breaking region is done by building in an electric field, effective fields, or both, and no mention was made of diffusion. We will demonstrate in concrete terms in this chapter that diffusion is relatively unimportant in charge separation and photovoltaic action. We will see it can be very important in charge collection. After we establish this point, we will then turn to the question of how solar cell structures are achieved in practice. We will examine the criteria for selecting an absorber, and then will focus on length scales. Our primary questions about scale are: Are there natural length scales that occur in photovoltaic materials and structures and, if so, what are they, and do they lie in the or nano- or microscale regime, or both?

3.2 BASIC STRUCTURES FOR PHOTOVOLTAIC ACTION

We determined in Section 2.5 that the mathematical system of equations describing the physics of solar cells points to (1) the presence of an electric field, (2) the presence of effective force fields (changes with position in affinities or equivalently in the LUMO or HOMO levels and densities of states), and (3) diffusion, as the three possible causes of photovoltaic action. To examine the relative importance of these potential mechanisms, we will look at a series of very basic structures, each designed to focus on one of these possibilities.

3.2.1 General comments on band diagrams

We will construct the band diagrams of these basic structures and use the system of mathematical equations and boundary conditions developed in Chapter 2 to analyze their behavior under illumination. By using the numerical approach to solve the mathematical system, we will not have to make any of the simplifying assumptions usually needed to obtain an analytical solution. Avoiding these assumptions will give us confidence that we have correctly captured all that is taking place.

Before we actually start assembling the band diagrams of the structures we will use to explore the electric field, effective field, and diffusion sources of photovoltaic action, we should first review the rules for constructing band diagrams for multi-component material systems in thermodynamic equilibrium (TE). To do that, we make use of Figure 3.1, which shows two semiconductors in various stages of forming a materials system in TE. The materials are isolated from one another and their band diagrams are lined up in Figure 3.1a with respect to the vacuum

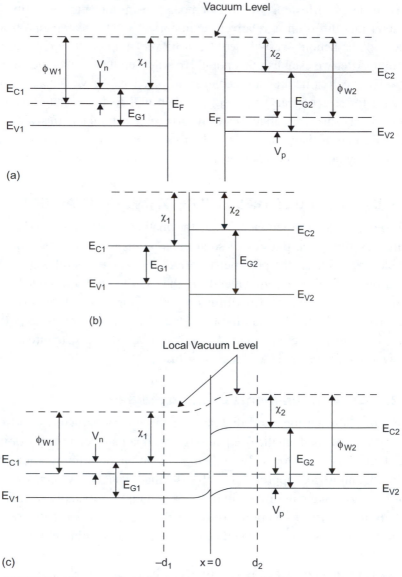

FIGURE 3.1 (a) Two semiconductors prior to contact, (b) the same two semiconductors during contact, and (c) the same two semiconductors after contact, in thermodynamic equilibrium.

(escape) level so that we can see the relative positions of the bands and energy gaps. Figure 3.1a also shows the positions of the Fermi levels in each material—again with respect to the common reference of the vacuum level—using the workfunctions ϕ_{W1} and ϕ_{W2}. As can be seen in

Figure 3.1a, material 1 is n-type and material 2 is p-type. The quantities V_n and V_p also locate the Fermi levels but they do not use a common reference; i.e., V_n locates the Fermi level of material 1 with respect to the conduction band edge of material 1 while V_p locates the Fermi level of material 2 with respect to the valence band edge of material 2. The other notation used in Figure 3.1a is first introduced in Chapter 2. We are using a semiconductor-semiconductor example in Figure 3.1, but the same rules we are about to review apply to metal-semiconductor, etc., combinations.

Figure 3.1b is part of a thought experiment in which we place these two materials together (actually one would be deposited or grown on the other) and watch them begin to come into thermodynamic equilibrium. Of course, TE is that very special situation for which a system of materials has one temperature and one Fermi level. Since they have not come into TE in Figure 3.1b, we do not draw the Fermi levels but, thanks to Figure 3.1a, we can certainly see how the band edges and vacuum level line up on contact. Since we are forming an abrupt interface in this example, there are abrupt steps in the electron affinity χ and hole affinity $E_G + \chi$ at the $x = 0$ position. These steps would be graded if one material were graded into the other.

Figure 3.1c shows the continuation of the thought experiment where we have allowed the materials system to come to TE, with one Fermi level and one temperature. As seen, far to the left of the junction region, the Fermi level has its old position with respect to the local vacuum level, as shown by ϕ_{W1}, and its old position in the band gap of material 1, as shown by V_n. Far to the right of the junction, the Fermi level also has its old position with respect to the local vacuum level, as shown by ϕ_{W2}, and its old position in the band gap of material 2, as shown by V_p. This must be the case, since V_n and V_p are dictated by doping far from a metallurgical junction and the affinities cannot change for a given material. Establishing one Fermi level for the system in TE is accomplished by the creation of an electrostatic potential energy between $x = -d_1$ and $x = d_2$ or, correspondingly, by the creation of a new potential energy component for electrons. Remembering that band diagrams show total energy levels for electrons, we see that this electron potential energy shifts the energy levels of material 2 up with respect to those of material 1 (or shifts the energy levels of material 1 down with respect to those of material 2) just enough to equate the Fermi level across the whole materials system. The electron potential energy being developed between $x = -d_1$ and $x = d_2$ is seen in the band bending in this region; i.e., the valence bands, conduction bands, and local vacuum level are all bending due to the presence of this newly created electron potential energy. The orientation of the electron potential energy difference across the junction, which is typically called the built-in potential, tells us that the materials system must have developed a negative charge on the right side of the interface and thus a positive charge on the left side of the interface. This can be seen by the conduction band's bending toward, and the valence band's bending away from, the Fermi level in material 2 (resulting in a net negative charge), and by the conduction band's bending away from, and the valence band's bending toward, the Fermi level in material 1 (resulting in a net positive charge).

In fact, the first derivative of the local vacuum level $E_{VL}(x)$ with position is the built-in electrostatic field creating the built-in electron potential energy. Through Poisson's equation, the derivative of the permitivity times the local vacuum level derivative is equal to the charge density ρ (or space charge) causing that field[1]; i.e.,

$$\frac{d\left[\varepsilon \dfrac{d}{dx}(E_{VL}(x))\right]}{dx} = \rho \tag{3.1}$$

Using our notation from Eq. 2.45 from Chapter 2 allows us to write, the following:

$$\frac{d\left[\varepsilon \dfrac{d}{dx}(E_{VL}(x))\right]}{dx} = \rho = e\left[p - n + \sum p_T - \sum n_T + N_D^+ - N_A^-\right] \tag{3.2}$$

Total derivatives are used in these equations since we will be assuming one-dimensional structures. Since the first derivative of the local vacuum level with x (i.e., the electric field) is seen to be zero in Figure 3.1c away from the junction in either direction, the charge distribution at the junction region must be a dipole. Equation 3.2 shows this dipole is developed across very short distances if accumulation (majority carrier population enhancement) or heavy doping is present.

The total variation in E_{VL} across the dipole (i.e., the band bending from $x = -d_1$ to $x = d_2$ in Fig. 3.1c) is the built-in potential V_{Bi}. It is the sum of V_{Bi1}, the band bending in material 1, plus V_{Bi2}, the band bending in material 2; i.e.,

$$V_{Bi} = V_{Bi1} + V_{Bi2} \qquad (3.3)$$

As shown in Figure 3.1a, the workfunction difference is what forces the creation of V_{Bi}. Consequently,

$$V_{Bi} = \phi_{w2} - \phi_{w1} \qquad (3.4)$$

If we define the built-in potential as the total band bending developed in TE from contact to contact, then there can be additional contributions to the built-in potential due to workfunction differences at the contacts to our materials of Figure 3.1. We will return to this point later.

Things can also get a bit more complicated when there are permanent dipoles present at the interface between materials 1 and 2. We postpone discussion of that detail until Chapter 5.

We have reviewed the rules for drawing band diagrams for a two-material system. We ignored contacts to the constituent materials and assumed these materials 1 and 2 of Figure 3.1 extended far to the left and far to the right from the junction. In practice, there can be a number of materials involved in a solar cell structure and the band bending may even extend from contact to contact. The same rules apply in such a situation: line up the components as in Figure 3.1b and then let charge flow back and forth until there is one Fermi level, as required for TE. However, solving Poisson's equation in this case may require a numerical analysis. We will examine a number of situations like that in the device chapters.

3.2.2 Photovoltaic action arising from built-in electrostatic fields

The structures we are about to explore are not purported to be optimized. We are exploring them to see if they can act as "charge separation engines" and give photovoltaic action. Our criterion for photovoltaic action is straightforward: is there an open-circuit voltage and a short-circuit current under illumination? In this section we focus on a built-in electric field in an absorber to see if it gives rise to significant photovoltaic action under illumination, which it most likely will, since an electric field pushes electrons one way and holes the opposite way. We will have the AM1.5G spectrum (see Fig. 1.1) impinge on the sample from the left and use an absorber material whose absorption process produces free electrons and holes. The absorption is modeled with the Beer-Lambert law, with reflection assumed at the back surface (see Section 2.4). The bulk recombination mechanism in the absorber is taken to be S-R-H recombination. Figure 3.2a shows the computer-calculated band diagram of our structure in TE and Table 3.1 gives its material parameters. It is clearly a very simple configuration, with a built-in electric field created by the workfunction difference between the two metal contacts. The doping density of the absorber is so low that the field penetrates across the whole absorber; i.e., this is an example of the band bending running from contact to contact.

To determine what happens when illumination impinges on this structure, we have the computer numerically solve the full set of mathematical equations and boundary conditions given in Section 2.4. The resulting light and dark J-V characteristics are presented in Figure 3.3. As seen from Figure 3.3, this very basic structure, with a simple built-in electric field, does give photovoltaic action. In fact, Figure 3.3 shows a short circuit current density of $J_{sc} \approx 13\,mA/cm^2$ and an open circuit voltage $V_{oc} = 0.37\,V$. These arise because, under illumination, the built-in field pushes photogenerated electrons to the right and photogenerated holes to the left (in Figure 3.2a), thereby forcing a segment of the device's J-V curve to lie in the power quadrant, as seen in Figure 3.3. In this segment of the J-V characteristic, the current is seen from the J-V plots of Figure 3.3 to emerge from the positive (left) electrode. If we change the impinging direction of the light and have it enter through the right contact, the sense of the photovoltaic action stays the same. It is interesting to note that the dark J-V characteristic, also given in Figure 3.3, demonstrates an asymmetry in voltage. The dark J-V behavior also arises from the fact that the built-in electric field inside the structure has made one direction different from the other.

Using the numerical analysis results, we can explore the contributions from electrons and holes to the total current density J (which must be a

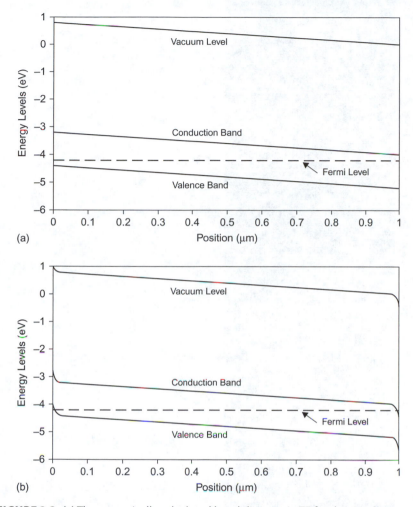

FIGURE 3.2 (a) The numerically calculated band diagram in TE for the simple structure of Table 3.1. Here $\phi_{WL} - \phi_{WR} < E_G$. (b) A schematic band diagram in TE for a case for which $\phi_{WL} - \phi_{WR} \geq E_G$. The Fermi level penetration into the bands in this schematic is exaggerated to call attention to the phenomenon. In both (a) and (b), the built-in field has been created by contact workfunction differences.

constant across this one-dimensional structure) as a function of position. The short-circuit (SC) condition (under illumination with V = 0) is particularly interesting for such an investigation because the maximum currents are flowing at SC. These contributions are seen in Figure 3.4. The figure shows that holes are carrying all the current as they emerge from the anode (left contact). As one moves toward the cathode, electrons and holes share the current; and, finally, at the cathode, electrons

Table 3.1 Parameters Used in Numerical Modeling of Photovoltaic Action Arising from a Built-in Electrostatic Field

Length	Band gap	Electron affinity	Absorption properties	Doping density	Front contact workfunction and surface recombination speeds	Back contact workfunction and surface recombination speeds	Electron and hole mobilities	Band effective densities of states	Bulk defect properties
1000 nm	$E_G = 1.12$ eV	$\chi = 4.05$ eV	Absorption data for Si used (See Fig. 3.19)	$N_A = 1.0 \times 10^{13}$ cm^{-3}	$\phi_W = 4.90$ eV, $S_n = 1 \times 10^7$ cm/s $S_p = 1 \times 10^7$ cm/s	$\phi_W = 4.25$ eV, $S_n = 1 \times 10^7$ cm/s $S_p = 1 \times 10^7$ cm/s	$\mu_n = 1350$ cm^2/vs $\mu_p = 450$ cm^2/vs	$N_C = 2.8 \times 10^{19}$ cm^{-3} $N_V = 1.04 \times 10^{19}$ cm^{-3}	Donor-like gap states from E_V to mid-gap $N_{TD} = 1 \times 10^{12}$ cm^{-3}eV^{-1} $\sigma_n = 1 \times 10^{-15}$ cm^2 $\sigma_p = 1 \times 10^{-17}$ cm^2 Acceptor-like gap states from mid-gap to E_C $N_{TA} = 1 \times 10^{12}$ cm^{-3}eV^{-1} $\sigma_n = 1 \times 10^{-17}$ cm^2 $\sigma_p = 1 \times 10^{-15}$ cm^2

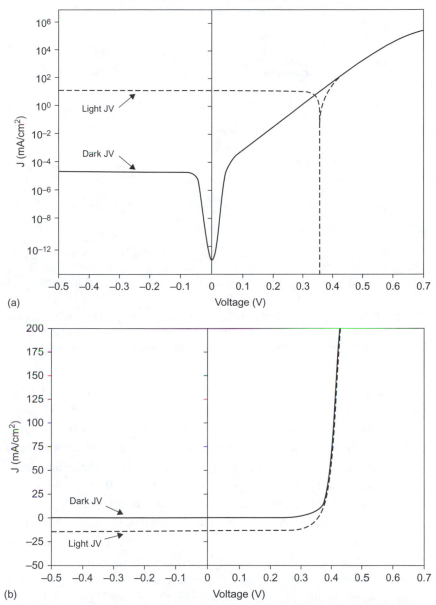

FIGURE 3.3 Numerically computed light and dark J-V characteristics of the structure of Figure 3.2 and Table 3.1 shown in (a) semi-log and (b) linear plots.

emerge carrying all the current. Figures 3.5 and 3.6 use the information contained in the numerical analysis to delve into the behavior of the current densities in much greater detail. Figure 3.5a shows the electron current density's drift and diffusion components at thermodynamic

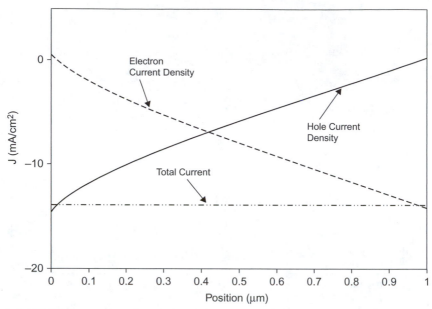

FIGURE 3.4 The numerically calculated electron, hole, and total current densities for the structure of Figure 3.2a and Table 3.1 as a function of position for the short-circuit condition. Negative conventional current density is defined as flow to the left in Figure 3.2a.

equilibrium, whereas Figure 3.5b shows these components at the SC condition. From Figure 3.5a, it is apparent that electron drift and diffusion currents are present inside the device even in TE—and they are relatively large in magnitude. However, they must exactly total to zero in thermodynamic equilibrium; i.e., $J_n \equiv 0$ at TE from the principle of detailed balance.[3] From Figure 3.5b it can be seen that, at the short-circuit condition, the electron drift and diffusion balance of Figure 3.5a has been upset and there is a net electron current density due to drift. This is caused by the photogenerated excess carriers (those present in addition to what exists in TE) experiencing the built-in electrostatic field. It is interesting to note that on the right side of the structure the net electron current density is much smaller in magnitude than the electron drift current density. There is still a net drift current density in this region because drift dominates over diffusion. This electron diffusion under the short-circuit condition existing in the vicinity of the right contact and is a residual effect from the high TE electron population there. It is seen to be heading in the wrong direction to contribute constructively to the net current at the electrodes. Figures 3.6a and 3.6b contain the same information for holes. Figure 3.6a

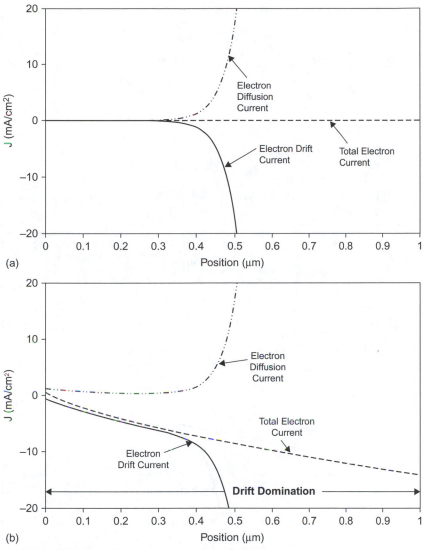

FIGURE 3.5 The numerically computed electron drift and diffusion components for the structure of Figure 3.2a as a function of position for (a) thermodynamic equilibrium and (b) short circuit. Electron drift is seen to dominate over diffusion across the whole structure in the short-circuit condition. Negative conventional current density is defined as flow to the left in Figure 3.2a.

demonstrates that hole drift and diffusion are present inside the device at TE but they too must exactly sum to zero in thermodynamic equilibrium; i.e., $J_p \equiv 0$ at TE from the principle of detailed balance. Figure 3.6b shows that, in the simple structure of Figure 3.2a, at the short-circuit

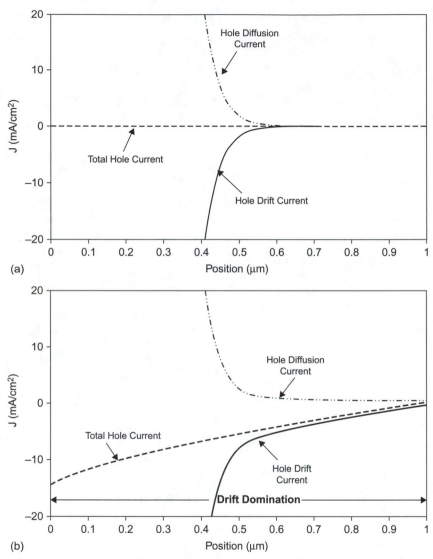

FIGURE 3.6 The numerically determined hole drift and diffusion components for the structure of Figure 3.2a as a function of position for (a) thermodynamic equilibrium and (b) short circuit. Hole drift is seen to dominate over diffusion across the whole structure in the short-circuit condition. Negative conventional current density is defined as flow to the left.

condition, the hole drift and diffusion balance of part (a) is also upset due to the impinging illumination. Now there is a net hole current density due to drift. On the left side of the structure the net hole current density is much smaller in magnitude than the hole drift current density but exists because of this hole drift. There is residual hole diffusion at short circuit (left over from TE—see Fig. 3.6a), but it is confined to the vicinity of the left contact and is heading in the wrong direction. This current density analysis underscores that, under illumination, the built-in field pushes photogenerated electrons to the right and photogenerated holes to the left in Figure 3.2a through the mechanism of drift, thereby giving rise to the photovoltaic action.

It can be seen from Figure 3.2a that there is an upper bound on how much splitting can occur between the Fermi level in the right metal contact and the Fermi level in the left metal contact in the power quadrant. This is very significant, since the splitting of the contact Fermi levels is the external voltage developed. The largest value of this contact Fermi level splitting (i.e., the magnitude of the difference between the Fermi level position in the right contact and that in the left contact) in the power quadrant is the open-circuit voltage V_{oc}. The largest V_{oc} possible can occur for Figure 3.2a when the splitting of the Fermi level of the right contact over that of the left contact has forced the bands to become flat in the absorber; i.e., when the splitting has totally wiped out the built-in electric field, which is the only "charge separating engine" in this cell. Thus, the biggest V_{oc} possible must be the built-in potential $V_{Bi} = \phi_{WL} - \phi_{WR}$. This is consistent with Eq. 2.47 of Chapter 2, since the integral from contact to contact of the electrostatic field present at TE is the built-in potential. If the splitting were to exceed V_{Bi}, the electric field in the device would reverse direction and the current would be leaving the anode; i.e., the device would be consuming, not producing, power.

This observation on the limit on V_{oc} for Figure 3.2a begs the following question: what happens if $\phi_{WL} - \phi_{WR} \geq E_G$? Or, in other words, can V_{oc} exceed E_G in this simple device? The answer for this structure is no, or, more accurately, it depends on the band densities of states distribution in energy but usually the upper bound on V_{oc} can not exceed E_G significantly for this simple structure of Figure 3.2b. The reason for this is as follows. If the contact workfunctions were such that $\phi_{WL} - \phi_{WR} \geq E_G$, then the band bending V_{Bi} must exceed E_G, since $V_{Bi} = \phi_{WL} - \phi_{WR}$. The additional band bending that is required in this case develops immediately adjacent to the contacts. This additional band bending (let us call it V_{BiR} at the right and V_{BiL} at the left) occurs at the contacts because, as the Fermi level begins to closely approach the band edge E_C at the right contact in Figure 3.2 (within a few kT where kT = 0.026eV at RT), Eq. D.1 from Appendix D shows that a huge electron population increase occurs immediately adjacent to this right contact. The same thing is occurring for the hole population immediately adjacent to the left contact, if the Fermi level there begins to closely approach the band edge E_V. As a consequence, the charge needed to develop V_{BiR} at the right contact interface and V_{BiL} at the left contact interface is easily set up within a very short distance of the absorber-contact interfaces. This is depicted in the sketch provided by Figure 3.2b for TE, with the Fermi level penetration into the bands exaggerated. Under illumination, the Fermi level position within the conduction band at the right contact in Figure 3.2b and the Fermi level position within the valence band at the left contact in Figure 3.2b would not be expected to shift much with the voltage since any change would cause huge differences in charge at the contacts. The usual terminology is to say that the Fermi level is pinned with respect to the absorber band bending at the contacts; i.e., not much happens to the band bending V_{BiR} and V_{BiL} at the contacts nor to the Fermi level positions with respect to the band edges near the contacts, under illumination and with the development of voltage[‡]. With the constraint that the Fermi level positions with respect to the band edges will be relatively constant at the contacts, it can be deduced from Figure 3.2b that the largest contact Fermi level splitting must be such that

$$V_{oc} \lesssim E_G \qquad (3.5)$$

While Eq. 3.5 sets upper bound on V_{oc} for simple built-in potential structures, loss kinetics—i.e., recombination in the bulk and at contacts—will cause the actual open-circuit voltage to be less than this. The stronger the recombination loss mechanism is, the more photovoltaic action is suppressed, since the photogenerated carriers do not live long enough to be separated and collected.

While the role of built-in electric fields in causing photovoltaic action is demonstrated here using a free electron-hole pair–producing absorber, we could use exciton-producing absorbers as well. In that case, to be

---

[‡]Fermi level pinning can occur at an interface if the Fermi level is forced into a region of energy with a high density of states. The situation can arise due to the proximity of band edges or due to a large density of gap states.

useful, the excitons must dissociate somewhere in the structure into free electrons and holes. If the excess carriers appear in a built-in field region, they will be separated and a photovoltage will arise. Our next examples show that there are two other sources of photovoltaic action, as we expected from Chapter 2. Only one of these is as important as a built-in electric filed. All three may exist in a structure simultaneously.

3.2.3 Photovoltaic action arising from diffusion

We now consider a structure with an absorber that is exactly like the preceding one. The absorber is taken to have the same material properties as that of Table 3.1 but the new structure has no built-in electric field and no effective force fields (i.e., no electron affinity, no hole affinity, and no density of states changes). There is no contact workfunction difference and both contacts have $\phi_W = 4.62\,\text{eV}$. The resulting band diagram is quite boring—everything is flat—so it is not shown. Since there are no electric and no effective force fields, the only candidate for causing photovoltaic action is diffusion of electrons and holes.

As may be seen in Figure 3.7, numerical simulation of this device's response when subject to an AM1.5G spectrum impinging from the left does show photovoltaic action, but it is extremely small. The short circuit current density can be determined from this figure to be $J_{sc} \approx 0.4\,\text{mA/cm}^2$ whereas the open circuit voltage is $V_{oc} = 0.003\,\text{V}$. This photovoltaic action arises from the fact that photogeneration is falling off exponentially from the left contact, which causes both electrons and holes to diffuse toward the right contact. Since electrons have a higher mobility than holes in this absorber (Table 3.1), they have a higher diffusion coefficient according to Eqs. 2.20 and 2.29. As a consequence, an electric field is created by the carriers themselves as they undergo unequal diffusion, and it is oriented to try to pull the electrons back toward the slower holes. This electric field is plotted for the open-circuit condition in Figure 3.8. As always, the integral across the device of the electrostatic field at the open circuit condition minus that present at TE is the open-circuit voltage, as discussed in Section 2.4; i.e., for the situation under discussion,

$$V_{oc} = \int_{\text{structure}} [\xi(x)]dx$$

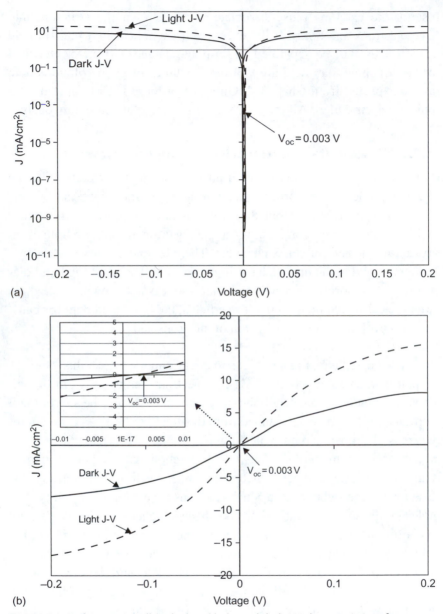

FIGURE 3.7 The numerically calculated light and dark J-V characteristics of a structure with neither a built-in electric field nor effective fields. J-V shown in (a) semi-log and (b) linear plots. The resulting V_{oc} is 0.003 V.

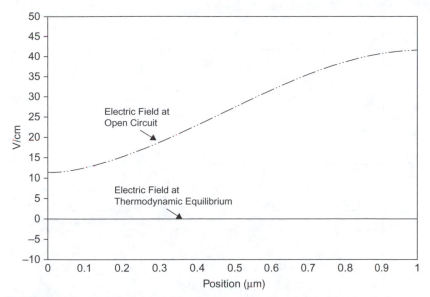

FIGURE 3.8 The electric field at TE and at open circuit under illumination in the structure with neither a built-in electric field nor effective fields.

Here the fact that $\xi_0(x) = 0$ for this structure has been used. The small photovoltage arising for this device from diffusion, or more precisely from diffusion differences, is called the Dember potential.[2] We note that V_{oc} is positive in this example according to our conventions. Had light come in the right contact, it would have been negative. It is interesting to note that Figure 3.7 shows no asymmetry in voltage for the dark J-V. This is to be expected because there is no inherent feature in the structure that makes one direction different from the other. The Dember potential is generally of minor importance in solar cells. One other comment: while diffusion is in itself not a significant generator of photovoltaic action, it can be very much a part of what is going on in a cell. We have seen a glimpse of this, in Figures 3.5 and 3.6.

3.2.4 Photovoltaic action arising from effective fields

To assess the impact of effective force fields on photovoltaic action, we need to create a structure in which there is no built-in electric field in TE but there is an effective force field for electrons, holes, or both. To conservatively gauge the potential of effective force fields, we use the heterostructure of Figure 3.9, which only has a graded electron affinity;

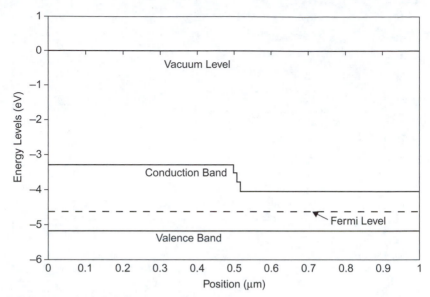

FIGURE 3.9 The band diagram in TE for the simple structure of Table 3.2. A graded electron affinity is present that gives rise to an electron effective force field.

i.e., there is no hole effective force field present. This grading is seen to occur step-wise over 20 nm. There is no band bending in the structure because $\phi_{WL} = \phi_{W1} = \phi_{W2} = \phi_{WR}$, where ϕ_{WL} is the workfunction of the left contact, ϕ_{W1} is the workfunction of material 1 (left absorber), ϕ_{W2} is the workfunction of material 2 (right absorber), and ϕ_{WR} is the workfunction of the right contact. The material parameters of this configuration are listed in Table 3.2. We anticipate that this change in the electron affinity seen in Figure 3.9 will make one direction different from the other—at least for photogenerated electrons—and thereby give rise to photovoltaic action.

Numerical simulation of this heterostructure subjected to an AM1.5G spectrum impinging from the left shows that the electron effective field existing in this example does cause photovoltaic action. A short circuit current density of $J_{sc} \approx 4 \, \text{mA/cm}^2$ and an open circuit voltage $V_{oc} \cong 0.06 \, \text{V}$ can be noted from the light J-V characteristics of Figure 3.10. This open-circuit voltage is an order of magnitude higher than a Dember potential and, like the photovoltaic action caused by a built-in electric field, it does not change direction if the light is switched to impinging from the right side. Careful inspection of the dark J-V shows it is essentially symmetric in voltage—this is to be expected because the Fermi level position in Figure 3.9 suggests that the device will act approximately like a

Table 3.2 Parameters Used in Numerical Modeling of Photovoltaic Action Arising from a Built-in Electron Effective Field

Parameter	Material 1	Material 2
Length	500 nm	500 nm
Band gap	$E_G = 1.90\,eV$	$E_G = 1.12\,eV$
Electron affinity	$\chi = 3.27\,eV$	$\chi = 4.05\,eV$
Absorption properties	Absorption data for Si used with cut-off at $E_G = 1.90\,eV$	Absorption data for Si used (See Fig. 3.19)
Doping density	$N_A = 7.0 \times 10^9\,cm^{-3}$	0.0
Front contact workfunction and surface recombination speeds	$\phi_{WL} = 4.62\,eV$, $S_n = 1 \times 10^7\,cm/s$, $S_p = 1 \times 10^7\,cm/s$	N.A.
Back contact workfunction and surface recombination speeds	N.A.	$\phi_{WR} = 4.62\,eV$, $S_n = 1 \times 10^7\,cm/s$, $S_p = 1 \times 10^7\,cm/s$
Electron and hole mobilities	$\mu_n = 1350\,cm^2/vs$ $\mu_p = 450\,cm^2/vs$	$\mu_n = 1350\,cm^2/vs$ $\mu_p = 450\,cm^2/vs$
Band effective densities of states	$N_C = 2.8 \times 10^{19}\,cm^{-3}$ $N_V = 1.04 \times 10^{19}\,cm^{-3}$	$N_C = 2.8 \times 10^{19}\,cm^{-3}$ $N_V = 1.04 \times 10^{19}\,cm^{-3}$
Bulk defect properties	Donor-like states falling off from E_V and acceptor-like states falling off from E_C as $10^{14} \times e^{-E'/0.01}$ with cross-sections $\sigma_n = 10^{-15}\,cm^2$ and $\sigma_p = 10^{-17}\,cm^2$ and $\sigma_n = 10^{-17}\,cm^2$ and $\sigma_p = 10^{-15}\,cm^2$, respectively	Donor-like states falling off from E_V and acceptor-like states falling off from E_C as $10^{14} \times e^{-E'/0.01}$ with cross-sections $\sigma_n = 10^{-15}\,cm^2$ and $\sigma_p = 10^{-17}\,cm^2$ and $\sigma_n = 10^{-17}\,cm^2$ and $\sigma_p = 10^{-15}\,cm^2$, respectively
Heterostructure interface light reflection	Neglected	
Back light reflection		Total reflection assumed

p-type resistor in the dark. As we will demonstrate in Chapter 5, if there are affinity steps at the interface for both electron and holes, the dark J-V becomes asymmetric and the V_{oc} under light becomes significantly larger. Figure 3.11 shows the electron, hole, and total current densities for the structure of Figure 3.9 for the short-circuit condition. Holes are carrying all the current as they emerge from the anode (left contact), share more of

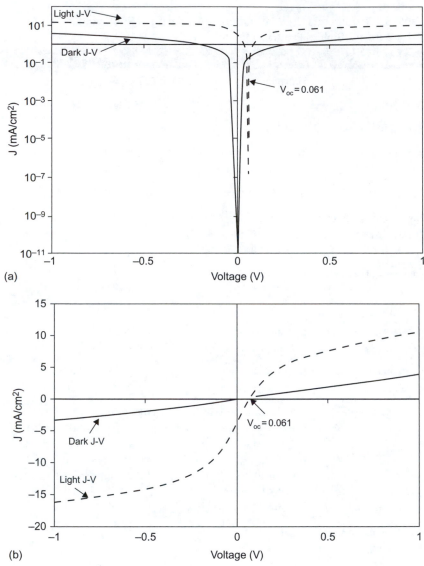

FIGURE 3.10 The numerically determined light and dark J-V characteristics of the structure of Figure 3.9. J-V shown in (a) semi-log and (b) linear plots. The resulting V_{oc} is 0.06 V.

the current with electrons with progress toward the cathode, and, finally, at the cathode, electrons emerge carrying all the current. Figure 3.12a looks at the drift and diffusion components of the electron current density at thermodynamic equilibrium, whereas Figure 3.12b looks at these components at the short-circuit condition. In Figure 3.12a, $J_n \equiv 0$ everywhere,

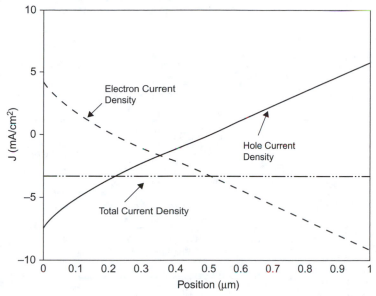

FIGURE 3.11 The numerically determined electron, hole, and total current densities for the structure of Figure 3.9 and Table 3.2 for the short-circuit condition.

as required by TE, but this is a very precarious balance at the heterostructure interface, since the large electron population gradient there is causing a large positive electron diffusion conventional current density that is balanced by the large effective force drift current arising from the affinity change. This happens at three planes because the grading occurs in three steps. Alternatively, this can also be viewed as a situation where electron emission from material 2 into material 1 is in balance with electron emission from material 1 into material 2 at three planes. The drift–diffusion formalism, which must include drift due to effective fields, as explained in Section 2.3.1, is more convenient, however, for mathematical analyses. This is especially true for graded changes like those seen in Figure 3.9. For the short-circuit condition depicted in Figure 3.12b, the electron current density's diffusion component plays a dominant role over electron drift everywhere in the device except at the electron-affinity graded interface. Diffusion disadvantageously carries electrons to the recombination sink of the front contact everywhere to the left of ~0.2 μm but, to the right of this plane in material 1, it advantageously carries electrons to the "separation engine" that is the affinity step region. At that region, drift due to the electron effective force field sweeps electrons across the interface

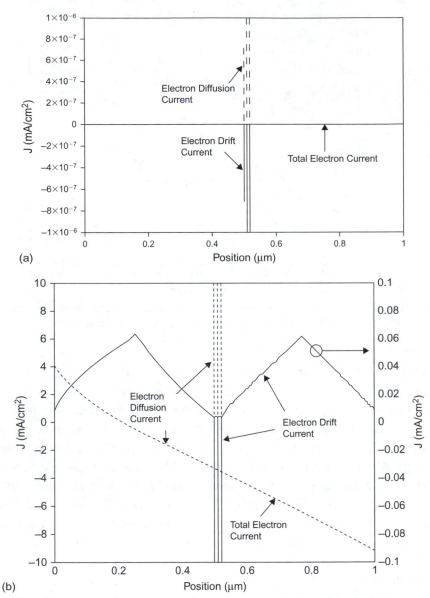

FIGURE 3.12 The numerically calculated electron drift and diffusion components for the structure of Figure 3.9 for (a) thermodynamic equilibrium and (b) short circuit. Drift component includes the effective force field contribution, which exists in the electron affinity grading region.

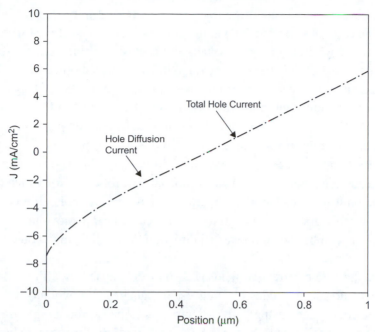

FIGURE 3.13 The numerically computed hole diffusion components for the structure of Figure 3.9 for short circuit. The hole drift component is much smaller and not shown.

toward material 2 at each of the three planes of the affinity steps. In material 2, diffusion takes over again, resulting in a net movement of electrons to the back (right) contact. The hole current components are not shown for TE, not only because $J_p \equiv 0$, as must be the case for TE, but also because the drift components and diffusion components are each zero for this heterostructure. Figure 3.13 shows that, at short circuit, diffusion is the dominant hole transport mechanism. It is advantageous in material 1, where it brings holes to the front (left) contact, but disadvantageous in material 2, where it directs them to the back contact hole recombination sink. There is no hole drift transport due to an effective force since such a force field does not exist for holes in this structure. There is a small component of electron drift current in Figure 3.12b and of hole drift current (not shown) due to the electric field that develops at short circuit. As we know from Eq. 2.47 of Chapter 2, the value of this field at open circuit integrated across the device gives the open circuit voltage. The calculation simplifies to this since there is no electric field at TE.

The above example shows the role of an electron affinity step (or in this specific case, the role of steps) in making one direction different from the other, thereby setting up photovoltaic action. Basically, the example shows how the existence of states in one direction and their absence in the other gives a favored transport direction under light. This need only happen for one carrier. Returning to the full definitions of electron and hole effective fields first given in Section 2.3.1.1, it can be seen that the gradients in the band effective densities of states can also give rise to photovoltaic action. Table 3.2 shows there were no changes in N_C and N_V in the example in this section. Numerical simulations (not shown) for cases for which there are no built-in fields, no electron affinity steps, and no hole affinity steps do show that gradients in N_C, N_V, or both can cause photovoltaic action. They, too, can be used to make one direction different from the other.

While the role of effective fields (i.e., barriers caused by affinity differences or density of states differences) in causing photovoltaic action is demonstrated here using a free electron-hole pair–producing absorber, exciton-producing absorbers can be used as well. In that case, to be useful, the excitons must dissociate somewhere in the structure (e.g., at the effective force field barrier in question). The resulting excess (i.e., more than are there in thermodynamic equilibrium) free electrons and holes must therefore encounter the effective field barrier. This barrier then makes one direction different from the other for these carriers, giving rise to photovoltaic action and carrier separation. (We pursue the exciton case in depth in Chapter 5.) To summarize the observations of this section: all three sources of photovoltaic action may exist simultaneously in a structure, but built-in electric field and effective fields are the major sources. They are the important "charge separating engines".

3.2.5 Summary of practical structures

We have established that built-in electric fields and built-in effective force fields (i.e., electron affinity, hole affinity, and densities of states changes) are the principal sources of photovoltaic action. One or both must be present in a viable cell structure. This conclusion applies whether a solar cell is based on an exciton-producing or free electron-hole–producing absorber. In either case, ultimately, absorber excitations must be turned into free electrons and holes and the solar cell must possess a structure that makes one direction different from the other for these carriers, thereby setting up their separation. Diffusion does not give rise to significant photovoltaic action because it does not make one direction inherently different from the other. However, as we saw in the last example, it can be instrumental in moving carriers to the "separation engines" provided by built-in electric field and effective field regions.

In designing solar cell structures we clearly want configurations with a built-in electrostatic field, built-in effective fields, or both. Figure 3.14 gives a compilation of various types of common solar cell structures. Some are seen to have built-in electrostatic fields, as signified by band bending regions, to have effective force regions (electron affinity changes, hole affinity changes, or both), or to have combinations of both sources of photovoltaic action. The regions of band bending due to electric fields are called electric field barrier regions since they use built-in electrostatic fields to inhibit carrier motion in one direction. The regions of change in electron affinity or conduction band density of states, hole affinity or valence band density of states, or both are termed effective field barrier regions, since they inhibit carrier motion in one direction using effective fields. Some of the structures in Figure 3.14 are used with exciton-producing absorbers; some are used with free electron-hole–producing absorbers. In the former case, the structure has an interface with an appropriate affinity change for exciton dissociation. The p–n and p–i–n homojunctions of Figure 3.14 rely on built-in electrostatic fields for carrier separation and photovoltaic action. They incorporate no effective field mechanism and are used with free electron-hole–producing absorbers. These cells are the subject of Chapter 4. Heterojunctions always involve effective force fields. The device Figure 3.14f only has this type barrier. The one depicted in Figure 3.14c has both a built-in electrostatic field and effective fields. Both, depending on design details (addressed in Chapter 5), can be used with exciton-producing or free electron-hole–producing absorbers. The electric field barrier region in the case of the device of Figure 3.14c can be designed to exist across the whole structure (not shown) analogous to the p–i–n homojunction. Schottky-barrier-type cells shown in Figure 3.14 have both the electric field and effective field mechanisms present. The former is relied upon in free electron-hole absorber cases while the latter may be involved also for excitonic absorbers. These devices are treated in depth in Chapter 6. The electrolyte-semiconductor cell, also discussed in Chapter 6, uses a built-in electrostatic field but can also use effective forces, depending on the electrolyte and absorber. The dye-semiconductor

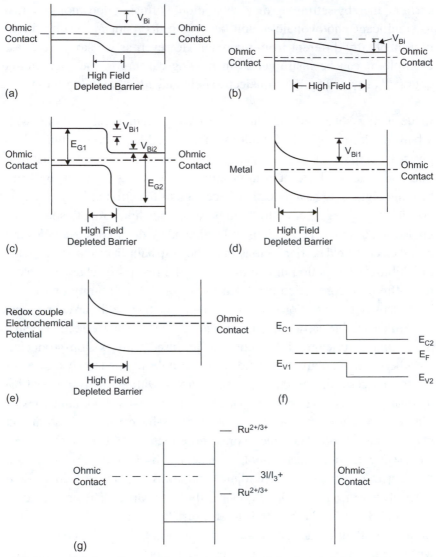

FIGURE 3.14 Solar cell types. Some only have a built-in electric field present as the "charge separation engine": (a) the p–n homojunction and (b) the p–i–n homojunction. Some have both a built-in electric field and effective fields present which may be used to varying degrees depending on the specific cell and the absorption mechanism: the heterojunction shown in (c), also (d) the Schottky-barrier-type cell, and (e) the semiconductor-electrolyte cell. Some rely solely on effective fields: (f) a heterojunction with no built-in electric field and (g) the dye-semiconductor (shown here for the case of Ru-based dye).

cell (usually called the dye-sensitized solar cell) uses effective forces arising from steps in the available states for separation. The absorber excitations in these cells are excitons, and these devices the subject of Chapter 7.

We note that all the built-in fields shown in the structures are created by
workfunction differences. It has been suggested that the built-in electric
fields arising from polarization in ferroelectrics, and not from the usual
workfunction (e.g., doping) differences, might also be exploited in solar
cells.[4]

3.3 KEY MATERIALS
3.3.1 Absorber materials

Absorber materials may be classified as semiconductors or dyes. There
are inorganic and organic absorbers. They can vary from single-element
materials (e.g., silicon) to polymers (e.g., poly(3-hexylthiophene)). As
noted earlier, they all share one attribute: they have absorption-caused
excitations (1) which match in energy with the photon-rich range of
the solar spectrum and (2) which are, or can be converted to, free electrons and holes. Figure 3.15 is a plot of the short-circuit current density
J_{sc} possible from an absorber as a function of the absorber's band gap
E_G. This is obtained by integrating the AM1.5G spectrum of Figure 1.1

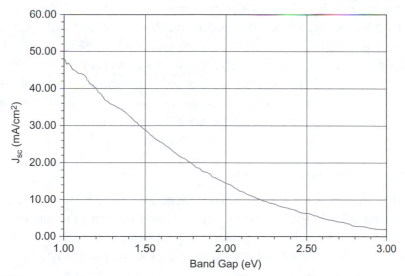

FIGURE 3.15 The potential short-circuit current density versus absorber band gap
for an AM1.5G spectrum.

over a wavelength range from 410 nm to hc/E_G. Figure 3.15 assumes all impinging photons enter the cell and every photon produces an excitation resulting in an electron in the external circuit. This ideal situation of no optical losses and no recombination losses means there is an external quantum efficiency[‡] (EQE) of 1. In reality, EQE < 1.0.

The optimum absorber, from an efficiency point of view, is not the material with the largest J_{sc} but the one with the largest

$$\eta = \frac{FF(J_{sc} V_{oc})}{P_{IN}} \tag{3.6}$$

where use has been made of Eq. 1.3 and the definition of fill factor. Section 3.2.2 suggested that V_{oc} follows E_G, at least for cells where the photovoltaic action arises from a built-in electric field. Figure 3.15 shows J_{sc} decreases with increasing E_G. Consequently we expect that there is some band gap value that optimizes the (FF $J_{sc} V_{oc}$) product for a given absorber and cell design. Studies of which absorber band gap maximizes the power conversion efficiency η give results that depend on the assumptions used. If one looks at the ultimate maximum efficiency permitted by thermodynamics, the answer is $E_G \approx 1.1 \, eV$ with $\eta = 44\%$.[5] This result is based on taking all photons with energy less than E_G as a loss, all photons with energy $\geq E_G$ as collected, but all photon energy $> E_G$ as lost. It takes the cell output voltage as E_G. If recombination losses are added and if the current-voltage features are taken into consideration, then the answer shifts to an E_G as large as $\approx 1.5 \, eV$ and the maximum η falls into the ~25% range or lower, with the value depending on the specific loss mechanisms assumed.[5,6] Selecting the ideal absorber band gap is, in reality, an even more complex issue than these analyses suggest because one has to factor in all the other variables involved in the overall optimization, such as material cost, manufacturing cost, operational lifetime, and environmental impact.

3.3.1.1 ABSORBER PROPERTIES

Figure 3.16 gives the absorption coefficient, as defined in Section 2.2.4.1 and Appendix A for a common organic absorber thin film and several inorganic thin-film absorber materials. The difference in $\alpha(E)$ behavior in Figure 3.16 between the organic absorber and the three inorganic absorbers

---

[‡]EQE measures if an impinging photon turns into an electron doing work in an external circuit.

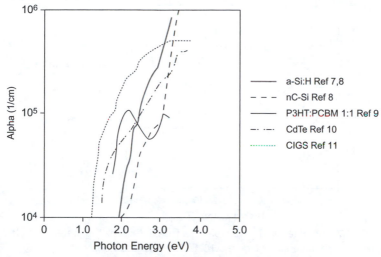

FIGURE 3.16 Absorption coefficients, as defined in Section 2.2.4.1 and Appendix A for an amorphous silicon-hydrogen (a-Si:H), a nanocrystalline silicon (nC-Si), a poly (3-hexylthiophene): phenyl buckyball butyric acid methyl ester mixture (P3HT:PCBM), CdTe, and copper-indium-gallium-selinide (CIGS) thin films.

is due to the organic material's having clusters of molecular orbitals (peaks in the density of states), whereas the inorganics have valence and conduction bands with increasing densities of states as E moves away from the band edges (see Fig. 2.10). Figure 3.17 gives the corresponding potential J_{sc} that these materials can produce as a function of film thickness for EQE = 1. The semiconductor CIGS ($CuIn_xGa_ySe$) has its $\alpha = \alpha(E)$ begin at the lowest energies in Figure 3.16 due to its having the lowest energy band gap (1.15 eV for the CIGS shown here[8,9]) among these absorbers. The compound semiconductor CdTe is a direct gap material, with E_G = 1.45 eV.[10,11] The organic P3HT:PCBM is a mixture of the absorber poly(3-hexylthiophene) (P3HT) and the electron transport material C_{61}-butyric acid methyl ester (PCBM). It is the only material among these particular examples for which the absorption process produces excitons. The specific a-Si:H absorption coefficient behavior shown is for a material with a band gap of ~1.8 eV. It is seen to have a strong absorption behavior since the **k**-selection rules do not apply. The particular silicon nanocrystalline material shown has a band gap of 1.12 eV (Its band gap has not been affected by quantum size effects) and strong absorption behavior due to the relaxation of the **k**-selection rules arising from the impact of confinement on photons, electrons, or both.

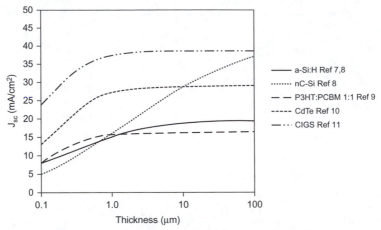

FIGURE 3.17 The potential J_{sc} available as a function of the log of thin-film absorber thickness for the materials of Figure 3.16. EQE = 1 is assumed.

Figure 3.17 is useful because it shows how thick these various materials of Figure 3.16 need to be to have a chance to collect all their potential J_{sc}. It is a conservative assessment since it assumes the absorption follows the Beer-Lambert law (Eq. 2.1); i.e., Figure 3.17 neglects the reflection, scattering, and interference effects which can be advantageously built into cells as discussed in Section 2.2.4.2. It is very interesting to note that log thickness is the abscissa in this plot, since the various $\alpha = \alpha(E)$ data for these materials enter exponentially into the J_{sc} versus thickness calculation. Whether or not we can collect the potential short-circuit currents shown in Figure 3.17 depends on cell design, carrier mobilities, diffusivities, and recombination.

As we discussed in Chapter 1, it is really the product

$$\frac{\text{energy conversion efficiency}}{\text{true costs}} \times \text{cell lifetime}$$

which is critical for wide-scale terrestrial applications of photovoltaics. Consequently, absorber thickness is very important since it enters into this product in two ways: through efficiency and through material and time of deposition (manufacturing) costs. Because of these two facts, the argument has been made that highly absorbing, robust, and inexpensive absorbers may give the best optimization, even though they may

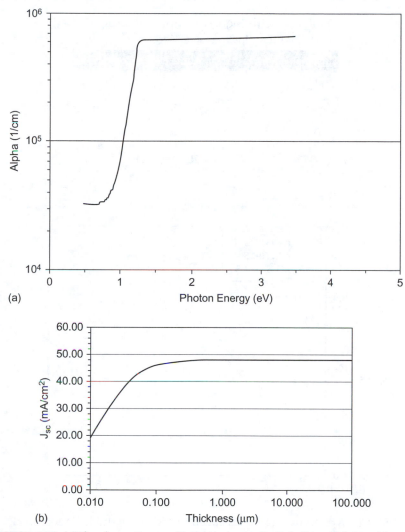

FIGURE 3.18 (a) The absorption coefficient behavior for FeS_2 (from Ref. 14, with permission) and (b) potential J_{sc} versus thickness for FeS_2. EQE $= 1$ is assumed.

not give the highest values for Eq. 3.6.[12,13] Iron pyrite (FeS_2) is an interesting example of such a material. It is a very abundant, inexpensive, strongly absorbing material. It has an indirect band gap at 0.95 eV, a direct gap at 1.03 eV, and the $\alpha = \alpha(E)$ behavior seen in Figure 3.18.[14] These α values may be compared to the absorption coefficients shown in Figure 3.16. Examples of some other highly absorbing, robust, and inexpensive absorbers are given in Table 3.3.[13]

Table 3.3 Some Highly Absorbing, Robust, and Potentially Inexpensive Materials

Material	Some References
Zn_3P_2	15–17
Cu_2S	16, 18
CuO	16, 19
Cu_2O	20, 21

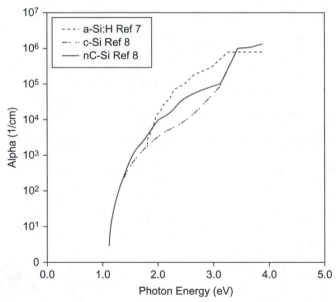

FIGURE 3.19 A comparison of the absorption coefficients, as defined in Section 2.2.4.1 and Appendix A, for three types of silicon absorbers: an amorphous silicon-hydrogen material (a-Si:H), single-crystal silicon (c-Si), and a nanocrystalline silicon (nC-Si).

Figure 3.19 presents absorption data for three different forms of silicon.[7,8,22] Data for a-Si:H and nanocrystalline Si data of Figure 3.16 together with $\alpha = \alpha(E)$ for single-crystal silicon (c-Si) are given. The latter material is an indirect band gap material with $E_G = 1.12\,\text{eV}$. The burden imposed on the absorption process in c-Si by the **k**-selection rules is apparent from a comparison of these plots. Figure 3.20, with an abscissa that is the log of thickness, shows how dramatically

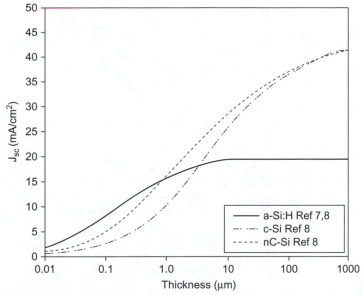

FIGURE 3.20 The potential J_{sc} available as a function of the log of absorber thickness for the materials of Figure 3.19. EQE = 1 is assumed.

the type of silicon material affects the J_{sc} potentially attainable and the thickness required to attain it. Figure 3.21 follows up on Figure 3.19 and takes a more in-depth look at the evolution of the absorption coefficient $\alpha = \alpha(E)$ as a function of grain size for several silicon polycrystalline materials produced by solid-phase crystallization of the same a-Si:H precursor.[23] Shown are nanocrystalline materials with grain sizes Δ of 10 nm and 150 nm and a microcrystalline material of grain size $\Delta = 1 \mu m$. Phonon confinement effects are observed in the 10-nm grain-size nC material.[23]

Dyes are a very interesting class of absorbers. They are molecules with absorption-caused excitations that (1) match in energy with the photon-rich range of the solar spectrum and (2) can be converted into free electrons and mobile cations, as seen in Figure 3.14g. The electron transfer from the dye molecule to the semiconductor conduction band sketched in that figure is somewhat similar to what occurs at the affinity step in Figure 3.9. Similarly, the hole transition from the molecule to the ion level seen in Figure 3.14g is similar to a hole affinity-driven separation. Since the light caused excitation in a dye is an exciton, the transfers just

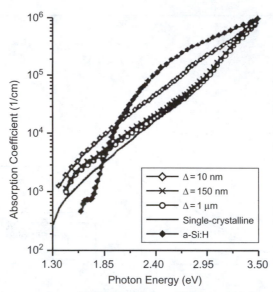

FIGURE 3.21 A comparison of the absorption coefficients, as defined in Section 2.2.4.1 and Appendix A, for two nanocrystalline silicon materials and one microcrystalline silicon material produced from the same precursor by solid-phase crystallization. Data for amorphous silicon-hydrogen material and single-crystal silicon are given for comparison. (From Ref. 23, with permission.)

described necessitate exciton dissociation first. A dye differs from a semiconductor absorber in that it only performs the absorption function and generally has no role in transport. For this reason, it is often referred to as a sensitizer. In the dye-semiconductor solar cell configuration, which is usually called the dye-sensitized solar cell (DSSC), the dye is present in a monolayer on the semiconductor surface. As is discussed in Chapter 7, the morphology of the cell structure leads to the overall absorption and J_{sc}. Since monolayers of the materials are used and since their absorption properties are generally evaluated in solutions, their absorption response is usually expressed as an absorbance A_{abs} (see Appendix A). Plots of A_{abs} for two ruthenium-based dyes ($C_{58}H_{86}N_8O_8RuS_2$ and $C_{42}H_{52}N_6O_4RuS_2$) are given in Figure 3.22.

3.3.2 Contact materials

3.3.2.1 METAL CONTACTS

An ohmic contact is one that ideally passes the current required without dropping any voltage. Metals are excellent for contacts due to their

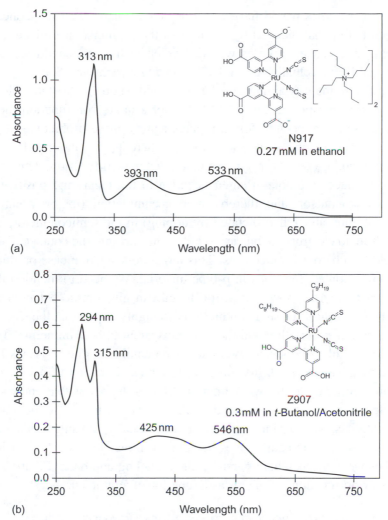

FIGURE 3.22 The absorbance for the ruthenium-based dyes (a) $C_{58}H_{86}N_8O_8RuS_2$ (N917) and (b) $C_{42}H_{52}N_6O_4RuS_2$ (Z907) for a 1-cm path length in solution. (From Ref. 24, with permission.)

low resistivity. As a general "rule of thumb," one wants to use a large workfunction metal as the contact to p-type semiconductor materials, whether inorganic or organic. This follows from our band diagram discussion of Section 3.2.1. If a metal with a workfunction less than that of a given p-type material were used to contact the p-type semiconductor, then a dipole oriented positive in the metal and negative in the semiconductor would have to result to equate the Fermi levels in TE.

This means holes would have to be depleted (majority carrier concentration below that dictated by doping) in the p-type semiconductor, an electrostatic barrier to holes would develop in the semiconductor, and we would have actually succeeded in making a rectifying Schottky barrier diode to the p-type material. There would be no electrostatic barrier to electron transport at the contact. Use of a large workfunction metal with a p-type semiconductors mitigates against these problems. To get the full benefits, one wants $\phi_{WM} \geq \phi_{Wp}$, where ϕ_{WM} is the metal workfunction and ϕ_{Wp} is the p-type semiconductor workfunction. This situation produces a dipole-oriented negative in the metal and positive in the semiconductor, and thereby hole accumulation (majority carrier concentration above that dictated by doping) in the semiconductor, and therefore no electrostatic barrier to hole transport into the contact. There would be a barrier for electrons. This large-workfunction case produces an ohmic contact for holes in p-type material. We need to mention that there is a way a low-workfunction metal can give ohmic behavior for p-type material: if the semiconductor is so highly doped that the electrostatic barrier is very thin and thereby transparent to hole tunneling. This is very useful if the ideal metal cannot be employed due to issues like cost, chemical compatibility, processing damage, or interdiffusion. The obvious corollary to this workfunction "rule of thumb" is that one wants to use a low-workfunction metal as the contact to n-type semiconductor materials, whether inorganic or organic. Ideally, the situation should be $\phi_{WM} \leq \phi_{Wn}$, but again issues like chemical reactions, cost, diffusion, or damage can necessitate pursuing the tunneling approach. Figure 3.23 shows workfunctions of metals as a function of atomic number.

3.3.2.2 HOLE TRANSPORT-ELECTRON BLOCKING AND ELECTRON TRANSPORT-HOLE BLOCKING LAYERS

When used in contact formation, wide band gap hole conductor and wide band gap electron conductor materials with appropriately aligned electron and hole affinities are often referred to as hole transport-electron blocking layers (HT-EBL) and electron transport-hole blocking layers (ET-HBL), respectively. While we focus on their contact role here, many such materials may also be used in heterojunction formation, as is discussed in depth in Chapter 5. In contact formation, these materials can perform two distinct functions, arising from exploiting the

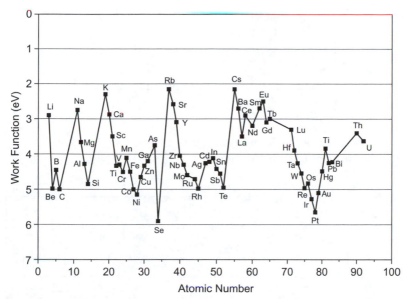

FIGURE 3.23 Workfunctions of metals as a function of their atomic number. (From Ref. 25, with permission.)

changes in the electron affinity, hole affinity, and densities of states (or, equivalently, by changes in the LUMO level and HOMO level). These functions are (a) selective contact formation and (b) exciton-blocking interfaces.

(a) Selective ohmic contacts

We define a selective ohmic contact as one that passes one carrier type only, ideally with no voltage drop. This is very advantageous in a solar cell because it aids in making one direction different from the other; i.e., it helps encourage photocarrier separation. If a contact is the anode of a cell, for example, then one does not want electrons leaving through this contact or recombining with the holes at the contact surface (see Section 2.3.2). An HT-EBL material can be very useful in this particular case. Figure 3.24 shows two such layers interfaced with an absorber. Figure 3.24a shows a graded interface, and Figure 3.24b shows an abrupt interface. Photogenerated holes in Figure 3.24 are able to easily traverse the HT-EBL material and enter the contact. Photogenerated electrons

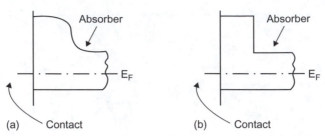

FIGURE 3.24 Use of an HT-EBL to form a selective ohmic contact for holes: (a) graded structure; (b) abrupt structure.

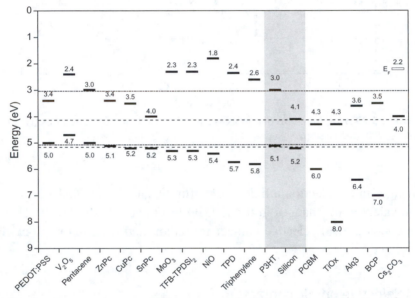

FIGURE 3.25 Band edges (LUMO-HOMO positions) for several HT-EBL and ET-HBL materials. The absorbers P3HT and Si are shown (gray stripe) for comparison. The HT-EBL materials are generally to the left of these two absorbers and the ET-HBL materials are to the right. In this band diagram, ZnPc = zinc phthalocyanine; CuPc = copper phthalocyanine; SnPc = tin phthalocyanine; TFB-TPDSi$_2$ is a cross-linked blend of poly[9,9-dioctylfluorene-co-N-[4-(3-methylpropyl)]-diphenylamine] and 4,4'-bis[(p-trichlorosilylpropylphenyl)phenylamino]biphenyl; TPD = N,N'-bis (3-methylphenyl)-N,N'-diphenylbenzidine; Alq3 = tris(8-hydroxyquinoline aluminum); and BCP = bathocuproine. (From Refs. 26–38.)

are seen to encounter an effective force barrier and are blocked from entering the contact. Several organic and inorganic HT-EBL and ET-HBL materials are shown in Figure 3.25. The band edges of the absorbers P3HT and silicon are shown for comparison.

(b) Exciton-blocking interfaces

If the absorber adjacent to the HT-EB of our example in Figure 3.24 is an exciton-producing material, then the positioning of an HT-EBL at the contact, as shown, can also serve to block the diffusion of excitons to the left and thereby encourage diffusion to an exciton-dissociation heterojunction to the right (not shown).

3.3.2.3 TRANSPARENT CONTACTS

Light obviously needs to enter solar cell structures, which often means transparent contacts are needed to allow light entry while still providing properly positioned electrodes. The materials utilized must have the high conductivities (e.g., $\geq 5 \times 10^2$ siemens/cm), high transparencies, band edge positions, and workfunctions needed to serve as ohmic contacts. Since absorption in these materials is by the free carrier intra-band process 1 of Figure 2.11, there is a trade-off between conductivity and light transmission arising from doping (actually, alloying at the concentrations used). There is also a trade-off between resistance and transmission with film thickness. Transparent conductive oxide (TCO) materials can meet these trade-off requirements. TCOs that have been examined for transparent contact purposes include indium oxide, tin oxide, indium tin oxide (ITO), zinc oxide, aluminum zinc oxide (AZO), gallium indium oxide, indium zinc oxide, gallium indium tin oxide, and zinc indium tin oxide. Among these, tin oxide, ITO, and zinc oxide are the most prevalent in actual use.[39] Energy band edge position and workfunction information for ITO materials are given in Figure 3.26, along with the corresponding information for several AZO materials. Illustrative transmission data (for the AZO materials) are presented in Figure 3.27.

3.4 LENGTH SCALE EFFECTS FOR MATERIALS AND STRUCTURES

3.4.1 The role of scale in absorption and collection

3.4.1.1 ABSORPTION LENGTH

There are a number of naturally occurring lengths in photovoltaics and their magnitudes can vary from the nano- to micrometer scales. One of these basic lengths is the absorption length. It can be formally defined as a function of λ, as in Section 2.2.4.1. However, this definition is not that useful when considering a spectrum of impinging light wavelengths. To

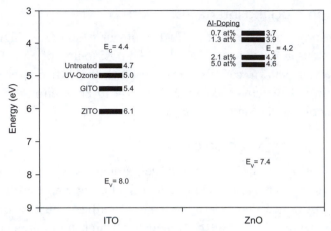

FIGURE 3.26 The band edge positions for ITO and ZnO. Fermi-level positions (stripes), which are seen to depend on processing and alloying (Ga and Zn for ITO; Al for ZnO), are also shown. The ZnO alloying data imply the band gap is changing. The notation at% stands for atomic percent. (From Ref. 40.)

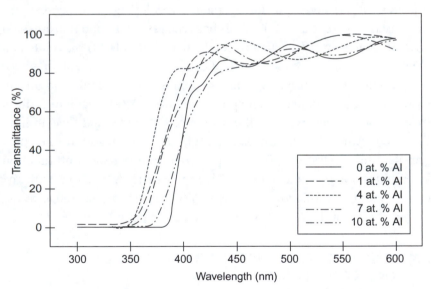

FIGURE 3.27 Transmission behavior for ZnO (0% Al) and several AZO transparent contact materials. The notation at. % stands for atomic percent. The undulations are due to interference effects. (From Ref. 41.)

have a more useful quantity, we define an absorption length L_{ABS} (using figures like Figures 3.17 and 3.20) as the thickness required to absorb 85% of the impinging light or equivalently required to develop 85% of the potential J_{sc} available. For the absorbers represented in Figures 3.17 and 3.20,

L_{ABS} is seen to vary from several hundred nm to tens of microns. Using these figures to determine how thick a material needs to be to get a certain degree of absorption is a conservative assessment since it assumes the absorption follows the Beer-Lambert law (Eq. 2.1); i.e., reflection, scattering, and interference effects, which can be built into cells using features ranging from simple reflection layers to plasmonic or photonic structures, are neglected in those figures.

3.4.1.2 EXCITON DIFFUSION LENGTH

There are also lengths, called collection lengths (L_C), that can be used to gauge the distances over which light absorption-caused excitations can be transported. In the case of excitons, the collection length can be determined by noting that transport can only be by diffusion. Using this fact gives the following equation for the exciton population:

$$D_E \frac{d^2 P_E}{dx^2} = G' - \frac{P_E}{\tau_E}$$

or

$$\frac{d^2 P_E}{dx^2} = \frac{G'}{D_E} - \frac{P_E}{D_E \tau_E} \tag{3.7}$$

where G' represents generation of excitons, D_p is the exciton diffusion coefficient, and τ_E is the exciton lifetime—a measure of the time an exciton exists before extinction. A dimensional analysis of Eq. 3.7 (i.e., looking at the denominator of the second term on the right) shows that there is a naturally arising length L_E^{Diff} characterizing exciton diffusion given by

$$L_E^{Diff} = [D_E \tau_E]^{1/2} \tag{3.8}$$

This collection length is called the exciton diffusion length and is a measure of how far an exciton can diffuse in an absorber before relaxing back to the ground state. Exciton diffusion lengths in conjugated polymer absorbers, for example, are generally believed to be in the 5 to 10 nm range although there is some evidence that suggests these lengths actually may be significantly longer.[42]

3.4.1.3 ELECTRON AND HOLE DIFFUSION AND DRIFT LENGTHS

When the light absorption-caused excitations are free electrons and holes, collection length L_C is controlled by diffusion, drift, or some mixture of the two. Examination of the equations of Sections 2.2.5.1 and 2.4 shows that there are naturally arising electron and hole collection lengths due to diffusion and due to drift. To be specific, if drift is neglected and a linearized recombination model is valid, then the equations of those sections reduce to

$$D_n \frac{d^2n}{dx^2} = G'' - \frac{n - n_0}{\tau_n} \tag{3.9}$$

for electrons where G'' is the generation term. A dimensional analysis of Eq. 3.9 shows that there is a naturally arising electron diffusion length L_n^{Diff} given by

$$L_n^{Diff} = [D_n \tau_n]^{1/2} \tag{3.10}$$

where D_n is the electron diffusion coefficient and τ_n is the electron minority carrier lifetime.

Similarly, if drift is neglected and a linearized recombination model is valid, then the equations of those sections reduce to

$$D_p \frac{d^2p}{dx^2} = G'' - \frac{p - p_0}{\tau_p} \tag{3.11}$$

for holes. A dimensional analysis of Eq. 3.11 shows that there is a naturally arising hole diffusion length L_p^{Diff} given by

$$L_p^{Diff} = [D_p \tau_p]^{1/2} \tag{3.12}$$

where D_p is the hole diffusion coefficient and τ_p is the hole minority carrier lifetime. These diffusion lengths given by Eqs. 3.10 and 3.12

have the meaning that, if a photocarrier species is being collected by diffusion to a "carrier separation engine" (i.e., a built-in electric field or an effective field region), then collection is expected to be effective in the absorbing region over a diffusion-length distance extending out from that "separation engine." As we have already seen, diffusion can play a major role in photocarrier collection in regions of an absorber that have neither built-in electric field nor built-in effective fields.

If the equations of Sections 2.2.5.1 and 2.4 are examined again, but this time with diffusion neglected, then drift collection lengths emerge. For electrons, the neglect of diffusion and the use of a linearized recombination model reduce the equations of those sections to

$$\mu_n \xi \frac{dn}{dx} = -G'' + \frac{n - n_0}{\tau_n} \tag{3.13}$$

where we have assumed a constant electric field over the collection region. A dimensional analysis of Eq. 3.13 shows that there is an electron drift length L_n^{Drift} given by

$$L_n^{Drift} = \frac{D_n \tau_n \xi}{kT} \tag{3.14}$$

Here the Einstein relation $D_n = kT\mu_n$ introduced in Appendix D has been used. For holes, the neglect of diffusion and the use of a linearized recombination model reduce the equations of those sections to

$$\mu_p \xi \frac{dp}{dx} = G'' - \frac{p - p_0}{\tau_p} \tag{3.15}$$

where again we have assumed a constant electric field over the collection region. A dimensional analysis of Eq. 3.13 shows that there is a hole drift length L_p^{Drift} given by

$$L_p^{Drift} = \frac{D_p \tau_p \xi}{kT} \tag{3.16}$$

This equation uses the Einstein relation $D_p = kT\mu_p$ from Appendix D. Equations 3.14 and 3.16 give the length over which a field of strength ξ can collect a photocarrier with the lifetime $\tau_{n,p}$.

3.4.1.4 ABSORPTION LENGTH AND COLLECTION LENGTH MATCHING ISSUES

There are absorbers with (1) absorption and collection lengths both in the nano-scale, (2) absorption lengths in the microscale and collection lengths in the nano-scale, and (3) absorption and collection lengths both in the microscale. This is captured in Table 3.4, which uses several example materials. Figure 3.28 shows the problems that arise due to the commonly encountered situation of absorption length > collection length. In the arrangement depicted in Figure 3.28a, the absorber has wasted material. In Figure 3.28b, the absorber volume has been reduced, thereby eliminating unutilized material but now light is lost. Light loss can be addressed with the use of back reflection, photonic structures, and plasmonic structures, as discussed in Chapter 2. Light loss can also be addressed by the use of tandem or triple cells arranged to be electrically

Table 3.4 Some Collection Lengths, Absorption Lengths, and Lateral Collection Electrode Spacing Distances

Material	Collection length	Absorption length	Electrode spacing scale (Fig. 3.28c)	Absorber thickness
μC poly-Si	2- to 5-μm range by diffusion[39]	~80 μm (From Fig. 3.20)	~5 μm	~80 μm
a-Si:H	~300 nm by drift[39]	~1 μm (From Fig. 3.20)	~300 nm	~1 μm
P3HT	Excitons ~5–10 nm[26,42]. Could be as high as 150 nm[42]. Holes ~200–300 nm by drift (estimated from cell film thicknesses)	300–400 nm (From Fig. 3.17)	5–150 nm if exciton splitting is done at electrode interface; ~10–300 nm if exciton splitting is done at an intermediate interface	~300 nm
FeS$_2$	<100 nm[44]	~60 nm (From Fig. 3.18)	<100 nm	~60 nm

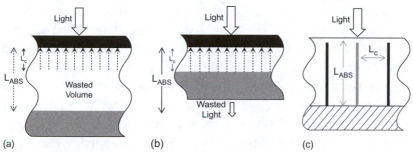

FIGURE 3.28 Three solar cells, all with the same absorption L_{ABS} and collection length L_C. The dashed arrows in (a) and (b) indicate the volume of the device that generates useful (i.e., collectable) current. (a) Conventional design with planar electrodes (top and bottom dark volumes) and with absorber layer thickness equal to the absorption length but greater than the collection length. (b) Conventional design with planar electrodes (top and bottom dark volumes) and with absorber layer thickness equal to the collection length. (c) The lateral collection design in which the absorption and collection lengths are decoupled and not parallel. The vertical anode and cathode electrodes alternate in position but connect to the their respective interconnect structures on the bottom substrate.

and optically in series; i.e., light passing out of the first cell passes through the second cell and so on. It can be seen from Figure 3.28 that these absorption length–collection length mismatch issues arise because, in the conventional cell configuration, these lengths are in parallel.

The lateral collection configurations such as that of Figure 3.28c have been proposed to address the mismatch problem.[43] This approach wastes neither light nor active layer material because L_C and L_{ABS} are arranged to be essentially perpendicular to each other. With this configuration, electrode spacing and device thickness can be varied independently and customized in principle to fit the properties of the absorbing material. Lateral collection can be accomplished by incorporating both anode and cathode elements into the volume of the absorbing material, as seen in Figure 3.28c, or by having one set of electrode elements in the absorber and the other electrode on the absorber. Lateral collection configurations have the effect of shortening the path that absorption-generated excitons, electrons, or holes must travel in the cell from the site of their creation. A lateral collection configuration is designed to aid the entity (exciton, electron, or hole) having the greatest difficulty in being collected. Since collection lengths and absorption lengths can be in the nano- or microscale length range depending on the absorber, the spacings and heights in structures represented by Figure 3.28c can be in the nano- or microscale. Table 3.4 also summarizes this situation for several example absorber materials. Two collection lengths are listed for the representative organic absorber P3HT, since it is an exciton-producing absorber and (1) excitons must be collected and dissociated and (2) the resulting electrons and holes must be collected. While the lateral collection configuration can lead to more current collection, its use must be examined on a case-by-case basis, since it may result in more recombination and therefore lower open-circuit voltage. This trade-off depends on the relative importance of bulk and contact recombination.

In excitonic cells, the excitons must diffuse to a region where they can dissociate. Figure 3.29 shows the heterojunction exciton-dissociation process that must take place to produce free electrons and holes in organic solar cells based on exciton-producing absorbers like P3HT. As seen, a properly designed interface with appropriate steps in the electron and hole affinities (or, equivalently, in the LUMO and HOMO levels) can provide the energy to dissociate the excitons, resulting in free electrons in the right-hand material conduction band (grouping of molecular orbitals) and free holes in the left-hand material

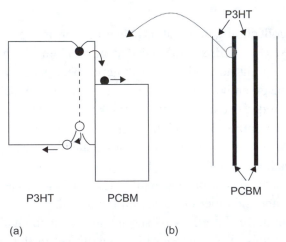

(a)   (b)

FIGURE 3.29 (a) Exciton dissociation at a heterojunction and subsequent electron and hole separation due to effective forces (i.e., affinity changes); (b) schematic of the device arrangement.

valence band (grouping of molecular orbitals) of Figure 3.29a. In the jargon of organic electronics and opto-electronics, the material receiving an electron from the exciton dissociation is the acceptor material and the material whose exciton dissociation produced the electron (and hole) is termed the donor. In the example of Figure 3.29, P3HT is the absorber and donor and PCBM is the heterojunction-former and acceptor. Figure 3.29b depicts a lateral collection structure carried out at the nano-scale for a P3HT-PCBM cell. Here the P3HT columns have a diameter as required by Table 3.4 for harvesting the excitons produced in the P3HT to the P3HT-PCBM interface for dissociation. The PCBM volume is minimized since this material is not a strong absorber but is an electron conductor. The PCBM columns then connect with a top planar conductor for electron collection and the P3HT columns connect to a transparent bottom conductor for hole collection. Studies of P3HT-PCBM mixtures show that they can undergo a phase separation on annealing, which can result in a structure approximating the features of Figure 3.28c and Figure 3.29b.[42] Organic heterojunction cells are discussed further in Chapter 5.

3.4.2 Using the nano-scale to capture lost energy

In solar cell operation, the extra energy of photons with $h\nu > E_G$ is lost to heat. The ideal situation would be to use those photons with $h\nu > 2E_G$ to generate multiple carriers. Furthermore, all the energy of photons with $h\nu < E_G$ is lost. The ideal situation would be to use several of these photons together to generate a carrier. If these objectives of making better use of photon energies could be attained meaningfully, then the ultimate efficiency analysis results discussed in Section 3.3.1 would be very favorably improved. Nano-scale materials engineering offers the possibility of attaining these goals.

As discussed in Chapter 2, semiconductor nanoparticle (quantum dot) structures in the sub-10-nm range open the door, through quantum confinement effects to band gap tuning and to the possibility of absorption that leads to multiple exciton generation (MEG). With MEG, one supra-band gap photon ($h\nu > 2E_G$) can produce more than one exciton. This is extremely attractive since it makes use of the otherwise wasted excess energy possessed by the hot species resulting from supra-band gap photon absorption. Quantum dots of PbSe, PbS, PbTe, and CdSe have been shown to exhibit MEG,[45] and, therefore, offer the possibility of free carrier multiplication (CM). As discussed in Chapter 2, the modeling of CM in quantum dots is based on a MEG process analogous to impact ionization; however, other models have also been put forward.[45,46] The minimum threshold for MEG found in this modeling is the expected $2E_G$, which means low band gap QD materials can be used to fully utilize the whole photon-rich region of the solar spectrum (see Fig. 1.1). Once multiple excitons have been created from one supra-band gap photon in a quantum dot, the issues are how the excitons can be dissociated and how the free electrons and holes can be collected. In other words, how do we get to the free carrier multiplication stage? The dissociation required is analogous to that undergone by the exciton produced by absorption at a dye molecule in the DSSC cell of Figure 3.14g. In that cell, the excited dye molecule's exciton relaxes by emitting an electron to the levels at the semiconductor conduction band edge and a hole to the anion level in the electrolyte. In quantum dots, the analogous collection of free carriers out of MEG appears to be hampered by QD surface states and short exciton lifetimes.[47] The use of QD absorbers for CM has been tried in several cell structures, including tandem cells, but with limited success so far, apparently due to these collection issues.[47]

In an effort to capture the lost energy of sub-band gap photons, there have been proposals (dating back to 1960) to introduce states in the band gap to allow two-photon generation processes.[48,49] As discussed in Chapter 2, two-photon processes based on delocalized intermediate bands (IB) in the gap are expected to be the most effective. Such bands, if properly designed, are believed to be able to support a two-step photon absorption-free carrier generation process without enhancing the competing loss mechanism of nonradiative recombination. Such IB levels are proposed to be attainable with quantum dot structures immersed in a wider band gap absorber.[50] An analysis of IB two-photon generation based on quantum dot structures for up to 1000 suns has shown its potential also for concentrator applications.[51]

3.4.3 The role of scale in light management

Light trapping has long been achieved in wafer-based solar cells using 2- to 10-μm pyramidal structures etched into the cell's surface. Using this micron-length scale technology is obviously out of the question in thin-film structures. Light trapping has long been achieved for thin film cells by forming wavelength-scale texturing on a substrate and then subsequently depositing the thin-film cell. This substrate-texturing approach also becomes problematic as cells are made thinner in an effort to save absorber material cost, since it affects cell morphology and back surface recombination.[52] Nanotechnology adds to these tools which are available for light management in solar cells. As discussed in Chapter 2, it makes possible two new approaches: photonic structures and plasmonics. Both have already been shown to improve cell performance, as discussed in Chapter 2. Both involve nano-scale structures and therefore can be compatible with the thinner devices.

REFERENCES

1. S. Sze, K.K. Ng, Physics of Semiconductor Devices, third ed., John Wiley & Sons, Hoboken, NJ, 2007, pp. 360–380.

2. H. Dember, Photoelectric E.M.F. in cuprous-oxide crystals, Phys. Z. 32 (1931) 554. 856 (1931); Kristallphotoeffekt in Klarer Zinkblende, Naturwissenschaften 20, 758 (1932).

3. W. Shockley, Electrons and Holes in Semiconductors, D. Van Nostrand Co., Princeton, NJ, 1956, pp. 254, 299, 491.

4. T. Choi, S. Lee, Y.J. Choi, V. Kiryukhin, S.W. Cheong, Switchable ferroelectric diode and photovoltaic effect in $BiFeO_3$, Science 324 (2009) 61.

5. W. Shockley, H. Queisser, Detailed balance limit of efficiency of p–n junction solar cells, J. Appl. Phys. 32 (1961) 510.

6. J. Loferski, Theoretical considerations governing the choice of the optimum semiconductor for photovoltaic solar energy conversion, J. Appl. Phys. 27 (1956) 777.

7. T. Searle, Properties of Amorphous Silicon and Its Alloys, INSPEC, 1998.

8. M. Vanecek, Optical properties of microcrystalline materials, J. Non-Crystalline Solids (1998), 227–230, 967–972.

9. Y. Kim, S.A. Choulis, J. Nelson, D.D.C. Bradley, S. Cook, J.R. Durrant, Composition and annealing effects in polythiophene/fullerene solar cells, J. Mater. Sci. 40 (2005) 1371–1376; V.D. Mihailetchi, H.X. Xie, B. de Boer, L.J.A. Koster, P.W.M. Blom, Charge transport and photocurrent generation in poly(3-hexylthiophene): Methanofullerene bulk-heterojunction solar cells, Adv. Funct. Mater. 16, 699–708 (2006).

10. A. Fahrenbruch, Modeling Results for CdS/CdTe Solar Cells, Colorado State University Technical Report, March 2000. Available on CSU Web site: http://www .physics.colostate.edu/groups/photovoltaic/PDFdocs.htm

11. M. Gloeckler, A.L. Fahrenbruch, J.R. Sites, Proc. of the IEEE Photovoltaic Energy Conversion, 3rd World Conference, Osaka, Japan, vol. 1, 491 (2003); AMPS Web site http://www.ampsmodeling.org/

12. A. Ennaoui, S. Fiechter, Ch. Pettenkofer, N. Alonso-Vante, K. Buker, M. Bronold, Ch. Hopfner, H. Tributsch, Iron disulfide for solar energy conversion, Sol. Energy Mater. Sol. Cells 29 (1993) 289.

13. C. Wadia, Y. Wu, S. Gul, S.K. Volkman, J. Guo, A.P. Alivastos, Surfactant-assisted hydrothermal synthesis of single phase pyrite FeS_2 nanocrystals, Chem. Mater. 21 (2009) 2568.

14. A. Ennaoui, S. Fiechter, H. Goslowsky, H. Tributsch, Photoactive synthetic polycrystalline pyrite (FeS_2), J. Electrochem. Soc. 132 (1985) 1579.

15. L. Bryja, K. Jezierski, M. Ciorga, A. Bohdziewicz, J. Misiewicz, Temperature dependence of energy gap of amorphous thin films of Zn_3P_2, Vacuum 50 (1–2) (1998) 5–7.

16. R. Clasen, P. Grosse, A. Krost, F. Levy, Condensed Matter Subvolume C: Non-Tetrahedrally Bonded Elements and Binary Compounds I (Lanboldt-Bornstein), Springer-Verlag, Berlin and Heidelberg, Germany, 1998.

17. J.M. Pawlikowski, Absorption edge of Zn_3P_2, Phys. Rev. B. 26 (8) (1982).

18. S.V. Bagul, S.D. Chavhan, R. Sharma, Growth and characterization of Cu_xS (x = 1.0, 1.76, and 2.0) thin films grown by Solution Growth Technique (SGT), J. Physics Chem. Solids 68 (2007) 1623.

19. P. Yu, N. Sukhorukov, N. Loshkareva, A.S. Moskin, V.L. Arbuzov, S.V. Naumov, Influence of electron irradiation on the fundamental absorption edge of a copper monoxide CuO single crystal, Tech. Phys. Lett. 24 (2) (1998).

20. K. Akimotot, S. Ishizuka, M. Yanagita, Y. Nawa, G.K. Paul, T. Sakurai, Thin film deposition of Cu_2O and application for solar cells, Solar Energy 80 (2006) 715.

21. L.C. Olsen, F.W. Addis, W. Miller, Experimental and theoretical studies of Cu_2O solar cells, Solar Cells 7 (1982) 247–279.

22. Optical Properties of Silicon, technical data sheet available on Virginia semiconductor website: http://www.virginiasemi.com/vsitl.cfm

23. A. Kaan Kalkan, S.J. Fonash, Control of Enhanced Absorption in Poly-Si., Proceedings of the Spring Materials Research Society Meeting, Amorphous and Microcrystalline Silicon Technology, Materials Research Society, vol. 467:415, 1997.

24. Aldrich Web site at http://www.sigmaaldrich.com/materials-science/organic-electronics/dye-solar-cells.html

25. Electron Work Function of the Elements, in CRC Handbook of Chemistry and Physics, 89th Edition (Internet Version 2009), D.R. Lide, ed., CRC Press/Taylor and Francis, Boca Raton, FL.

26. Z. Xu, L. Chen, M. Chen, G. Li, Y. Yang, Energy level alignment of poly (3-hexylthiophene): [6,6]-phenyl C_{61} butyric acid methyl ester bulk heterojunction, Appl. Phys. Lett. 95 (2009) 013301.

27. V. Shrotriya, G. Li, Y. Yao, C. Chu, Y. Yang, Transition metal oxides as the buffer layer for polymer photovoltaic cells, Appl. Phys. Lett. 88 (2006) 073508.

28. Y. Kinoshita, T. Hasobe, H. Murata, Controlling open-circuit voltage of organic photovoltaic cells by inserting thin layer of Zn–Phthalocyanine at pentacene/C_{60} interface, Jpn. J. Appl. Phys. 47 (2) (2008) 1234.

29. P. Peumans, S.R. Forest, Very-high-efficiency double-heterostructure copper phthalocyanine/C_{60} photovoltaic cells, Appl. Phys. Lett. 79 (1–2) (2001) 126.

30. B.P. Rand, J. Xue, F. Yang, S. Forrest, Organic solar cells with sensitivity extending into the near infrared, Appl. Phys. Lett. 87 (2005) 233508.

31. A.W. Hains, T.J. Marks, High-efficiency hole extraction/electron-blocking layer to replace poly(3,4-ethylenedioxythiophene):poly(styrene sulfonate) in bulk-heterojunction polymer solar cells, Appl. Phys. Lett. 92 (2008) 023504.

32. M.D. Irwin, D.B. Buchholz, A.W. Hains, R.P.H. Chang, T.J. Marks, p-type semiconducting nickel oxide as an efficiency-enhancing anode interfacial layer in polymer bulk-heterojunction solar cells, PNAS 105 (8) (2008) 2783; L. Ai, G. Fang, L. Yuan, N. Liu, M. Wang, C. Li, Q. Zhang, J. Li, X. Zhao, Influence of substrate temperature on electrical and optical properties of p-type semitransparent conductive nickel oxide thin films deposited by radio frequency sputtering, Appl. Surf. Sci., 254, 2401 (2008).

33. J. Cui, A. Wang, N.L. Edleman, J. Ni, P. Lee, N.R. Armstrong, T.J. Marks, Indium tin oxide alternatives—high work function transparent conducting oxides as anodes for organic light-emitting diodes, Adv. Mater. 13 (19) (2001) 1476.

34. P. Destruel, H. Bock, I. Séguy, P. Jolinat, M. Oukachmih, E. Bedel-Pereira, Influence of indium tin oxide treatment using UV-ozone and argon plasma on the photovoltaic parameters of devices based on organic discotic materials, Polym. Int. 55 (2006) 601.

35. M.C. Scharber, D. Mühlbacher, M. Koppe, P. Denk, C. Waldauf, A.J. Heeger, C.J. Brabec, Design rules for donors in bulk-heterojunction solar cells – towards 10% energy-conversion efficiency, Adv. Mater. 18 (6) (2006) 789.

36. S.H. Park, A. Roy, S. Beaupre, S. Cho, N. Coates, J.S. Moon, D. Moses, M. Leclerc, K. Lee, A. Heeger, Bulk heterojunction solar cells with internal quantum efficiency approaching 100%, Nature Photonics 3 (5) (2009) 297.

37. K.L. Wang, B. Lai, M. Lu, X. Zhou, L.S. Liao, X.M. Ding, X.Y. Hou, S.T. Lee, Electronic structure and energy level alignment of $Alq_3/Al_2O_3/Al$ and Alq_3/Al interfaces studied by ultraviolet photoemission spectroscopy, Thin Solid Films 363 (1–2) (2000) 178.

38. J. Huang, Z. Xu, Y. Yang, Low-work-function surface formed by solution-processed and thermally deposited nanoscale layers of cesium carbonate, Adv. Funct. Mater. 17 (12) (2007) 1966.

39. G. Beaucarne, Silicon thin-film solar cells, Adv. OptoElectronics 10 (2007) 1155.

40. X. Jiang, F.L. Wong, M.K. Fung, S.T. Lee, Aluminum-doped zinc oxide films as transparent conductive electrode for organic light-emitting devices, Appl. Phys. Lett. 83 (9) (2003) 1875; T.W. Kim, D.C. Choo, Y.S. No, W.K. Choi, E.H. Choi, High work function of Al-doped zinc-oxide thin films as transparent conductive anodes in organic light-emitting devices, Applied Surface Science 253, 1917–1920 (2006); P. Ravirajan, A.M. Peiró, M.K. Nazeeruddin, M. Graetzel, D.D.C. Bradley, J.R. Durrant, Nelson, Hybrid polymer/zinc oxide photovoltaic devices with vertically oriented ZnO nanorods and an amphiphilic molecular interface layer, J. Phys. Chem. B 110 (15) 7635 (2006).

41. J.G. Lu, Z.Z. Ye, Y.J. Zeng, L.P. Zhu, L. Wang, J. Yuan, B.H. Zhao, Q.L. Liang, Structural, optical, and electrical properties of (Zn,Al)O films over a wide range of compositions, J. Appl. Phys. 100 (2006) 073714.

42. J.S. Moon, J.K. Lee, S. Cho, J. Byun, A.J. Heeger, Columnlike structure of the cross-sectional morphology of bulk heterojunction materials, Nano Lett. 9 (1) (2009) 230; A.I. Ayzner, C.J. Tassone, S.H. Tolbert, B.J. Schwartz, Reappraising the need for bulk heterojunctions in polymer-fullerene photovoltaics: The role of carrier transport in all-solution-processed P3HT/PCBM bilayer solar cells, J. Phys. Chem. C (2009) 113, 20050–20060.

43. U.S. Patents 6399177, 6919119, and 7341774.

44. P. Altermatt, T. Kiesewetter, K. Ellmer, H. Tributsch, Specifying targets of future research in photovoltaic devices containing pyrite (FeS_2) by numerical modelling, Sol. Energy Mater. Sol. Cells 71 (2002) 181.

45. M.C. Beard, K.P. Knutsen, P. Yu, J.M. Luther, Q. Song, W.K. Metzger, R.J. Ellingson, A.J. Nozik, Multiple exciton generation in colloidal silicon nanocrystals, Nano Lett. 7 (2007) 2506.

46. V.I. Rupasov, V.I. Klimov, Carrier multiplication in semiconductor nanocrystals via intraband optical transitions involving virtual biexciton states, Physical Review B 76 (2007) 125321.

47. S.J. Kim, W.J. Kim, A.N. Cartwright, P.N. Prasad, Carrier multiplication in a PbSe nanocrystal and P3HT/PCBM tandem cell, Appl. Phys. Lett. 92 (2008) 191107.

48. M. Wolf, Proc. IRE 48 (1960) 1259.

49. A. Luque, A. Marti, Increasing the efficiency of ideal solar cells by photon induced transitions at intermediate levels, Phys. Rev. Lett. 78 (1997) 5014.

50. A. Luque, A. Marti, A.J. Nozik, Solar cells based on quantum dots: multiple exciton generation and intermediate bands, MRS Bulletin 32 (2007) 236.

51. S.P. Bremner, M.Y. Levy, C.B. Honsberg, Limiting efficiency of an intermediate band solar cell under a terrestrial spectrum, Appl. Phys. Lett. 92 (2008) 171110.

52. K.R. Catchpole, A. Polman, Plasmonic solar cells, Optics Express 16 (26) (2008) 21793.

Chapter | Four

Homojunction Solar Cells

4.1 INTRODUCTION

We now begin our detailed examination of specific photovoltaic structures, and the first to be explored is the p–n and p–i–n homojunction solar cell class. The lineage of this solar cell type can be traced back to the work by Ohl[1] in 1941 in which he demonstrated a grown Si p–n junction photovoltaic device. About 12 years later, a 6% efficient, diffusion-formed single-crystal Si p–n junction device was demonstrated[2] and by 1958, 14% efficient single-crystal silicon devices were available using the diffused junction technology. While the Si cells continued to

DOI: 10.1016/B978-0-12-374774-7.00004-2

develop, efforts began on p–n homojunction cells based on other single-crystal semiconductor materials, such as GaAs.[3] A shallow junction GaAs homojunction device with $\eta = 22\%$ (under one sun conditions) was demonstrated,[4] as was a (p)Ga$_y$Al$_{1-y}$ As/(p)GaAs/(n)GaAs device with $\eta = 22\%$(under essentially one sun conditions).[5] The wide gap (p)Ga$_y$Al$_{1-y}$As layer in this structure reduced photogeneration near the front surface and served as a selective ohmic contact. These Si and III–V compound semiconductor p–n junction technologies have continued to be refined over the succeeding years. Thin-film p–n homojunction structures, attractive because of the potential material cost savings, also quickly emerged; they initially used semiconductors such as polycrystalline CuInS$_2$,[6] polycrystalline CuInSe$_2$,[7] and hydrogenated amorphous silicon (a-Si:H).[8] Interestingly, homojunction cells have all been based on electron-hole producing absorber materials. Homojunction cells based on exciton-producing absorbers have not yet been realized.[9] The consensus is that they may not be possible[10–12] due to (1) the inability of built-in cell fields to dissociate excitons[13–16] and consequently due to (2) the need for a semiconductor heterojunction or metal–semiconductor interface for exciton dissociation and charge separation.

Figure 4.1 presents band diagrams for some of the various homojunction junction structures possible. The principal symmetry-breaking, charge-separation region in all of them utilizes an electric field built-in via a p–n or p–i–n homojunction. Several of these examples have ancillary effective field regions arising from HT-EBL/absorber electron affinity changes. One also has an ancillary effective field region arising from ET-HBL/absorber hole affinity change. Figure 4.1e has an ancillary electric field region at the back contact due to an n-n$^+$ "high-low" junction. All the different types of ancillary regions seen in these cells are being used to create selective ohmic contacts—either by effective fields or by an electric field. In Figures 4.1b–4.1d, the front wide-band-gap regions also have the advantage of stopping photocarrier generation near this contact, where the photon flux is most intense. This stops photoelectron losses to recombination at the contact. The graded affinity region in Figure 4.1b is termed a heteroface structure in p–n homojunction terminology. The ancillary electric field region in Figure 4.1e is termed a back surface field region. The built-in electrostatic potential is seen to be located in and about the p–n metallurgical junction region in

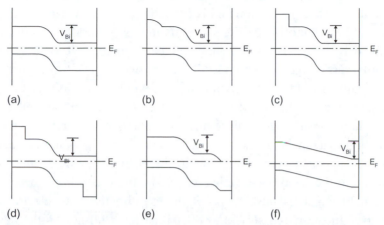

FIGURE 4.1 Some homojunction solar cell configurations with light entering from the left: (a) p–n absorber structure with no ancillary fields, (b) p–n absorber structure with a graded ancillary effective field region forming a selective ohmic contact, (c) p–n absorber structure with an abrupt ancillary effective field region forming a selective ohmic contact, (d) p–n absorber structure with abrupt ancillary effective field regions forming selective ohmic contacts at the front and back, (e) p–n absorber structure with an abrupt ancillary electric field region forming a selective back ohmic contact, and (f) a p–i–n structure with no ancillary fields. In these structures, V_{Bi} is the built-in electron electrostatic potential energy developed across the device in TE. In (a)–(d) this built-in electrostatic potential is seen to reside at the p–n junction region. In (e) it is seen to be developed in two regions while in (f), this built-in potential is developed across the whole p–i–n cell. Since the illumination comes from the left, the structures (b–d) that suppress front surface recombination can be very useful.

Figures 4.1a–d whereas in Figure 4.1e it is seen to have a contribution also from the back of the cell and in Figure 4.1f the built-in potential is developed across the whole p–i–n cell. Whether or not there are ancillary field regions present, the principal "charge separating engine" in these devices lies in the same semiconductor in its various p, n, or even i regions. Hence, as noted, this whole class is referred to as homojunction solar cells.

In the p–n configurations in Figure 4.1 there are flat band regions adjacent to the p–n junctions which have no built-in field. We know there is no built-in electric field present in these regions since Appendix D establishes that the negative of the derivative of the local vacuum level and of either band edge in a homojunction is always the electrostatic field. In these flat band regions there is a majority carrier whose population is established by doping. This population is generally so large it is difficult to modify by illumination, at least under one sun illumination conditions. The other carrier has a much lower population and is termed the minority carrier in these regions. This minority carrier's population is generally significantly modified by illumination. In such flat band regions, it is the excess minority carriers generated under illumination that need to be brought to the built-in electrostatic, symmetry breaking region as can be seen by examining the flat band regions in this figure. Consequently, diffusion must be used to collect minority photocarriers to the charge-separation machinery of the built-in electric field, p–n junction region. Photocarrier drift is very important in the barrier regions of these p–n cells where the band edge derivatives show there is a built-in electrostatic field which is pushing holes in one direction and electrons in the other. In the p–i–n structure, drift is utilized all across the device for collection and separation. The homojunction designs of Figures 4.1a–e are suitable for absorber materials that have carrier diffusion lengths which are capable of getting minority photocarriers out of the flat band regions. The p–i–n design of Figure 4.1f is ideal for the situation where the doping adjacent to the contacts, the contact workfunctions, or both set up a field across the structure such that L_n^{Drift} and L_p^{Drift} are both $\approx L_{ABS}$ and the cell width is $\approx L_{ABS}$.

4.2 OVERVIEW OF HOMOJUNCTION SOLAR CELL DEVICE PHYSICS

4.2.1 Transport

We begin our more in-depth examination of how things move around inside the homojunction class of solar cells by considering the simple p–n structure seen in Figure 4.2. Our ultimate goal in this chapter is to understand the origins of the functional relationship between current density J and the voltage V across the cell. If we understand that, we can analyze devices and design better ones. From Figure 4.2, we see that photogenerated free electrons and holes produced by the impinging light are subject to the carrier loss mechanisms of (a) bulk recombination (mechanisms 1, 4, and 5), (b) recombination at the top contact (mechanisms 2 and 3), and (c) recombination at the back contact (mechanisms 6 and 7). While all this recombination is taking place, the lucky photogenerated electrons that escape the top layer,[†] without recombining, will do so by diffusing to the electrostatic-field

---

[†]In p–n homojunction usage, the top layer is commonly referred to as the emitter layer and the bottom layer is commonly termed the base.

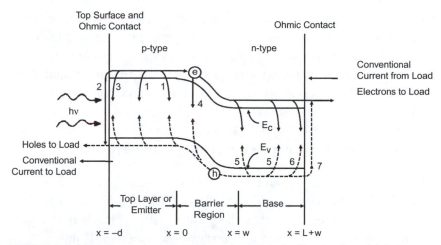

FIGURE 4.2 A simple p–n homojunction solar cell under illumination with light entering from the left. Shown are the bulk photocarrier loss mechanisms 1, 4, and 5 as well as the contact recombination loss mechanisms 2, 3, 6, and 7. Particle current flow is shown. The distinction between mechanisms 2 and 3 and between 6 and 7 has been discussed in depth in Section 2.3.2.9.

(barrier) region. There they will join photogenerated electrons which have been produced in this barrier region and together they will be swept out by drift to the bottom material, if they avoid recombination in the electric field (barrier) region. Once in the bottom material, the electrons are still subject to recombination but become majority carriers moving to the back contact. They are principally moving by drift in this layer but in a very small electric field—such a field can give rise to a significant electron current here, since electrons are the majority carrier in the bottom material. Correspondingly, photogenerated holes in the bottom layer will be principally diffusing to the electrostatic-field barrier region, where they will join holes generated there and together will be swept out by drift toward the top material, all while hopefully avoiding recombination. Once in the top material the holes become majority carriers moving to the front contact, principally as a majority-carrier drift current set up by a very small electric field.

Electrons that succeed in avoiding the loss mechanisms provided by processes 1–7 emerge from the right contact (particle flow) of Figure 4.2 and cause current flow through the external circuit. They traverse the external circuit, do work, and return to the left contact, where they annihilate exactly the same number of emerging holes. These holes too have survived the loss mechanisms in the cell. Figure 4.2 makes it clear that to increase the photovoltaic action, it is necessary to reduce the photocarrier loss effects due to processes 1–7. Note that when the cell of Figure 4.2 is under illumination, the left contact is positive with respect to the right contact at any point on the J-V characteristic in the power quadrant. Note also that in this power-producing mode, conventional current is moving in the minus-x direction of Figure 4.2. While light is depicted impinging onto the left contact in Figure 4.2, none of this description of the cell operation would change if it were impinging onto the right contact.

The upper bound on V_{oc} can be seen from Figure 4.2 to be the built-in electrostatic potential energy V_{Bi} for this simplest of p–n homojunctions. If V_{oc} were to exceed V_{Bi}, then the electric field in the barrier region would change sign, as would the current flow direction. If this were to happen, the current flow direction change would move the device J-V into the power-consumption mode. While V_{Bi} establishes the upper limit on V_{oc}, we stress that it is the kinetics of loss mechanisms 1–7 that establish the actual open-circuit voltage V_{oc}. The more processes 1–7 can be suppressed, the more the open-circuit V_{oc} approaches its limit V_{Bi}. The quantity V_{oc} can be viewed as the voltage necessary to reduce the barrier electrostatic field enough, and, therefore, to drive the electron and hole quasi-Fermi levels apart enough, that loss exactly equals generation.

The net result of all this activity seen under illumination in Figure 4.2 can be expressed in very general mathematical terms by picking any plane[‡] and doing some accounting. This accounting must determine the conventional current density J crossing that plane when the cell is developing a voltage V. Taking the plane $x = L + W$ as the one where we will do this "bean-counting" using electrons as the "beans," it can be seen that the net number of electrons leaving the device per time per area at $x = L + W$ (actually leaving at an x just to the right of processes 6 and 7, as discussed in Section 2.3.2.9) causes the conventional current density J flowing into the device at $x = L + W$ from the load. The bean-counting says that J must arise from $\int_{-d}^{L+W} \int_{\lambda} G_{ph}(\lambda, x)d\lambda dx$, the electrons (and holes) photogenerated per area per time between $-d$ and $L + W$, minus $\int_{-d}^{L+W} \mathcal{R}(x)\, dx$, which counts the electrons (and

---

[‡]The net current density is the same at any plane in the cell since a one-dimensional structure is assumed.

holes) lost per area per time to bulk recombination between $-d$ and $L + W$. The quantities $-J_{ST}(-d)$ and $-J_{SB}(L + W)$, which count the electron (and hole) charge lost per area per time to recombination at the contacts also enter into this accounting. Putting this all together gives

$$
J = -\left[e \int_{-d}^{L+W} \int_{\lambda} G_{ph}(\lambda, x) d\lambda dx \right.
$$

$$
\left. -e \int_{-d}^{L+W} \mathscr{R}(x) \, dx - J_{ST}(-d) - J_{SB}(L + W) \right| \quad (4.1a)
$$

Equation 4.1a is the continuity concept from Section 2.3.3 in its integral form. Here $G_{ph}(\lambda, x)$ is the free carrier photogeneration function introduced in Section 2.2.6 and its λ integration in Eq. 4.1a is over the impinging spectrum. Loss mechanisms 1, 4, and 5 are all accounted for by the integral $\int_{-d}^{L+W} \mathscr{R}(x) \, dx$. The integrand $\mathscr{R}(x)$ represents band-to-band, S-R-H, or Auger net recombination taking place at some point x under the impinging light spectrum. These loss mechanisms are discussed in detail in Section 2.2.5.1. The quantity $J_{ST}(-d)$ accounts for the electrons (and consequently holes) lost at the top (front, light-entering) surface through loss mechanisms 2 and 3, and $J_{SB}(L + W)$ accounts for the electrons (and consequently holes) lost at the back surface through mechanisms 6 and 7 under the impinging light spectrum. Before we proceed any further, we need to look at the sign convention used in Eq. 4.1a. Since we have picked the power quadrant to be the fourth quadrant in Chapter 1 and since J is in the negative x-direction in Figure 4.2 when the device is producing power, the minus sign in front of all the terms in Eq. 4.1a is present in order to be consistent with our sign choices for the x-coordinate and the power quadrant.

It follows from Eq. 4.1a that the short circuit current density J_{sc} is given by

$$
J_{sc} = -\left[e \int_{-d}^{L+W} \int_{\lambda} G_{ph}(\lambda, x) d\lambda dx \right.
$$

$$
\left. -e \int_{-d}^{L+W} \mathscr{R}^{SC}(x) \, dx - J_{ST}^{SC}(-d) - J_{SB}^{SC}(L + W) \right| \quad (4.1b)
$$

Here the superscripts SC denote that the loss terms are evaluated at short circuit. It also follows from Eq. 4.1a that the open circuit voltage occurs when

$$
\begin{aligned}
0 = - \Bigg[& e \int_{-d}^{L+W} \int_{\lambda} G_{ph}(\lambda, x) d\lambda dx \\
& - e \int_{-d}^{L+W} \mathscr{R}^{OC}(x)\, dx - J_{ST}^{OC}(-d) - J_{SB}^{OC}(L+W) \Bigg]
\end{aligned} \qquad (4.1c)
$$

where the superscripts OC denote that the loss terms are evaluated at open circuit.

Now, if we could evaluate the various terms in Eq. 4.1a, we would have the functional relationship we are looking for; i.e., we would have the J-V characteristic. Evaluating these terms takes a great deal of effort and is postponed until Section 4.3, where we use a numerical analysis approach, and Section 4.4, where we use an analytical approach. For now, we have Eqs. 4.1a–4.1c. They are a very general, "overall concepts" type of statements and they apply to all the structures seen in Figure 4.1. Since we have Eq. 4.1a, let us see if we can learn more from it. The loss terms in Eq. 4.1a [i.e., $e \int_{-d}^{L+W} \mathscr{R}(x)\, dx - J_{ST}(-d) - J_{SB}(L+W)$] depend on voltage, a fact we used in Eqs. 4.1b and c. They also depend on illumination. To be specific, they depend on the impinging spectrum distribution and intensities, since the latter two spectrum attributes will affect carrier populations. If there is no illumination, Eq. 4.1a reduces to

$$
J = e \int_{-d}^{L+W} \mathscr{R}^{D}(x)\, dx + J_{ST}^{D}(-d) - J_{SB}^{D}(L+W) \qquad (4.2)
$$

where the superscript D is utilized to stress that the terms $e \int_{-d}^{L+W} \mathscr{R}(x)\, dx - J_{ST}(-d) - J_{SB}(L+W)$ are functions of voltage which must now be evaluated in the dark. Equation 4.2 is usually called the dark current density J-V characteristic and we denote it as $J_{DK}(V)$.

We can also use Eq. 4.1a to understand a helpful measure of cell performance called the quantum efficiency. This quantity measures how many electrons show up doing work in the external circuit for each impinging photon of wavelength λ. Actually, there are two quantum efficiencies: the external quantum efficiency (EQE), which we have already utilized in Chapter 3 and which measures the response to the testing photons $\Phi(\lambda)\Delta\lambda$ in bandwidth $\Delta\lambda$ that are impinging on a cell; and the internal quantum efficiency (IQE), which measures the response to the testing photons $\Phi_C(\lambda)\Delta\lambda$ in bandwidth $\Delta\lambda$ that have entered the device, where $\Phi_C(\lambda)\Delta\lambda$ has been corrected for the photons that were reflected or absorbed before entering the actual cell. As can be discerned from their definitions, both of these quantum efficiencies are ≥ 0. In general, they depend on the voltage across the cell (the bias) and the light spectrum impinging on the cell (the light bias). Most often, quantum efficiencies are reported for no voltage bias (short-circuit condition) and no light bias. However, quantum efficiencies obtain under voltage bias can be useful if charging, and therefore severe electrostatic field redistribution, is occurring under biasing. Similarly light bias can be quite informative if the large photocarrier population present under a light biasing is charging traps and rearranging the electric field in the device. Unless there is carrier multiplication present (see Section 2.2.6), both IQE and EQE are ≤ 100, if reported as a percentage, or ≤ 1, if based on unity. Since EQE is inherently positive, it follows from Eq. 4.1a that it, in the absence of a bias spectrum, is given by

$$
EQE = \Bigg[\int_{-d}^{L+W} G_{ph}^T(\lambda, x)dx\Delta\lambda
$$
$$
- \int_{-d}^{L+W} \mathscr{R}^T(\lambda, x)dx - J_{ST}^T(\lambda, -d) - J_{SB}^T(\lambda, L+W) \Bigg] \Big/ \Phi(\lambda)\Delta\lambda
$$

$$(4.3)$$

where the superscript T means the testing photons $\Phi(\lambda)\Delta\lambda$ are impinging on the cell. The definition of IQE follows by adjusting to having the photons per time per area for the bandwidth $\Delta\lambda$ in the denominator of Eq. 4.3 be the quantity $\Phi_C(\lambda)\Delta\lambda$. This discussion of quantum efficiencies not only applies to homojunctions but is very general and applicable to all of the various classes of solar cells. This generality of Eq. 4.3

comes from its simply being an expression of generation minus loss for a cell under test illumination conditions.

Equation 4.1a can also be used to give us insight into the superposition assumption. This assumption is commonly employed to construct the J-V characteristics for all types of solar cells. It assumes that J can be written as the difference between a photocurrent density, which is independent of voltage but depends on illumination, and a current density term which depends on voltage but is independent of the presence or absence of illumination. Since the latter component is always present, it must be the dark current density $J_{DK}(V)$ discussed earlier. Since a cell's dark current density is zero at short circuit, the photocurrent density must be the short circuit current density J_{sc}. With this superposition assumption, the sought-after relationship between J and V can be then written as

$$J = -[J_{sc} - J_{DK}(V)] \qquad (4.4)$$

The minus sign in front of all the terms on the right hand side again appears because of our sign conventions. When valid, Eq. 4.4 says a homojunction device (or any cell for which it is valid) has the electrical engineering diagrammatic interpretation of a constant current generator shunted by a voltage-dependent element whose J-V characteristic is the same under illumination or in the dark. Equation 4.4 also says these elements are oriented in opposition to one another. From our discussion on symmetry breaking in Ch. 3, one would expect $J_{DK}(V)$ characteristic for the structure of Figure 4.2 to ideally be that of a diode.

Substituting Eqs. 4.1b and 4.2, which are always valid, into the superposition assumption expressed by Eq. 4.4 shows that superposition assumes that J is

$$J = -e\left\{ \int_{-d}^{L+W} \int_{\lambda} G_{ph}(\lambda, x)d\lambda dx - e \int_{-d}^{L+W} \mathscr{R}^{SC}(x)\, dx - J_{ST}^{SC}(-d) \right.$$

$$\left. - J_{SB}^{SC}(L+W) - e \int_{-d}^{L+W} \mathscr{R}^{D}(x)\, dx - J_{ST}^{D}(-d) - J_{SB}^{D}(L+W) \right\}$$

$$(4.5)$$

In other words, it assumes the loss terms in Eq. 4.1a are given at any voltage by the sum of $e \int_{-d}^{L+W} \mathscr{R}^{SC}(x) \, dx + J_{ST}^{SC}(+d) + J_{SB}^{SC}(L+W)$ and $e \int_{-d}^{L+W} \mathscr{R}^{D}(x) \, dx + J_{ST}^{D}(-d) + J_{SB}^{D}(L+W)$. The question that obviously arises is how valid is superposition? Strictly speaking, for superposition to be valid, the system of equations describing what is taking place inside a cell must be linear, but a glance at this mathematical system (see Section 2.4) shows that the system is far from linear, unless a number of phenomena are not playing a major role.

We explore when the features of homojunction light (under illumination) and dark J-V characteristics in detail in Sections 4.3 and 4.4. In Section 4.3, we use the equations of Section 2.4 to numerically evaluate Eq. 4.1a without making any assumptions to get the J-V characteristics. Ideally, this will give us the opportunity to learn more about homojunction cell design and operation. Along the way, we will determine how well Eq. 4.4 matches what we see. In Section 4.4, we use an analytical approach to obtaining the J-V characteristics. This will give us the opportunity to examine in detail all the assumptions needed for Eq. 4.4 to be rigorously valid. Hopefully that, too, will give us further insight into homojunction behavior and performance optimization.

4.2.2 The homojunction barrier region

The electrostatic barrier region is the "charge separation engine" of homojunction cells. It is either of the p–n type seen in Figures 4.1a–e or of the p–i–n type seen in Figure 4.1f. These barriers break symmetry and make one direction different from the other, thereby causing charge separation and current flow. As may be noticed from the band diagrams of the p–n homojunction devices of Figures 4.1a–e, a characteristic feature of p–n cells is that they all have a barrier region that is constrained in its extent and flat band (no built-in electric field) regions on both sides of the electrostatic barrier in TE. Under illumination and current-flow conditions, our numerical analysis (Section 4.3) will show an electric field does develop in these flat band regions of a p–n cell but it usually tends to be small. When the electric field is small outside the barrier under light and non-existent in TE, the regions cannot have any significant charge density present under illumination. Because of this, they are often quasi-neutral regions. "Quasi" is Italian for "almost," so we can have quasi-neutral regions and quasi-Fermi levels (defined in Appendices C and D) in solar cell device physics. Since the electric field is small in these "almost" neutral regions of p–n homojunctions, we expect the minority-carrier collection to the barrier region to be dominated by diffusion when the cell is under illumination. Drift should be less important for these carriers in the quasi-neutral region because it would involve the product of a minority population and a very small electrostatic field. The p–i–n device of Figure 4.1f is the other extreme. It has a built-in electrostatic field, and therefore an electrostatic field barrier, that extends across the cell at TE and there are no flat band regions. Under operation, such cells have been designed to have collection of photogenerated carriers accomplished everywhere by drift. In many situations, analytical solutions to Poisson's equation (Eq. 2.45 in Section 2.3.4) yielding $E_C(x)$, $E_V(x)$, and $E_{VL}(x)$ and the electric field $\xi(x)$ can be obtained for p–n and p–i–n barrier regions as a function of voltage. Discussions of this analytical approach can be found in standard device physics books.[17] Of course, we can find $E_C(x)$, $E_V(x)$, and $E_{VL}(x)$ and $\xi(x)$ for both the quasi-neutral and barrier regions in all situations using numerical solution techniques.

4.3 ANALYSIS OF HOMOJUNCTION DEVICE PHYSICS: NUMERICAL APPROACH

We will now use numerical analysis to solve the equations of Section 2.4 exactly, and thereby determine the electrostatic field, recombination, and currents in p–n and p–i–n homojunction solar cells in the dark and under illumination. This will allow us to develop their dark and light J-V relationships. One sun illumination (specifically AM1.5G) will be used to determine the light J-V behavior. The power of the numerical approach is that one does not have to make the extensive assumptions that are necessary with the analytical approach to obtain the required J-V characteristics. All the rich device physics is retained. Numerical analysis also allows us to peel open a cell and watch, in great detail, what is going on inside. By using numerical analysis, we can thoroughly gauge the impact of HT-EBL and ET-HBL contact materials or the use of back surface high-low doping structures. All the devices that we explore in Sections 4.3 have absorption which produces only free electron-hole pairs. The examples from Sections 4.3.1 to 4.3.4 are designed to principally utilize minority-carrier diffusion collection in the cell flat band regions; i.e., diffusion is used to collect the photogenerated minority carriers to the electrostatic field barrier regions of these cells. Sections 4.3.5 and 4.36 examine the alternative p–i–n cell approach of using drift carrier collection everywhere. Although Section 3.4.1.4 showed that two- and even three-dimensional solar cell structures can be advantageous, one-dimensional configurations are used here to keep things relatively simple and to concentrate on the device physics.

4.3.1 Basic p–n homojunction

The first homojunction to be considered for numerical analysis is a Figure 4.2–type structure. It is seen in Figure 4.3 and is described by the parameters in Table 4.1. The material parameters employed in this table are typical of silicon. We call this simple cell our baseline device and start our numerical analysis study of homojunctions with it. There has been no attempt at optimization for this cell. As can be seen from Figure 4.3, the contacts play no role in creating the built-in potential. The built-in potential V_{Bi} is 0.62 eV and it is entirely developed by the workfunction difference between the p and n regions. The absorber is seen from Table 4.1 to have gap defect states present at a density level of $\sim 10^{14} \text{cm}^{-1}$. These states are spread across the band gap and act as S-R-H recombination centers. Transport at the contacts is modeled with the surface recombination speeds listed in the table.

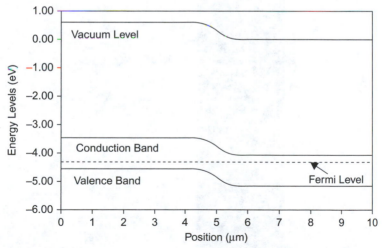

FIGURE 4.3 The band diagram of a very simple, hypothetical p–n homojunction in thermodynamic equilibrium. The built-in potential V_{Bi} is 0.62 eV. Under illumination, light enters from the left side.

Table 4.1 Basic p–n Homojunction Parameters

Overall length (p-region length) (n-region length)	Band gap	Electron affinity	Absorption properties	Doping density for p-region (N_A) and for n-region (N_D)	Front-contact workfunction and surface recombination speeds	Back-contact workfunction and surface recombination speeds	Electron and hole mobilities	Band effective densities of states	Bulk defect properties
10,000 nm (5000 nm) (5000 nm)	$E_G = 1.12\,\text{eV}$	$\chi = 4.05\,\text{eV}$	Absorption data for Si used (See Fig. 3.19)	$N_A = 1.0 \times 10^{15}\,\text{cm}^{-3}$, $N_D = 1.0 \times 10^{15}\,\text{cm}^{-3}$	$\phi_W = 4.93\,\text{eV}$, $S_n = 1 \times 10^7\,\text{cm/s}$, $S_p = 1 \times 10^7\,\text{cm/s}$	$\phi_W = 4.31\,\text{eV}$, $S_n = 1 \times 10^7\,\text{cm/s}$, $S_p = 1 \times 10^7\,\text{cm/s}$	$\mu_n = 1350\,\text{cm}^2/\text{vs}$, $\mu_p = 450\,\text{cm}^2/\text{vs}$	$N_C = 2.8 \times 10^{19}\,\text{cm}^{-3}$, $N_V = 1.04 \times 10^{19}\,\text{cm}^{-3}$	Donor-like gap states from E_V to mid-gap $N_{TD} = 1 \times 10^{14}\,\text{cm}^{-3}\,\text{eV}^{-1}$ $\sigma_n = 1 \times 10^{-14}\,\text{cm}^2$ $\sigma_p = 1 \times 10^{-15}\,\text{cm}^2$ Acceptor-like gap states from mid-gap to E_C $N_{TA} = 1 \times 10^{14}\,\text{cm}^{-3}\,\text{eV}^{-1}$ $\sigma_n = 1 \times 10^{-15}\,\text{cm}^2$ $\sigma_p = 1 \times 10^{-14}\,\text{cm}^2$

Using numerical techniques to simultaneously solve the whole system of equations in Section 2.4 for the structure of Table 4.1 and Figure 4.3, without any assumptions, gives the J-V characteristics under light and in the dark presented in Figure 4.4. This figure conveys this information as linear and semi-log J-V plots. The J_{sc} is seen to be $11.8\,mA/cm^2$, the $V_{oc} = 0.43\,V$, the $FF = 0.74$, and the efficiency is 3.8% under

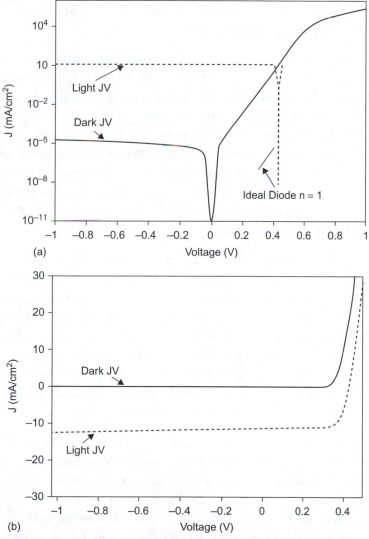

FIGURE 4.4 The numerically determined light and dark J-V behavior for the cell of Table 4.1 and Figure 4.3 depicted using a (a) semi-log and (b) linear plot. As is our custom, the sign convention for the linear plot has the fourth quadrant as the power quadrant.

the AM1.5G. These plots show that superposition predicts the open-circuit voltage V_{oc} well, since it is seen that $J_{DK}(V_{oc}) = J_{sc}$. The plots of the dark J-V that come from the numerical analysis show it to have the $e^{V/nkT}$ voltage dependence expected of a diode where n is the diode quality or n-factor. From the n = 1 line on this figure, it can be seen that the voltage dependence of the dark current is actually that of an ideal diode with n = 1. This behavior is seen to persist until the dark J-V enters into the series-resistance-limited region found at the higher voltages of the first quadrant.

Numerical analysis permits a detailed examination of how this cell is actually working. Considering TE first, we know that $J_n \equiv 0$ and $J_p \equiv 0$ in TE according to the principle of detailed balance (see Appendices B and C). However, there clearly are drift and diffusion currents flowing at TE due to the electric field and the population changes across the barrier region. These currents are seen in Figure 4.5a for electrons and in Figure 4.5b for holes. The electron drift and diffusion current densities as well as the hole diffusion and drift current densities must be in an exact point-by-point balance at TE, as these plots show. Under light, bias, or both this delicate balance is upset.

Now let us examine in some detail what the numerical analysis says about the cell of Figure 4.3 under light. In particular, let us focus on the origins of J_{sc}. Looking at Figure 4.6 it can be seen that in the region between ~2 μm and ~8 μm, both electrons and holes are carrying J_{sc}. After ~8 μm electrons are the sole carrier responsible for the photocurrent and carry it while also compensating for the holes' motion, which is in the wrong direction in this part of the cell. Between the front contact and ~1.8 μm, holes are carrying J_{sc} and compensating for the electrons, which are moving in the wrong direction.

The electron and hole diffusion and drift components at short circuit, which are presented in Figures 4.7 and 4.8, give even more details and are very insightful. For example, that unhelpful behavior of electrons in the region between the front contact and ~1.8 μm is seen in Figure 4.7 to be due to their diffusing (as particles) to the front contact. This is arising because the front contact is a very attractive recombination sink for electrons due to its large S_n value in Table 4.1. From ~1.8 μm to the left-hand edge of the electrostatic field–caused barrier region (easily seen since it is where the

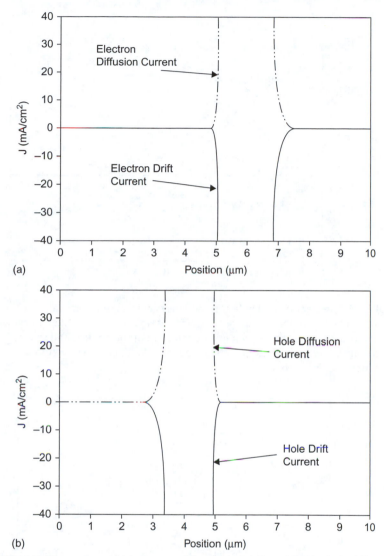

FIGURE 4.5 The computed (a) electron diffusion and drift and (b) the hole diffusion and drift current density components at TE for the device of Figure 4.3 and Table 4.1. The sign convention in these spatial plots, here and throughout the book, is tied to the x-coordinate system and consequently the positive direction for conventional current densities is left to right.

electron drift current becomes important), the electrons are again moving by diffusion but now in the useful direction of helping to carry J_{sc}. They are diffusing toward the barrier because it too is a sink for photogenerated electrons. This is the diffusion collection of minority excess carriers which

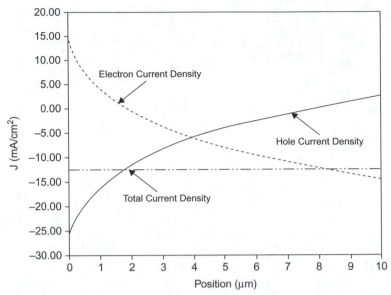

FIGURE 4.6 The electron and hole current density components produced by the numerical analysis at short circuit for the device of Figure 4.3 and Table 4.1. The sign convention in these spatial plots, here and throughout the book, is tied to the x-coordinate system and consequently the positive direction for conventional current densities is left to right.

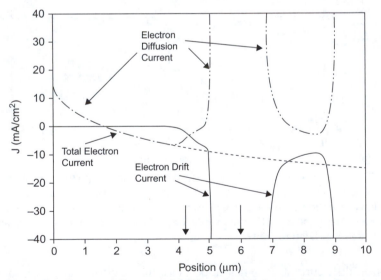

FIGURE 4.7 The electron diffusion and drift current density components at short circuit for the device of Figure 4.3 and Table 4.1. The broad arrows indicate the edges of the electrostatic-field barrier region. The sign convention is that the positive direction for conventional current densities is left to right.

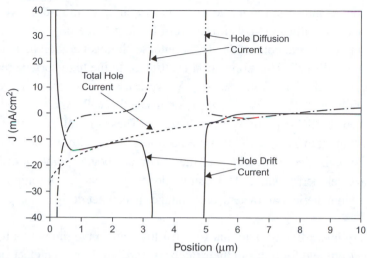

FIGURE 4.8 The computed hole diffusion and drift current density components at short circuit for the device of Figure 4.3 and Table 4.1. The sign convention is that the positive direction for conventional current densities is left to right.

we have mentioned many times. It occurs because the barrier's drift field whisks electrons into the n-material side. This important role of drift in the barrier region is seen in the plot of the electron drift current density given in Figure 4.7. The delicate balance seen in the barrier region for TE has been upset and now drift dominates over back diffusion in the barrier region, giving a net electron current density helping to carry J_{sc}. To the right of the right-hand edge of the space-charge region, the electrons are majority carriers having a net motion controlled by drift. This drift has to compensate in a micron thick region at the cathode for back diffusion from the back contact due to the contact's being somewhat of a bottleneck to exiting electrons. The same detail on the flow of the short-circuit current is provided for holes in Figure 4.8. The results are seen to be analogous to those for electrons. In the n-material region, where holes are the minority carrier, diffusion wins over drift in a collection region adjacent to the barrier and causes hole flow to the barrier. There is hole diffusion to the sink of the back contact but, as one moves closer to the sink of the barrier region, diffusion to the barrier region takes over. Drift and back diffusion compete in the barrier region but drift dominates, giving the barrier region its hole-sink property, and seeping holes to the p-material. In the p-material, holes are the majority carrier and move by drift. The symmetry with the electron motion of Figure 4.7 is lost, however, in the vicinity of the front contact. Unlike electron motion at the back contact, hole motion in the immediate vicinity of the front contact is dominated by diffusion. This arises from the large hole population just to the right of the front contact due to the intense photogeneration there as the light enters the cell. This population is attracted to the high surface recombination speed of the front contact. As we can see from these numerical simulation results, there is a lot going on inside this cell. And we have only examined the SC condition in detail so far. These complicated flow patterns set up in this cell at SC that we have just observed in Figures 4.6–4.8 must be optimized, particularly at the maximum power point, in order to optimize cell performance. This can be done with changes in region sizes, changes in material parameters, and changes in contact design.

It is very interesting to note that for drift to occur in the directions it does for electrons and for holes in their majority-carrier regions, an electrostatic field must develop in what had been the flat band regions in TE. If we consider how this electric field must be oriented to cause these majority carrier conventional currents, we can see that it gives rise to a voltage in the opposite sense to the voltage direction desired in the power quadrant. This last statement may be understood from the general expression for the voltage V across the electrodes (Eq. 2.47); i.e., $V = \int_{structure} [\xi(x) - \xi_0(x)]dx$, where $\xi(x)$ is the electric field at some operating point (J,V) and $\xi_0(x)$ is the built-in field at TE. This electrostatic field developed in what had been the field-free flat band regions of TE, if precipitable in the J-V characteristics, can appear as a phenomenological series resistance.

One way of viewing the voltage developed by the cell of Figure 4.3 when under AM15G illumination is depicted in the band diagram of Figure 4.9 for the case of open circuit. The fact that a voltage is being developed is seen in the electron and hole quasi-Fermi level split. As we noted as early as in Chapter 2, the open circuit voltage V_{oc} is specifically given by the magnitude of the difference in the Fermi level positions in the contacts. Interestingly, these numerical simulation results show that the quasi-Fermi levels are both flat almost all the way across the device for the open circuit condition. We will utilize that when we undertake our analytical analysis. From Appendix D, we know that the statements $J_n = en\mu_n(dE_{Fn}/dx)$ and $J_p = -ep\mu_p(dE_{Fp}/dx)$ include diffusion and drift conventional currents. Consequently, the noticeable variation in the electron quasi-Fermi level at the front contact is caused by electron particle diffusion to the

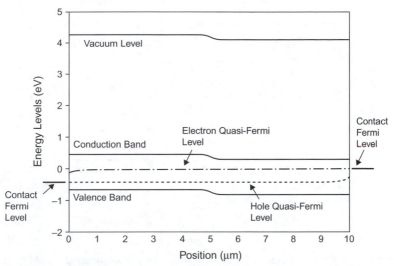

FIGURE 4.9 Band diagram of the device in Figure 4.3 numerically computed for open circuit. The computed flat electron and hole quasi-Fermi levels are seen.

front-contact electron recombination sink. This is, of course, is matched at open circuit by holes coming to the front contact. Only the variation in the electron quasi-Fermi level is noticeable here near the contact, since $p \gg n$ in this region. An analogous assessment applies to the hole quasi-Fermi level behavior seen at the back contact. As we said, the Fermi level split between the two contacts is always the voltage across the electrodes and, at open circuit, it must be $V_{oc} = \int_{structure} [\xi(x) - \xi_0(x)]dx$, where $\xi(x)$ is now the field in the cell at open circuit. Since the negative of the derivative of the local vacuum level is always the field (see Appendix D), a comparison of Figures 4.3 and 4.9 shows where the open-circuit voltage is being developed in the cell. This comparison shows V_{oc} is essentially completely arising due to a reduction in the barrier-region electrostatic field in this cell, and therefore due to a reduction in the barrier height. We reiterate that V_{oc} occurs when recombination equals generation. Another way of looking at this is to say V_{oc} occurs when the quasi-Fermi level splitting causes enough recombination to balance the generation.

4.3.2 Addition of a front HT-EBL

We will now utilize numerical analysis to explore how modifying the cell contacts affects cell operation. The p–n homojunction structure to be explored next is the device of Figure 4.3 but with a 40-nm front HT-EBL

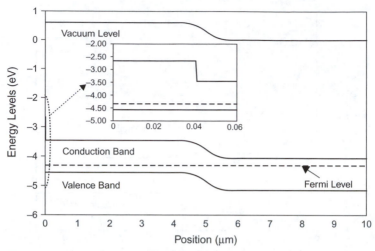

FIGURE 4.10 The computed TE band diagram of the p–n homojunction of Figure 4.3 with the addition of a 40-nm HT-EBL at the front contact. The built-in potential V_{Bi} is still 0.62 eV. Under illumination, light enters from the left side.

added. All the absorber material parameters are the same for this cell and the contact workfunctions are also the same as those of the cell of Figure 4.3. The added HT-EBL material has the doping density and hole affinity of the absorber but a band gap of 1.90 eV. The resulting band diagram in TE is given in Figure 4.10. The inset in the figure shows the HT-EBL region in more detail. As seen in the computed linear and semilog J-V plots of Figure 4.11, the impact of the addition of a front HT-EBL is dramatic. The performance parameters are now $J_{sc} = 25.9$ mA/cm^2, $V_{oc} = 0.48$ V, FF = 0.79, and $\eta = 9.9\%$ under AM1.5G. Superposition is seen to predict the open-circuit voltage (i.e., $J_{DK}(V_{oc}) = J_{sc}$) and the dark diode characteristic is seen to be very close to having n = 1.

The numerically computed current density analysis results given in Figures 4.12 and 4.13 for the short-circuit condition show the origins of the dramatic impact of the front HT-EBL. First, it can be seen that the general features of the electron-hole sharing of the short-circuit current carrying duties seen in Figure 4.6 for the simple p–n homojunction are reproduced in Figure 4.12, except now J_{sc} is much higher. However, unlike the case in Figure 4.6, in Figure 4.12 the electron current density at the front-contact region never becomes positive (never flows counter to the direction of J_{sc}). Actually, the insert in the figure shows the electron current is reduced almost to zero in the HT-EBL. Figure 4.13 makes it

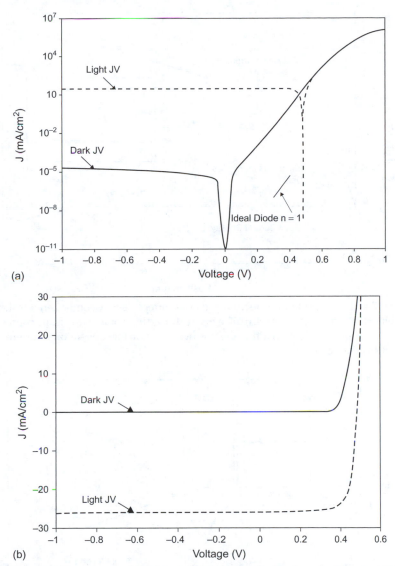

FIGURE 4.11 The computed light and dark J-V behavior for the cell of Figure 4.10 depicted using a (a) semi-log and (b) linear plot. The sign convention for the linear plot has the fourth quadrant as the power quadrant.

clear that the reason for this is that there is no electron diffusion to the front-contact recombination plane when the HT-EBL is present. Electrons that used to be lost to recombination at the front contact (Fig. 4.7) are now able to move (as particles) to the right and contribute to J_{sc}. In other words, electrons photogenerated in the absorber cannot get to the

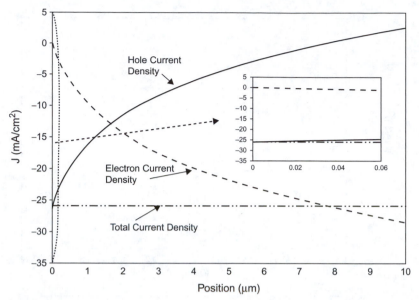

FIGURE 4.12 The computed electron and hole current density components at short circuit for the device of Figure 4.10. The insert shows the region of the front-contact HT-EBL. The sign convention is that the positive direction for conventional current densities is left to right.

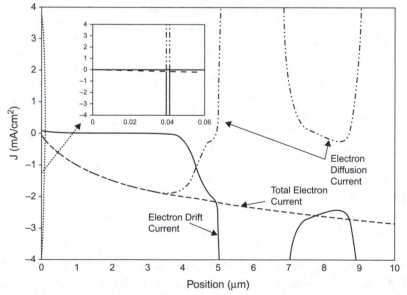

FIGURE 4.13 The computed electron diffusion and drift current density components at short circuit for the device of Figure 4.10. The insert shows the region of the front-contact HT-EBL. The competition between effective field drift and diffusion at the HT-EBL/absorber interface is seen in this insert. The sign convention is that the positive direction for conventional current densities is left to right.

front-contact recombination sink due to the barrier presented by the affinity step at the HT-EBL/absorber interface. The HT-EBL structure has effectively made the electron recombination speed S_n at the front contact equal to zero. This front electrode now is a selective ohmic contact and passes holes without any electron-hole losses. Since this selective ohmic contact has succeeded in essentially eliminating front-contact recombination losses, the open-circuit voltage has to increase to get bulk and back-contact losses equal to the photogeneration.

It is very interesting to go back to the insert in Figure 4.13 and to note the diffusion-drift competition at the HT-EBL/absorber interface. First of all, it is critical to realize that the drift seen at the HT-EBL interface with the absorber is not electrostatic field drift seen elsewhere in the cell but it is the effective field drift of Chapter 2. This drift and diffusion must also be present in TE, since the huge electron population difference and the affinity step at the interface are always there. From detailed balance we know they exactly cancel each other at TE. At short circuit, this effective force drift is seen to win at the interface and to carry a very small electron current present from limited photogeneration in the wide-gap HT-EBL into the absorber. Again we emphasize that this electron drift current at the HT-EBL boundary is not caused by an electrostatic field but, due the affinity step, is caused by an electron effective force field that only acts on electrons. The hole equivalent to Figure 4.13 is not shown since the flow pattern does not change much for holes from that seen for the basic p–n cell.

4.3.3 Addition of a front HT-EBL and back ET-HBL

Since the front HT-EBL structure proved to be so effective in enhancing the performance of the basic p–n homojunction of Figure 4.3, we now assess the impact of using the same HT-EBL at the front together with an ET-HBL at the back. We again stress that we are not trying to do a full performance optimization for this cell but we are exploring how to impact homojunction operation. This back ET-HBL that is being added to the cell has the same electron affinity as the absorber, but its band gap is 1.90 eV. The resulting, computed band diagram in TE is given in Figure 4.14. The insets in the figure show the front HT-EBL and back ET-HBL regions in more detail. The impact of having both of these structures is seen in the computer generated linear and semi-log J-V plots of Figure 4.15. The performance parameters are now $J_{sc} = 28.4 \, \text{mA/cm}^2$, $V_{oc} = 0.55 \, \text{V}$, FF = 0.79, and $\eta = 12.4\%$ under AM1.5G. Superposition is seen to be slightly off for this structure,

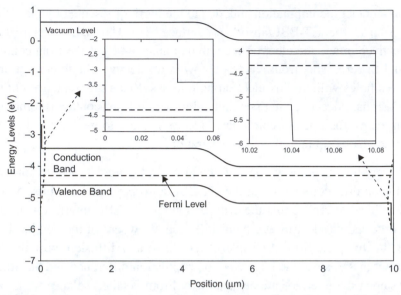

FIGURE 4.14 The computed TE band diagram of the p–n homojunction of Figure 4.3 but with the addition of a 40-nm HTL-EBL at the front contact (left insert) and a 40-nm ET-HBL at the back contact (right insert). The built-in potential V_{Bi} is still 0.62 eV. Under illumination, light enters from the left side.

with J_{sc} not exactly predicting V_{oc}. Interestingly, the dark diode characteristic has an n-diode factor that is a function of voltage and greater than unity. The non-linear aspects of the full equation set of Section 2.4 are showing up in the light and dark J-V behavior of this device.

Figure 4.16 shows that the short-circuit electron current density continues to be zero at the front contact and now the hole current density is zero at the back contact, with the inclusion of the ET-HBL. The details for the electron diffusion and drift components are not presented because the behavior is similar to that of Figure 4.13. The details for the hole short-circuit diffusion and drift components are seen in Figure 4.17. The figure shows that hole diffusion to the back-contact recombination plane has been stopped and holes, which used to be lost to recombination at the back contact (Fig. 4.8), are now able to move to the left and contribute to J_{sc}. In other words, holes photogenerated in the absorber cannot get to the back-contact recombination sink due to the barrier presented by the hole affinity step at the ET-HBL/absorber interface. The ET-HBL structure has effectively made the hole recombination speed $S_p = 0$ at this back contact. The back electrode now has a selective ohmic contact and passes

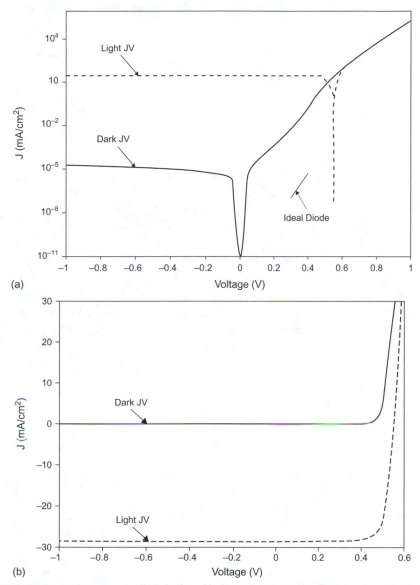

FIGURE 4.15 The numerically calculated light and dark J-V behavior for the cell of Figure 4.14 depicted using a (a) semi-log and (b) linear plot. Note that the n-factor is a function of voltage and larger than unity.

electrons without any electron-hole losses. Since this selective ohmic contact has succeeded in essentially eliminating back-contact recombination losses, the open-circuit voltage has to increase to get the only recombination mechanism remaining, bulk losses, to equal the photogeneration.

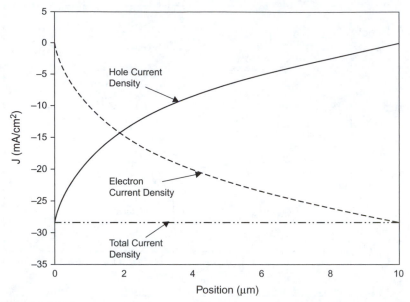

FIGURE 4.16 The computed electron and hole current density components at short circuit for the device of Figure 4.14. The sign convention is that the positive direction for conventional current densities is left to right.

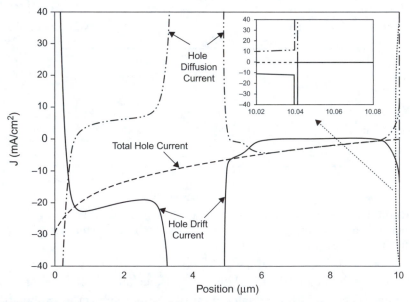

FIGURE 4.17 The computed hole diffusion and drift current density components at short circuit for the device of Figure 4.14. The insert shows the details near and in the ET-HBL. The sign convention is that the positive direction for conventional current densities is left to right.

It can be noted that there is diffusion-drift competition at the ET-HBL/ absorber interface. The drift, in this case, is due to the hole effective field at the interface. These two components must also be present in TE, since the things causing them, the huge hole population difference and the hole affinity step at the interface, are always there. From detailed balance we know they exactly cancel each other at TE, and, as seen in Figure 4.17, they come close to summing to zero at short circuit too. As noted, the hole drift current at the interface is not caused by an electrostatic field but, due the hole affinity step, is caused by a hole effective force field that acts only on those particles. Figure 4.17 shows that immediately to the left of this interface there is an electrostatic hole drift current but this becomes overwhelmed by hole diffusion to the left as the electrostatic barrier region is approached.

4.3.4 Addition of a front high-low junction

We now turn to the approach of using a high-low junction to form a HT-EBL and consider the device seen in Figure 4.18. In our simulation of this structure we will assume a front layer of 40 nm of heavily doped

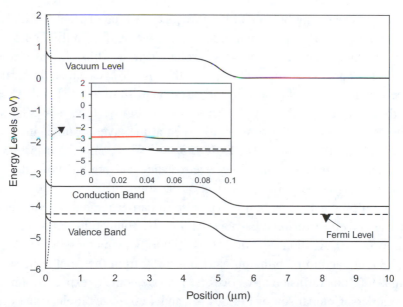

FIGURE 4.18 The computed TE band diagram of the p–n homojunction of Figure 4.3 but with the addition of a 40-nm heavily doped layer causing a high-low junction-type HT-EBL at the front contact. The band bending across the p–n junction is still 0.62 eV but V_{Bi}, the potential energy difference developed from contact to contact, is now 0.84 eV. Under illumination, light enters from the left side.

($N_A = 1.0 \times 10^{19} cm^{-3}$) p-type material has been added to the basic cell of Figure 4.3. The influence of the resulting high-low junction is seen in the insert of Figure 4.18 to extend out into the absorber. The heavily doped layer has all the same material parameters as the absorber, including the same affinities as the absorber. The front-contact workfunction has been changed to 5.15 eV to give the band diagram for this cell in TE seen in Figure 4.18. The structure is a very different form of an HT-EBL from those that we have tried so far. It uses an electric field and an electrostatic barrier, rather than an effective force field and an affinity step barrier. It creates electrostatic forces that act on both carriers. (The previous HT-EBL and ET-HBL structures each had effective forces that acted on one carrier type only.) The impact of this type of HT-EBL structure on device performance is captured in the linear and semi-log J-V plots of Figure 4.19 and is seen to be very similar to that of the front HT-EBL formed by an electron affinity step. The performance parameters for the device in Figure 4.18 are $J_{sc} = 25.6 mA/cm^2$, $V_{oc} = 0.48 V$, FF = 0.79, and $\eta = 9.8\%$ under AM1.5G. Superposition is seen to predict the open-circuit voltage V_{oc} and the dark diode characteristic is seen to be that of an ideal diode; i.e., n = 1.

We begin examining how this high-low junction-type of front HT-EBL is functioning by looking at Figure 4.20. First, we note from the figure that there is competition between very strong electron diffusion and electric field drift currents at the right side of the high-low electrostatic barrier. The beginning of this is seen in the insert at ~40 nm and the end is seen at the right side of the high-low electrostatic barrier (i.e., at ~800 nm). This is analogous to the competition seen at the layer/absorber interface in the effective force-type HT-EBL structure. The competition in Figure 4.20 happens over the high-low electrostatic field region. For effective field cases, it happens over the extent of the affinity step. As we have seen many times, this competition is typical of a barrier region (whether electrostatic or effective force barrier). At TE, the competition must balance out to $J_n \equiv 0$. At short circuit, drift is seen to be winning between 40 nm and 800 nm and to be bringing electrons away from the front surface. At about 1000 nm, diffusion takes over, as we have come to expect for minority carriers. Comparison of Figures 4.7 and 4.20 makes it clear that once again the success of a front HT-EBL, even the high-low junction-type, lies in its blocking of electron diffusion out of the absorber region to the front-contact recombination plane ($S_n = 1 \times 10^7 cm/s$). Electrons from

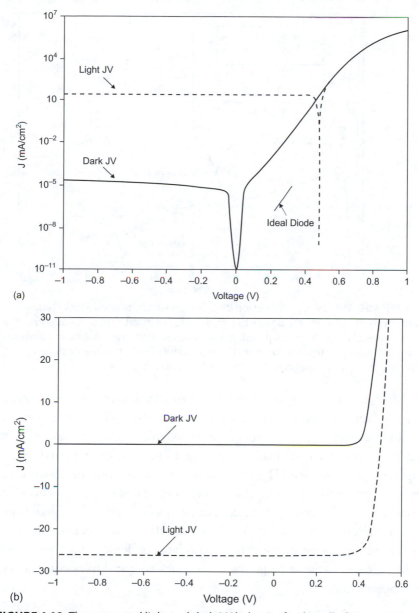

FIGURE 4.19 The computed light and dark J-V behavior for the cell of Figure 4.18 depicted using a (a) semi-log and (b) linear plot.

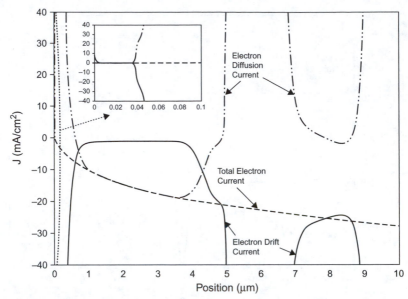

FIGURE 4.20 The computed electron diffusion and drift current density components at short circuit for the device of Figure 4.18. The insert shows the region of the front-contact HT-EBL high-low junction. The competition between drift and diffusion at the high-low junction can be seen. The sign convention is that the positive direction for conventional current densities is left to right.

the absorber that used to be lost to recombination at the front contact (Fig. 4.7) are now able to move as particles to the right and contribute to J_{sc}. In other words, electrons photogenerated in the absorber cannot get to the front-contact recombination sink due to the barrier presented by the high-low junction. The HT-EBL structure has effectively made the electron recombination speed S_n at the front contact equal to zero. This electrode now has a selective ohmic contact and passes holes with significantly reduced electron-hole losses. Since this selective ohmic contact has succeeded in supressing front-contact recombination losses, the open-circuit voltage has to increase to get bulk and back-contact losses equal to the photogeneration.

This electrostatic field type of HT-EBL is different from its effective force counterpart (Section 4.3.2) in one regard: in its effect on hole transport at the front surface. This is made clear by comparison of Figure 4.21 with Figure 4.17. In the latter figure, which applies to an effective force type front HT-EBL, the sink provide by the front contact hole recombination speed S_p is seen to set up a hole diffusion current

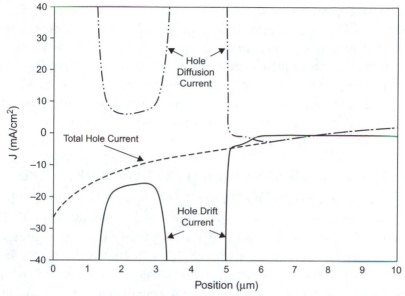

FIGURE 4.21 The computed hole diffusion and drift current density components at short circuit for the device of Figure 4.18. The sign convention is that the positive direction for conventional current densities is left to right.

which takes over from hole drift very near the front contact and brings the holes to the front electrode. In fact, this is the same behavior seen in Figure 4.8 which applies to the case of no HT-EBL of any kind. In the case of the electrostatic field type HT-EBL, Figure 4.21 shows that hole drift takes over the task of pushing holes all the way to the front contact and hole diffusion no longer has a front contact role. This change has occurred because the electrostatic field type of HT-EBL affects both electron and hole transport, since electrostatic fields, as opposed to effective fields, necessarily act on both carrier species.

Obviously, a high-low ET-HBL could also be added to the back of the p–n homojunction cell of Figure 4.18 and additional benefits would accrue. In general, combinations of electrostatic and effective field selective contacts, including those with graded affinities, can be used with p–n homojunctions. In our numerical simulations we added such layers without addressing a lot of processing and materials compatibility issues. Among them are whether or not the inclusion of effective field type selective ohmic contacts causes interface states that could act as additional recombination conduits, whether or not the use of the heavy doping needed for the electrostatic field type of selective ohmic contacts creates defects that provide more recombination conduits, and whether or not the benefits of these contacts justifies the added processing complexity. We could explore these questions. For example, we could plug in defects that could arise from the additional processing, give them various energy level and spatial distributions, and study their impact. This is all left to the interested reader.

4.3.5 A p–i–n cell with a front HT-EBL and back ET-HBL

We now return to the device of Figure 4.14, a p–n homojunction cell having selective ohmic contacts due to a front electron affinity step HT-EBL and a back hole affinity step ET-HBL. As we recall, this cell gives the performance seen in Figure 4.15; i.e., $J_{sc} = 28.4\,\text{mA/cm}^2$, $V_{oc} = 0.55\,\text{V}$, FF = 0.79, and $\eta = 12.4\%$ for AM1.5G. We now take the absorber from this cell and see how it would perform if it were placed in a p–i–n, instead of a p–n, homojunction. We also keep the selective ohmic contacts of Figure 4.14 but dope them, as described below. The computed band diagram of this p–i–n cell is given in Figure 4.22 for TE. All the material parameters are the same as those of the Figure 4.14 cell except for the doping

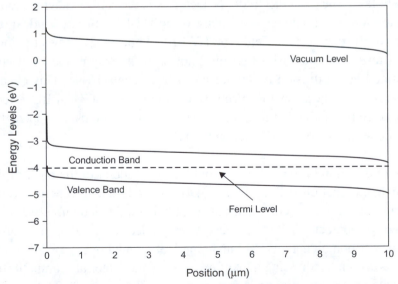

FIGURE 4.22 The computed TE band diagram of the p–i–n homojunction corresponding to the p–n homojunction of Figure 4.14. Under illumination, light enters from the left side.

and the electrode workfunctions. The absorber now is undoped and both selective ohmic contact regions are degenerately doped, as Figure 4.22 shows. These have p^+ and n^+ doping densities of $5 \times 10^{19} \, \mathrm{cm}^{-3}$. The electrode workfunctions have been appropriately modified to give flat bands at the contacts. Figure 4.22 shows that the changes have resulted in having a built-in potential of about 1 V developed over the 10-μm absorber.

Although a built-in potential of $\sim$1V exists across this cell, the electrostatic field in much of the absorber is seen to be lower than the $\sim 10^4 \, \mathrm{V/cm}$ that would be expected. The reduction of the field in the absorber is due to the gap defect states listed in Table 4.1. These are storing charge that shields some of the electrostatic field emerging from the heavily doped selective contact regions; i.e., the charge developed in these gap states blocks electric flux lines originating in the heavily doped layers from penetrating into the i-layer. This can be seen from the absorber's band-bending behavior adjacent to the doped layers in Figure 4.22. The result of this shielding is a field in the interior of the absorber that is suppressed to $\sim 5 \times 10^3 \, \mathrm{V/cm}$. Figure 4.23 presents the performance for this p–i–n in terms of the light and dark J-V characteristics. As seen, the p–i–n has $J_{sc} = 28.4 \, \mathrm{mA/cm^2}$, $V_{oc} = 0.55 \, \mathrm{V}$, FF = 0.72, and $\eta = 11.3\%$. Comparing Figures 4.15 and 4.23 and the efficiency, etc., shows that the performance of the p–n device is slightly better than the corresponding p–i–n cell for this particular set of material parameters and absorber thickness. Details for the electron and hole current density components for this p–i–n cell are given for the short-circuit condition in Figures 4.24 and 4.25. Figure 4.24 shows that, as expected, the electron current density is flowing by drift across the entire p–i–n cell (electron drift dominates over diffusion everywhere). Correspondingly, Figure 4.25 shows that, as expected, the hole current density is flowing by drift across the entire p–i–n cell (drift dominates over diffusion everywhere). A comparison of the hole current density components, for example, between Figures 4.17 and 4.25 shows the changes in the roles of the hole drift and diffusion components between the p–n and the p–i–n configurations.

4.3.6 A p–i–n cell using a poor $\mu\tau$ absorber

Some guidance on when to employ the p–i–n configuration becomes apparent if one looks at absorbers with poorer collection capabilities than those that have been considered so far. We know that the diffusion length, a measure of the effectiveness of collection by diffusion (see Section 3.4.1.3),

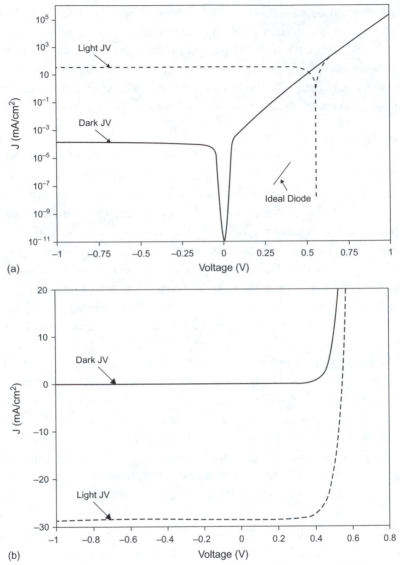

FIGURE 4.23 The computed light and dark J-V behavior of the p–i–n cell of Figure 4.22 depicted using a (a) semi-log and (b) linear plot. Superposition is seen to work well. The n-factor for the dark J-V is larger than unity indicating a less than ideal-diode-like dark J-V.

is proportional to $(\mu\tau)^{1/2}$ and that the drift length, a measure of the effectiveness of collection by drift, is proportional to $(\mu\tau\xi)$. Consequently, poor collection capability is synonymous with a small $(\mu\tau)$ product. Although the carrier lifetime concept is not being used in this section (i.e., we have

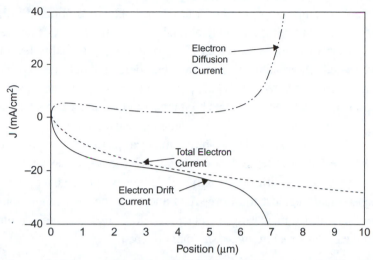

FIGURE 4.24 The numerically determined electron diffusion and drift current density components at short circuit for the p–i–n device of Figure 4.22. Electron drift dominates everywhere. The sign convention is that the positive direction for conventional current densities is left to right.

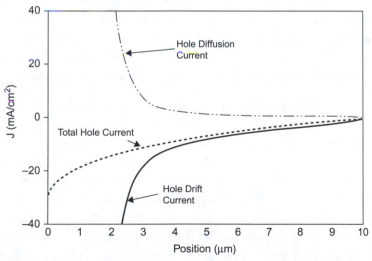

FIGURE 4.25 The computed hole diffusion and drift current density components at short circuit for the p–i–n device of Figure 4.22. Hole drift dominates everywhere. The sign convention is that the positive direction for conventional current densities is left to right.

the freedom of not being forced to utilize a linear recombination picture since we are employing a numerical analysis), it still follows that we can think of carrier lifetime τ values as a way of assigning a quality level to an absorber; i.e., materials with many gap-state defects would be expected to have low values of τ. With this $(\mu\tau)$ picture in mind, we create an absorber with poorer collection capabilities by increasing the defect gap states above what we have been using up to now while decreasing the mobilities below what we have been using. The resulting material parameters for this poorer absorber are listed in Table 4.2. While we have lowered the $(\mu\tau)$ product and the effect on the diffusion length will follow, we notice from $(\mu\tau\xi)$ that we can mitigate against the lower $(\mu\tau)$ and save the drift length, at least to some extent, by increasing the electric field ξ. It is this realization that makes the p–i–n structure of considerable interest.

To overcome poor collection capabilities, we need a short device. Shorter device lengths allow the built-in potential of a p–i–n structure to exist over shorter distances, which means an advantageous higher built-in electrostatic field can be attained. This higher field can then be used to at least partially compensate for a lower $(\mu\tau)$ product through drift collection. The need for a short device length requires that a material with poor collection capabilities must be a strong absorber, if it is to have any usefulness, and the hypothetical poor collection capability absorber of Table 4.2 has strong absorption properties. (The absorption coefficients of a-Si are used—see Table 4.2 and Figure 3.19.) Since the absorption length is short for the material of Table 4.2, we will try using this poor $\mu\tau$ absorber in devices whose length has been reduced to 430 nm (see Figure 3.20 in considering this choice).

We now can compare the performance of a p–n structure and a p–i–n structure, both based on the absorber of Table 4.2, using numerical analysis. The material and device parameters for these two structure types are listed in the table. Neither device has selective ohmic contacts. Figures 4.26 and 4.27 present the TE band diagram and performance results for the p–n cell. As seen, this device has $J_{sc} = 2.20 \text{ mA/cm}^2$, $V_{oc} = 0.87 \text{ V}$, FF = 0.75, and $\eta = 1.4\%$. Figure 4.28 is a plot of the p–n junction total electron and total hole current density components at short circuit. As may be seen, in the front of this cell the hole flow has to overcome that of the electrons which is in the wrong direction adjacent to the front contact. In the back, the hole flow is essentially negligible and electrons must carry the entire current density. Obviously the back of this particular cell is not effective.

Table 4.2 Some Parameters for Modeling Poorer Collectors

Device configuration	Overall length (p-region length) (n-region length)	Band gap	Electron affinity	Absorption properties	p-type doping density (N_A) and n-type doping density (N_D)	Back contact work function and surface recombination speeds	Electron and hole mobilities	Band effective densities of states	Bulk defect properties	Front-contact workfunction and surface recombination speeds
p–n structure	430 nm (p-region 215 nm) (n-region 215 nm)	$E_G = 1.80$ eV	$\chi = 3.80$ eV	Absorption data for a-Si (See Fig. 3.19)	$N_A = 3.0 \times 10^{18}$ cm^{-3} for 215 nm; $N_D = 8.0 \times 10^{18}$ cm^{-3} for 215 nm	$\phi_W = 4.95$ eV, $S_n = 1 \times 10^7$ cm/s; $S_p = 1 \times 10^7$ cm/s	$\mu_n = 20$ cm^2/Vs; $\mu_p = 2$ cm^2/Vs	$N_C = 2.5 \times 10^{20}$ cm^{-3}; $N_V = 2.5 \times 10^{20}$ cm^{-3}	Donor-like gap states from E_V to mid-gap: $N_D = 1 \times 10^{16}$ cm^{-3}eV^{-1}; $\sigma_n = 1 \times 10^{-15}$ cm^2; $\sigma_p = 1 \times 10^{-17}$ cm^2. Acceptor-like gap states from mid-gap to E_C: $N_A = 1 \times 10^{16}$ cm^{-3}eV^{-1}; $\sigma_n = 1 \times 10^{-17}$ cm^2; $\sigma_p = 1 \times 10^{-15}$ cm^2	$\phi_W = 5.20$ eV, $S_n = 1 \times 10^7$ cm/s; $S_p = 1 \times 10^7$ cm/s
p–i–n structure	430 nm, including a 25-nm contact p-layer and a 25-nm contact n-layer	Same as the p–n	Same as the p–n	Same as the p–n	$N_A = 3.0 \times 10^{18}$ cm^{-3} for 25 nm; $N_D = 8.0 \times 10^{18}$ cm^{-3} for 25 nm	Same as the p–n	Same as the p–n	Same as the p–n	Same as the p–n	Same as the p–n

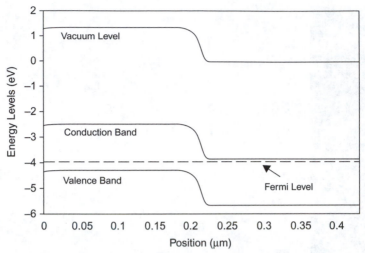

FIGURE 4.26 The computed TE band diagram for the p–n device of Table 4.2.

Figure 4.29 presents the carrier lifetimes for the short-circuit and open-circuit conditions for this cell. These are computed, in the case of electrons, by using the linear recombination model $\mathscr{R} = (n - n_0)/\tau$ from Chapter 2 and plugging in the actual (non-linearized) recombination $\mathscr{R}(x)$ and carrier density $n(x)$ given by the numerical analysis to thereby determine the lifetime. The corresponding procedure is followed for holes. The resulting value of τ_n seen for the p-side and of τ_p seen for the n-side are in Figure 4.29. From these plots, it can be determined that our absorber has a $(\mu\tau)$ product of $\tau_p\mu_p \sim 10^{-9}\,\mathrm{cm^2/v}$ for electrons and of $\tau_p\mu_p \sim 10^{-10}\,\mathrm{cm^2/v}$ for holes. Although Figure 4.29 plots lifetimes across the cell structure, it follows from our discussion in Section 2.2.5 that the electron lifetime, as we are deducing it, only has meaning in the quasi-neutral p-region. Correspondingly, the hole lifetime only has meaning in the quasi-neutral n-region. Importantly, these plots show that within the regions of their applicability, the behavior of these lifetimes is very close to being constant with position and biasing conditions. This turns out to be fairly true for p–n cells in general, and we use that observation in our analytical approach (Section 4.4), since, to make that analysis tractable, we will have to use the linearized recombination approximations and assume spatially independent lifetimes.

Figure 4.30 gives the computed band diagram at TE for the corresponding p–i–n structure while Figure 4.31 shows its dark and light J-V characteristics. The p–i–n design, which pushes the doped regions to

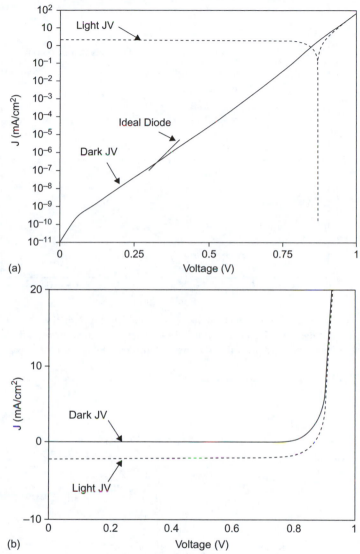

FIGURE 4.27 The computed light and dark J-V behavior of the p–n cell of Figure 4.26 depicted using a (a) semi-log and (b) linear plot. Note that the n-factor for the dark J-V is larger than unity indicating a less than ideal dark current behavior.

adjacent to each contact, is seen to be much better than the corresponding p–n device, with performance parameters of $J_{sc} = 11.23 \, \text{mA/cm}^2$, $V_{oc} = 0.91 \, \text{V}$, FF = 0.79, and $\eta = 8.1\%$. Figures 4.32a and b give the electron and hole current density components at short circuit for this p–i–n cell. As expected, drift is the dominant collection mechanism for both carriers, except for electrons very near the front contact due to that

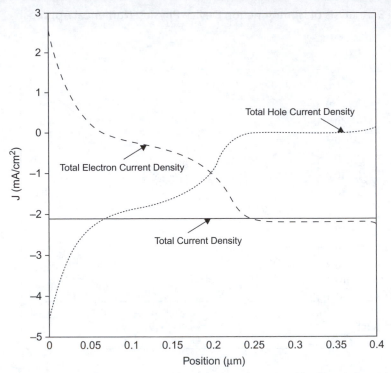

FIGURE 4.28 The computed electron, hole, and total current densities at short circuit for the p–n device of Figure 4.26 and Table 4.2. The sign convention is the usual: the positive direction for conventional current densities is left to right.

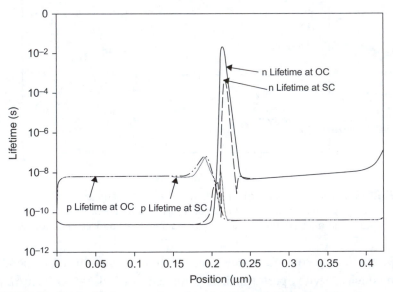

FIGURE 4.29 Carrier lifetimes as a function of position obtained by using the numerical analysis output for recombination and the carrier densities in linearized expressions of the form $\mathscr{R} = (n - n_0)/\tau$. Shown are the lifetimes for the open-circuit (OC) and short-circuit (SC) conditions. As discussed in the text, these lifetimes are meaningless outside the respective minority carrier regions.

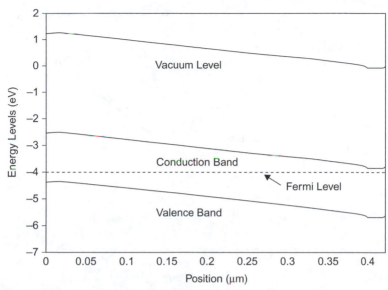

FIGURE 4.30 The computed TE band diagram for the p–i–n device of Table 4.2.

plane's photocarrier sink effect. As seen holes are involved in carrying the short circuit current now even in the back region of the cell. The carrier lifetimes, obtained in the same manner as those of Figure 4.29, are plotted versus position in Figure 4.33. One may argue that the electron lifetime has meaning on the left side of the structure and the hole lifetime has meaning on the right side of the structure where they are minority carriers. Even in those regions, they now vary considerably with position and biasing condition, and they are larger than those seen in Figure 4.29, although the absorber and its defects are the same. Of course, they must be larger since there is more current collection in the p–i–n and therefore less loss. These observations underscore carrier lifetime is not necessarily a true, fixed material parameter and can even depend on cell design.

As we have seen, homojunction cells can be designed to have a spread-out, built-in electrostatic-field collecting current across the device by drift and be of the p–i–n type. They can also be designed to have regions with flat bands in TE adjacent to a confined electrostatic-field barrier region and rely principally on diffusion to bring photogenerated carriers to the charge-separating barrier and be of the p–n type. In determining which approach is better, consideration must be given to absorption properties, carrier mobilities, the recombination processes, the gap-state distribution in space and energy, and the doping profiles. Consideration must also be given to processing issues, such as whether or not doping

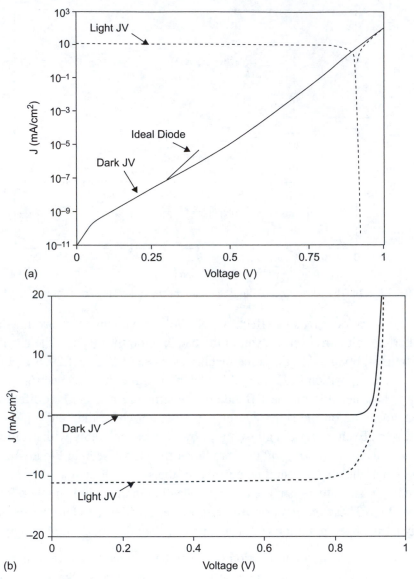

FIGURE 4.31 The computed light and dark J-V behavior of the p–i–n cell of Figure 4.30 depicted using a (a) semi-log and (b) linear plot. Note that the n-factor for the dark J-V is larger than unity.

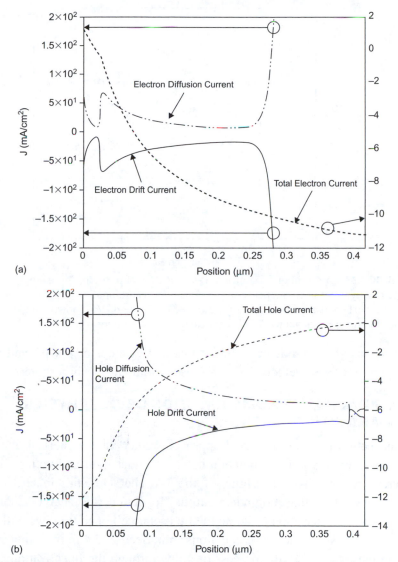

FIGURE 4.32 Computer results giving: (a) The electron diffusion and drift current density components as well as (b) the hole diffusion and drift current density components at short circuit for the p–i–n device of Figure 4.30 and Table 4.2. Note that different scales are used for the current components and the total carrier currents. As usual, the sign convention is that the positive direction for conventional current densities is left to right.

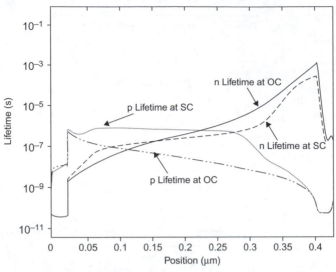

FIGURE 4.33 Carrier lifetimes computed as a function of position. These are obtained by using the numerical analysis output for recombination and the carrier densities in linearized expressions of the form $\mathscr{R} = (n - n_0)/\tau$. Shown are the lifetimes for the open-circuit (OC) and short-circuit (SC) conditions.

of the absorber introduces defect states and whether or not p^+ and n^+ doping introduces defects at the p^+-absorber and n^+-absorber interfaces.

4.4 ANALYSIS OF HOMOJUNCTION DEVICE PHYSICS: ANALYTICAL APPROACH

This section uses an analytical approach, rather than numerical analysis, to describe p–n homojunction behavior under light and voltage and thereby to describe the origins of the J-V characteristics in terms of equations rather than computer output. To obtain this analytical one-dimensional description of homojunction solar cell operation, we will be forced to linearize the set of equations from Section 2.4 to try to make it more manageable. In the process of going through the linearization, we will have the opportunity to examine the approximations behind Eq. 4.4, an expression that we have found works well in many cases for homojunctions (Section 4.3). The analytical approach is not capable of capturing all the details that we explored with numerical analysis. For example, it cannot handle the rich interplay in a region between drift and diffusion. The best it can do for the presence of selective ohmic contacts is to simply set the minority-carrier recombination speed equal to zero. We also need to note that we will not address p–i–n structures, since an analytical approach is complex and of doubtful value for that configuration. One

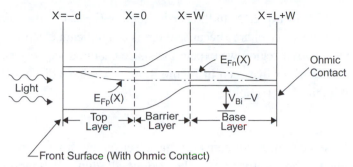

FIGURE 4.34 Schematic of a p–n homojunction solar cell configuration with light impinging at x = −d.

obvious problem, for example, is seen in Figure 4.33: there are no spatially independent and operating-conditions independent lifetimes to use in a linear, lifetime recombination model.

4.4.1 Basic p–n homojunction

Figure 4.34 gives the band diagram and coordinate direction used for our analytical approach to p–n homojunctions. The p-type portion of the absorber has been placed in the back of the cell for variety. This means that the back contact is now going to be the anode. The device is shown under illumination and, as a consequence, a voltage V is being developed. A portion, V_1, of the voltage is developed in the n-side; the remainder, V_2 (where $V = V_1 + V_2$), is developed in the p-side. Our objective is to find an analytical expression for the current density J being produced under illumination at a voltage V for this structure. Figure 4.34 assumes flat quasi-Fermi levels across the total barrier region and much of the cell, except for where minority-carrier bulk and surface recombination (at the contacts) are dominating. The numerical analysis results of Figure 4.9 show that, at least for open circuit, this type of quasi-Fermi level pattern is actually what is present.

We could hope to use Eq. 4.1 to get an analytical expression for the J–V characteristic. This would require expressions for the carrier populations everywhere across the device in order to evaluate the $\int_{-d}^{L+W} \mathscr{R}(x)\,dx$ term. Instead, we will use a current density formulation to obtain the J–V characteristic. It has the benefit of allowing us to go through the approximations involved in the analytical approach systematically. To begin this, let us pick a plane in Figure 4.34 to evaluate J. Our selection is x = 0, the plane just outside the space-charge region in the n-type portion of the absorber. As we have noted, the term space charge is the name given to the right hand side of Eq. 2.45. In the barrier region, Eq. 2.45 says there must be a significant space charge because there is a significant electric field but outside this barrier region we expect the space charge to be very small for cell designs such as that seen in Figure 4.34, as we have discussed previously. With this x = 0 choice,

$$J = J_n(0) + J_p(0) \qquad (4.6)$$

where these quantities are conventional electron and hole current densities defined to be positive, as we have done all along, if flowing left to right in Figure 4.34. This means that, for the cell orientation of Figure 4.34, we will have to use $-J$ for any plots in the power quadrant.

The plane x = 0 is an advantageous one to select, and of course this is why we picked it, since $J_p(0)$ can be obtained from:

$$J_p(0) = -eD_p \frac{dp}{dx}\Big|_{x=0} \qquad (4.7)$$

This recipe for $J_p(0)$ is so simple because we are making the standard assumptions employed in analytical analyses of p–n homojunctions.[17] The assumptions are that (1) in the n-type side of the electrostatic field barrier of Figure 4.32, the holes remain the minority carrier even under illumination, and (2) the absorber is a lifetime semiconductor (see Appendix E) and indeed the region x ≤ 0 is very close to being space charge neutral (i.e., it is quasi-neutral). These assumptions say that the mobility times the hole population times the electric field is small at x = 0 (and for all x ≤ 0) compared to $D_p(dp/dx)$ and that any electric field in x ≤ 0 is small and has a negligible effect on minority carriers. While we did not make any of these assumptions in the numerical analysis, Eq. 4.7 is consistent with what we saw in the numerical analyses results of the previous section; i.e., for homojunction designs represented by Figure 4.34, minority-carrier holes come out of the n-type absorber portion and arrive at the edge of the electrostatic-field barrier region by diffusion.[†]

[†]The n-type absorber portion (or the p-type absorber portion or both) can have an advantageously oriented constant electric field built in by well-designed doping variation. We did not examine this in Section 4.3, Its presence would give rise to a drift current also being present but balanced by diffusion in TE. This case can be handled easily in analytical analysis using minor adjustments. A full analytical discussion of this was first given by Wolf,[18] who showed that the incorporation of a constant electric field in the space-charge-neutral region into the analysis has the effect of aiding diffusion collection to the barrier region. Obviously, its addition in the numerical analysis approach is straightforward.

If we knew the functional form of the hole number density $p = p(x)$ in this $x \leq 0$ region, then we would have $J_p(0)$ and be well on our way to obtaining the cell's J-V characteristics. We can develop the function $p(x)$ for the quasi-neutral region $x < 0$ of the n-type portion of the absorber, if we use Eqs. 2.48b and d from Section 2.4 with the assumptions that holes are the minority carrier and their transport is by diffusion only and add new assumptions 3 and 4. Assumption 3 is that recombination for holes in the n-portion of the absorber follows the linear recombination lifetime model $\mathcal{R}(x) = (p - p_{n0})/\tau_p$ from Section 2.2.5.1, where τ_p is a constant hole minority-carrier lifetime. The appropriateness of a lifetime recombination model for formulating uni-molecular and bi-molecular recombination processes for such a region is discussed in Chapter 2 and Section 4.3. Assumption 4 is that contacts can be characterized with surface recombination speeds. One may argue that this is not an assumption limited to the analytical approach, since we used it in the numerical analysis. However, there is a huge difference. In the numerical analysis, the recombination-speed boundary condition could be made to have no influence, if we put layers beside the contacts. The layers, when present, dictated the conditions at the contacts. We do not have the flexibility to easily do that in the analytical approach. The contacts must be characterized by S_n and S_p to obtain a relatively simple analytical formulation.

Under these assumptions $p(x)$ satisfies

$$\frac{d^2p}{dx^2} - \frac{p - p_{n0}}{L_p^2} + \int_\lambda \frac{\Phi_0(\lambda)}{D_p} \alpha(\lambda) e^{-\alpha(x+d)} d\lambda = 0 \qquad (4.8)$$

subject to the boundary conditions

$$\frac{dp}{dx}\Big|_{x=-d} = \frac{S_p}{D_p}[p(-d) - p_{n0}] \qquad (4.9a)$$

$$p(0) = p_{n0} e^{E_{Fp}(0)/kT} \qquad (4.9b)$$

Here the hole quasi-Fermi level E_{Fp} is being measured positively down from the Fermi-level position in energy at the metal contact where holes are the minority carrier; i.e., it is being measured down in this case from the metal-contact workfunction position at x = −d in Figure 4.34. Eq. 4.8 assumes that absorption in this absorber material, over the wavelengths of interests (the solar spectrum), only produces free electron-hole pairs, follows the Beer-Lambert model, and generates free electron-hole pairs according to

$$\int_\lambda G_{ph}(\lambda, x)d\lambda = \int_\lambda G(\lambda, x)d\lambda = \int_\lambda \Phi_0(\lambda)\alpha(\lambda)e^{-\alpha(\lambda)(x+d)}d\lambda$$

as discussed in Section 2.2.4.1. In this equation, the spectrum of light of intensity $\Phi_0(\lambda)$ at wavelength λ enters into the structure from the left in Figure 4.34.

The solution to Eqs. 4.8, 4.9a, and 4.9b is worked out in Appendix F and is used in Eq. 4.7 to obtain

$$J_p(0) = e\int_\lambda \Phi_0(\lambda)\left\{\left[\frac{\beta_2^2}{\beta_2^2 - \beta_1^2}\right]\left[\frac{(\beta_3\beta_1/\beta_2) + 1}{\beta_3\sinh\beta_1 + \cosh\beta_1}\right] - \left[\frac{\beta_2^2 e^{-\beta_2}}{\beta_2^2 - \beta_1^2}\right]\right.$$

$$\left.\left[\left(\frac{\beta_3\cosh\beta_1 + \sinh\beta_1}{\beta_3\sinh\beta_1 + \cosh\beta_1}\right)\left(\frac{\beta_1}{\beta_2}\right) + 1\right]\right\}d\lambda$$

$$-\left\{\frac{eD_pP_{n0}}{L_p}(e^{V/kT} - 1)\right\}\left\{\frac{\beta_3\cosh\beta_1 + \sinh\beta_1}{\beta_3\sinh\beta_1 + \cosh\beta_1}\right\} \qquad (4.10)$$

As we have noted, Figure 4.34 assumes the hole quasi-Fermi level to be flat across the region where holes are the majority carrier and across the electrostatic-field barrier region. We take the flat hole quasi-Fermi level picture to be valid across the whole cell operating range in our analytical analysis; i.e., Eq. 4.10 uses $E_{Fp}(0) = V$. This is the fifth major assumption we have made in establishing the hole contribution to Eq. 4.6. The expression for $J_p(0)$ given by Eq. 4.10 is seen to depend on the incoming light, as it must, and on the voltage V developed across the cell. This piece of the J-V characteristic definitely displays super-position, as would be expected, since it comes from a linearized set of equations (Eqs. 4.8, 4.9a, and 4.9b).

Table 4.3 Top Region Beta Parameters

β Quantity	Definition	Physical significance
β_1	d/L_p	Ratio of n-portion quasi-neutral region length to hole diffusion length. Captures the physics of hole collection in the n-material by diffusion, while subject to recombination. Need $d \leq L_p$.
$\beta_2(\lambda)$	$d\alpha(\lambda)$	Ratio of n-portion quasi-neutral region length to the absorption length for light of wavelength λ. Captures the physics of absorption in the n-material quasi-neutral region. Want $d + W + L \cong 1/\alpha(\lambda)$.
β_3	$L_p S_p/D_p$	Ratio of top-surface hole carrier recombination velocity S_p to hole diffusion-recombination velocity D_p/L_p in the n-portion. Captures the physics of hole contact recombination versus hole recombination while diffusing. Need $D_p/L_p > S_p$.

The definitions of the β quantities appearing in Eq. 4.10 are listed in Table 4.3. These β quantities are useful because they obviously save a lot of writing but, much more importantly, because they nicely show the interplay between the various material parameters characterizing the n-portion of the absorber. Using a numerical analysis, as we did in the last section, gives a powerful tool for exploring complex p–n homo-junction solar cell structures with all sorts of contacts, defects, doping profiles, and band-gap grading. Using an analytical approach, as we are now doing, does not give this flexibility but it does capture the essence of p–n homojunction solar cell behavior. It also gives us these concise, insightful, dimensionless β quantities.

We handled $J_p(0)$ first because it was easy to get an analytical expression for this term by making the assumptions we discussed. Unfortunately, evaluating $J_n(0)$ is not so straightforward. Analytical evaluation of $J_n(0)$ requires our knowing both the electron diffusion and drift components at x = 0. We need both because electrons are the majority carrier and, although the charge density and therefore the electric field may be small in the quasi-neutral region of the n-material, the electron popu-lation times the electric field product may not be small. We also have the advantage of having used numerical analysis to look in detail in Figure 4.7 at how majority-carrier electrons are transported in the quasi-neutral region of the n-type material of this simple p–n homojunction cell. What we saw in that figure was that electron drift does dominate at what we are calling x = 0 in Figure 4.34. Our final assessment, therefore, is that drift is very important in determining $J_n(0)$ and we would need to know the electric field ξ in the region $x \leq 0$ to evaluate this current. That would necessitate numerical analysis, and we have done all that in the previous section. Since we are committed to analytical techniques at the moment, we use the continuity concept in its integral form to sidestep evaluation of the electron current density J_n at x = 0. In using continuity, we can transfer the problem of evaluating J_n at x = 0 to evaluating J_n at x = W.

Using the continuity concept allows $J_n(0)$ to be expressed as

$$J_n(0) = e \int_0^W \int_\lambda G_{ph}(\lambda, x) d\lambda dx - e \int_0^W \mathscr{R}(x)\, dx + J_n(W) \quad (4.11)$$

The advantage offered by Eq. 4.11 is that electrons are the minority carrier for the quasi-neutral region $x \geq W$ of the p-portion of the p–n homojunction we are analyzing. This fact permits us to employ the same approach in evaluating $J_n(W)$ that we took in evaluating $J_p(0)$; i.e., transport by diffusion dominates, as substantiated by the numerical analysis of Section 4.3 and, in particular, by Figure 4.7. This allows us to write

$$J_n(W) = eD_n \left.\frac{dn}{dx}\right|_{x=W} \quad (4.12)$$

The sign in Eq. 4.12 has been adjusted to make J positive when it is flowing in the +x direction. Now we also have to address the generation $\int_0^W \int_\lambda G_{ph}(\lambda, x) d\lambda dx$ as well as recombination $\int_0^W \mathscr{R}(x)\, dx$ in the electrostatic-field barrier region of width W. We will return to these terms, but first let us obtain $J_n(W)$.

Obtaining $J_n(W)$ from Eq. 4.12 necessitates determining the function n = n(x) for the region $x \geq W$. The required n(x) satisfies

$$\frac{d^2n}{dx^2} - \frac{n - n_{p0}}{L_n^2} + \int_\lambda \frac{\Phi_0(\lambda)}{D_n} e^{-\alpha(d+W)} \alpha(\lambda) e^{-\alpha(x+d)} d\lambda = 0 \quad (4.13)$$

which follows from Eqs. 2.48a and 2.48c along with application of the first four assumptions discussed above in the context of holes. The boundary conditions imposed on the solution to Eq. 4.13 are analogous to Eqs. 4.9a and 4.9b; i.e.,

$$n(W) = n_{p0}\, e^{E_{Fn}(W)/kT} \tag{4.14a}$$

and

$$\frac{dn}{dx}\bigg|_{x=W+L} = -\frac{S_n}{D_n}[n(W+L) - n_{p0}] \tag{4.14b}$$

The quantity $E_{Fn}(W)$ in Eq. 4.14a is the electron quasi-Fermi level position at $x = W$. This quasi-Fermi level is being measured positively up from the metal-contact Fermi energy position at $x = W + L$.

The $n(x)$ that results from solving the set of Eqs. 4.13, 4.14a, and 4.14b for the $x \geq W$ region is found in Appendix F. Using this function, we find from Eq. 4.12 that

$$
\begin{aligned}
J_n(W) = e\int_{\lambda} \Phi_0(\lambda) & \left\{ \left[\frac{\beta_6^2 e^{-\beta_4}}{\beta_5^2 - \beta_6^2} \right] \left[\frac{[(\beta_7\beta_5/\beta_6) - 1]e^{-\beta_6}}{\beta_7 \sinh\beta_5 + \cosh\beta_5} \right] \right. \\
& \left. + \left[\frac{\beta_6^2 e^{-\beta_4}}{\beta_5^2 - \beta_6^2} \right] \left[1 - \left(\frac{\beta_5}{\beta_6} \right) \left(\frac{\beta_7 \cosh\beta_5 + \sinh\beta_5}{\beta_7 \sinh\beta_5 + \cosh\beta_5} \right) \right] \right\} d\lambda \\
& - e \frac{D_n n_{p0}}{L_n} \left[e^{V/kT} - 1 \right] \left[\frac{\beta_7 \cosh\beta_5 + \sinh\beta_5}{\beta_7 \sinh\beta_5 + \cosh\beta_5} \right] \tag{4.15}
\end{aligned}
$$

This equation employs the assumption of a flat electron quasi-Fermi level across the quasi-neutral p-portion of the cell and across the electrostatic-field barrier region; i.e., Eq. 4.15 uses $E_{Fn}(W) = V$. This assumption

Table 4.4 Bottom Region Beta Parameters

β Quantity	Definition	Physical significance
$\beta_4(\lambda)$	$(d + W)\alpha(\lambda)$	Ratio of the absorber thickness up to the beginning of the quasi-neutral region in the p-portion to absorption length for light of wavelength λ. (This ratio depends on λ.) Captures the physics of absorption prior to the light's entry into the p-material quasi-neutral region. Desired value of this ratio depends on how much the p-portion is being depended on for absorption.
β_5	L/L_n	Ratio of p-portion quasi-neutral region length to electron diffusion length. Captures the physics of electron collection by diffusion, while subject to recombination, from the p-material quasi-neutral region. Need $L \leq L_n$.
$\beta_6(\lambda)$	$L\alpha(\lambda)$	Ratio of the p-portion quasi-neutral region length to absorption length for light of wavelength λ. (This ratio depends on λ.) Captures the physics of absorption in the p-material quasi-neutral region. Want $d + W + L \cong 1/\alpha(\lambda)$.
β_7	$L_n S_n/D_n$	Ratio of back-surface electron carrier recombination velocity S_n to the electron diffusion-recombination velocity D_n/L_n. Captures the physics of electron recombination at the back contact versus electron recombination while diffusing. Need $D_n/L_n > S_n$.

is analogous to the fifth assumption used for holes (discussed above). The dimensionless β quantities appearing in Eq. 4.15 are the p-region equivalents of those appearing in Eq. 4.10. Their definitions are given in Table 4.4. These quantities show the interplay among the various parameters characterizing the p-region of the absorber. $J_n(W)$ is seen to depend on light absorption, as we would expect, and on the voltage V developed across the cell. The piece of the J-V characteristic given by Eq. 4.15 is the product of a linear set of equations (Eqs. 4.13, 4.14a, and 4.14b) and, as a consequence, it is seen to obey superposition.

Now we can assemble our long-sought-after analytical J-V expression for a simple p–n homojunction solar cell. We do so by substituting Eq. 4.15 into Eq. 4.11 and then substitute the resulting expression into Eq. 4.6, together with substituting Eq. 4.10 into Eq. 4.6. This gives the full analytical J-V characteristic for a simple p–n solar cell; i.e.,

$$
J = e \int_\lambda \Phi_0(\lambda) \left\{ \left[\frac{\beta_2^2}{\beta_2^2 - \beta_1^2} \right] \left[\frac{(\beta_3 \beta_1 / \beta_2) + 1}{\beta_3 \sinh \beta_1 + \cosh \beta_1} \right] \right.
$$

$$
\left. - \left[\frac{\beta_2^2 e^{-\beta_2}}{\beta_2^2 - \beta_1^2} \right] \left[\left(\frac{\beta_3 \cosh \beta_1 + \sinh \beta_1}{\beta_3 \sinh \beta_1 + \cosh \beta_1} \right) \left(\frac{\beta_1}{\beta_2} \right) + 1 \right] \right\} d\lambda
$$

$$
- \left\{ \frac{e D_p p_{n0}}{L_p} (e^{V/kT} - 1) \right\} \left\{ \frac{\beta_3 \cosh \beta_1 + \sinh \beta_1}{\beta_3 \sinh \beta_1 + \cosh \beta_1} \right\}
$$

$$
+ e \int_\lambda \Phi_0(\lambda) \left\{ \left[\frac{\beta_6^2 e^{-\beta_4}}{\beta_5^2 - \beta_6^2} \right] \left[\frac{[(\beta_7 \beta_5 / \beta_6) - 1] e^{-\beta_6}}{\beta_7 \sinh \beta_5 + \cosh \beta_5} \right] \right.
$$

$$
\left. + \left[\frac{\beta_6^2 e^{-\beta_4}}{\beta_5^2 - \beta_6^2} \right] \left[1 - \left(\frac{\beta_5}{\beta_6} \right) \left(\frac{\beta_7 \cosh \beta_5 + \sinh \beta_5}{\beta_7 \sinh \beta_5 + \cosh \beta_5} \right) \right] \right\} d\lambda
$$

$$
- e \frac{D_n n_{p0}}{L_n} \left[e^{V/kT} - 1 \right] \left[\frac{\beta_7 \cosh \beta_5 + \sinh \beta_5}{\beta_7 \sinh \beta_5 + \cosh \beta_5} \right]
$$

$$
+ \int_\lambda \Phi_0(\lambda)(e^{-\beta_2} - e^{-\beta_4}) d\lambda - e \int_0^W \mathscr{R}(x)\, dx \tag{4.16}
$$

where

$$
\int_0^W \int_\lambda G_{ph}(\lambda, x)\, d\lambda dx = \int_\lambda \Phi_0(\lambda) e^{-\alpha d} (1 - e^{-\alpha W}) d\lambda
$$

$$
= \int_\lambda \Phi_0(\lambda)(e^{-\beta_2} - e^{-\beta_4}) d\lambda
$$

has been utilized. This analytical expression is definitely complex. By looking at the various terms of this equation, we can say with certainty that all of them, except for the last one, are linear in light intensity and voltage and will not cause the failure of superposition, so long as the assumptions upon which they are based are valid. We cannot make such a statement about the last term in Eq. 4.16, since we do not have analytical expression for it.

Obtaining an analytical expression for the electrostatic-field barrier region recombination term $e \int_0^W \mathscr{R}(x)\, dx$ in Eq. 4.16 presents a difficult task, since we cannot linearize the recombination model in that region and there is a strong electric field there that must be considered in the continuity equation for each carrier. However, this recombination term is often observed experimentally to have a behavior of the form

$$\int_0^W R(x)\, dx = \left\{ J_{SCR} \left(e^{(V/n_{SCR})kT} - 1 \right) \right\} \tag{4.17}$$

where, in reality, there can be voltage and also light dependence in the prefactor J_{SCR} and also in the n-factor n_{SCR}. J_{SCR} and n_{SCR} can depend on the kinetics of the loss paths, the carrier populations, the illumination level, and the electric field in the barrier region. Plugging Eq. 4.17 into Eq. 4.16 gives the sought-after general analytical J-V expression for a simple p–n solar cell under illumination:

$$J = e \int_\lambda \Phi_0(\lambda) \left\{ \left[\left[\frac{\beta_2^2}{\beta_2^2 - \beta_1^2} \right] \left[\frac{(\beta_3\beta_1/\beta_2) + 1}{\beta_3 \sinh\beta_1 + \cosh\beta_1} \right] - \left[\frac{\beta_2^2 e^{-\beta_2}}{\beta_2^2 - \beta_1^2} \right] \right. \right.$$

$$\left. \left[\left(\frac{\beta_3 \cosh\beta_1 + \sinh\beta_1}{\beta_3 \sinh\beta_1 + \cosh\beta_1} \right)\left(\frac{\beta_1}{\beta_2} \right) + 1 \right] \right\} d\lambda$$

$$+ e \int_\lambda \Phi_0(\lambda) \left\{ \left[\left[\frac{\beta_6^2 e^{-\beta_4}}{\beta_5^2 - \beta_6^2} \right] \left[\frac{[(\beta_7\beta_5/\beta_6) - 1]e^{-\beta_6}}{\beta_7 \sinh\beta_5 + \cosh\beta_5} \right] + \left[\frac{\beta_6^2 e^{-\beta_4}}{\beta_5^2 - \beta_6^2} \right] \right. \right.$$

$$\left. \left[1 - \left(\frac{\beta_5}{\beta_6} \right)\left(\frac{\beta_7 \cosh\beta_5 + \sinh\beta_5}{\beta_7 \sinh\beta_5 + \cosh\beta_5} \right) \right] \right\} d\lambda$$

$$+ e \int_\lambda \Phi_0(\lambda)(e^{-\beta_2} - e^{-\beta_4}) d\lambda - \left[\frac{eD_p p_{n0}}{L_p} \left(e^{V/kT} - 1 \right) \right]$$

$$\left[\frac{\beta_3 \cosh\beta_1 + \sinh\beta_1}{\beta_3 \sinh\beta_1 + \cosh\beta_1} \right] - e \frac{D_n n_{p0}}{L_n} \left[e^{V/kT} - 1 \right]$$

$$\left[\frac{\beta_7 \cosh\beta_5 + \sinh\beta_5}{\beta_7 \sinh\beta_5 + \cosh\beta_5} \right] - \left[J_{SCR} \left(e^{V/n_{SCR}kT} - 1 \right) \right] \tag{4.18}$$

This equation has been arranged with the photocurrent components first and then the opposing, voltage-dependent currents next. Equation 4.18 is clearly of the form of Eq. 4.4, if the electrostatic-field barrier region recombination is negligible or if there is no light dependence in the prefactor J_{SCR} or the n-factor n_{SCR}. The analytical approach has succeeded in obtaining a J-V characteristic that obeys superposition—and in establishing all the approximations and assumptions behind this property. Interestingly, if we evaluate Eq. 4.18 for the short-circuit condition (V = 0) and compare the result with Eq. 4.5, it becomes very clear why J_{sc} is not just simply $e \int_{-d}^{L+W} \int_{\lambda} G_{ph}(\lambda, x) d\lambda dx$. The expression for J_{sc} resulting from such a use of Eq. 4.18 is seen to contain β factors that characterize loss (i.e., β_1, β_3, β_5, and β_7). This analytical analysis emphasizes the point that we made earlier in Section 4.2.1: in the linearized mathematical picture, J_{sc} contains losses but all the voltage-dependent loss is in $J_{DK}(V)$.

This linearized model of p–n homojunction cell behavior is valid as long as the assumptions that led to it are valid. One necessary but not sufficient criterion for the validity of a linear model is the existence of superposition. If Eq. 4.4 does not correctly capture the physics of a device's operation, then the analysis we have just completed shows that its failure in the p–n junction case must be due to: (1) the electrostatic-field barrier region recombination loss mechanism; (2) the equation linearization utilized in the quasi-neutral regions; (3) the presence of physics not accounted for in the analytical model, such as trapping and electric field modification; (4) failure of the flat quasi-Fermi level assumptions, or (5) some combination of these. Equation 4.18 is complex, but in deriving it we learned a lot about the assumptions behind Eq. 4.4. Through the dimensionless β factors, we also learned a lot about the interplay of material properties in determining p–n homojunction cell performance.

If the barrier region recombination term in Eq. 4.18 is not dependent on illumination, then Eq. 4.18 has the diagrammatic, electrical engineering representation of a current generator element (all the light dependent terms) in parallel with a voltage dependent element (all the voltage dependent terms). If we multiply all the terms by the cell area, then the resulting equation and diagram represents the relationship between the cell's terminal current I and its terminal voltage V. This is a convenient formulation for adding in the effects of series resistance, which may arise from contacts, grids, or both, and of shunting resistance, which

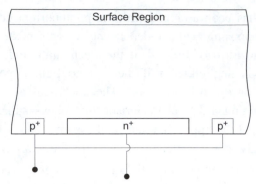

FIGURE 4.35 An interdigitated–back-contact cell.

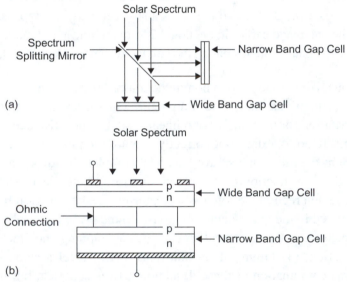

FIGURE 4.36 Two approaches to high-efficiency photovoltaic concentrator systems: (a) spectrum splitting and (b) cell cascade.

may arise from defects, pin holes, etc. For example, if series resistance R_S is present, it adds the resistor R_S in series with the parallel arrangement of the two basic elements. This means that mathematically it can be accounted for by simply replacing V by $V + IR_S$ everywhere in the I-V equation. If shunting resistance R_{SH} is present, it adds R_{SH} in parallel with the two basic elements. Mathematically, it requires subtracting from the I-V relationship the term $(V + IR_S)/R_{SH}$. Examination of the effects of series resistance shows it can lower I_{sc}, the fill factor, and decrease the slope of the I-V curve at V_{oc}. Examination of the effects

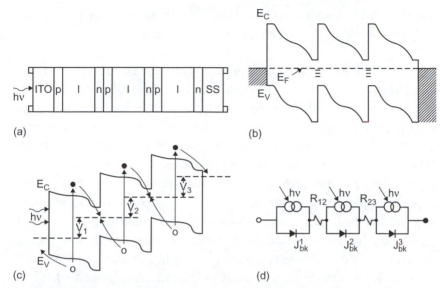

FIGURE 4.37 Illustration of: (a) use of three p–i–n structures in a homojunction unit tandem cell, with all units having the same band gap; (b) the band diagram in the dark; (c) the band diagram under illumination; and (d) the equivalent circuit. Here ITO = an indium tin oxide front transparent electrode and SS = the back stainless steel electrode and reflector.

of shunting resistance shows it can decrease the fill factor and V_{oc}. Of course, as we saw in Section 4.3, short circuit current, fill factor, and open circuit voltage can also be affected by much more fundamental causes than series and shunt resistances.

4.5 SOME HOMOJUNCTION CONFIGURATIONS

There actually are many different types of homojunction configurations. To underscore this point, some examples are seen in Figures 4.35–4.37. Figure 4.35 gives an example of a p–i–n homojunction that is definitely not one-dimensional. As seen, all the junction and contact regions are at the back of an intrinsic absorber. Light enters such structures through the contact-free "surface region." This front is passivated (i.e., processing is done to reduce gap states, which can support recombination, store charge, or both) to suppress carrier losses. Reducing gap state densities at this front surface can be very advantageous since this is where light enters and is most intense. The region among the back junctions is similarly passivated. These cells have several advantages. Among them are no contact-blocking of light entry into the cell and an esthetically pleasing front surface free of contacts and interconnects. Various types of back-contact cells for one sun and concentrator applications are reviewed in Ref. 18.

Figure 4.36 shows various homojunction cell arrangements that are used to address the problem of the wasted energy of supraband-gap photons in one-band-gap structures. In Chapter 2 we discussed using carrier multiplication to address this problem; here we see more straightforward approaches using spectrum-splitting and cell-cascade arrangements. Since these structures are relatively complex, they are generally only used in concentrator applications. As may be noted, spectrum-splitting approaches physically direct the lower-energy photons to a narrow-band-gap cell and the higher-energy photons to a wide-band-gap cell. Cascade approaches have the solar cells optically in series with light first imping-ing on the wide-band-gap cell to absorb the higher-energy photons and then on the narrow-band-gap cell. Cell-cascade arrangements have the cells in series electrically, as seen in Figure 4.36, and must make use of transparent ohmic contacts for the connections. In a series arrange-ment like this, the cells obviously must be designed to be producing the same currents at the maximum power point of the overall device. While Figures 4.36a and 4.36b show two cells in each case, in principle, mul-tiple cells could be used. However, cost is an issue and, as noted, these cells usually are only justifiable in concentrator applications.[19]

Figure 4.37 shows an approach to the collection length versus absorp-tion length issue we discussed in Section 3.4.1. Figure 4.37a shows the use of three p–i–n structures; hence, collection is designed to be by drift in each cell. All three cells have the same band gap and all three are arranged to be in tandem (physically integrated into one structure). The idea here is to not waste power in the impinging solar spectrum but to collect from the whole absorption length, which, in this case, must be about three device thicknesses; i.e., three times the drift col-lection length being used. The band diagram of Figure 4.37b shows the photogenerated electrons from the top cell must meet photogenerated holes from the middle cell and fill in these holes for current continu-ity. Correspondingly photogenerated electrons in the middle cell must meet photogenerated holes from the bottom cell and fill in those also for current continuity. The band diagram of Figure 4.37c shows that cur-rent matching is being accomplished by recombination in this specific example. This is not a good situation because it means energy is being lost. The internal energy loss at the two internal interfaces can be represented phenomenologically by the series resistors seen in Figure 4.37d. The problem of losing energy to attain current continuity in tandem structures can be overcome by using tunnel junctions at the n–p interfaces seen in Figure 4.37a, but it must be done without incurring deleterious optical effects. This goal can be achieved, for example, by utilizing high-low junctions with heavily doped high regions at the n–p interfaces. A discussion of such approaches and results are presented in Ref. 20. A tandem structure similar to that of Figure 4.37 could also be constructed so that it is composed of three homojunction cells with three different band gaps. The top (widest band gap) cell would then be designed to absorb the highest energy photons, the middle (middle band gap) cell would be designed to absorb the middle energy photons, and the bottom (lowest band gap) cell would be designed to absorb the lowest energy photons. These would be the photons with energies at or above its band gap but lower than the band gap of the middle cell.

REFERENCES

1. R.S. Ohl, U.S. Patent 2,402,662 (1941).

2. D.M. Chapin, C.S. Fuller, G.L. Pearson, A new silicon p–n junction photocell for converting solar radiation into electrical power, J. Appl. Phys. 25 (1954) 676.

3. D.A. Jenny, J.J. Loferski, P. Rappaport, Photovoltaic effect in GaAs p–n junctions and solar energy conversion, Phys. Rev. 101 (1956) 1208.

4. J.C.C. Fan, C.O. Bozler, R.C. Chapman, Applied Phys. Lett. 32 (1978) 390.

5. J. Woodall, H. Hovel, *Applied Phys. Lett.* 30 (1977) 492.

6. L.L. Kazmerski, G.A. Sanborn, J. Appl. Phys. 48 (1977) 3178.

7. L.L. Kazmerski, Ternary Compounds 1977, Conf. Ser.-Inst. Phys. 35 (1977) 217.

8. W.E. Spear, P.G. LeComber, S. Kinmond, M.H. Brodsky, Applied Phys. Lett. 28 (1976) 105.

9. B.A. Gregg, Excitonic solar cells, J. Phys. Chem. B. 107 (2003) 4688.

10. B.A. Gregg, R.A. Cormier, Doping molecular semiconductors: n-type doping of a liquid crystal perylene diimide, J. Am. Chem. Soc. 123 (2001) 7959.

11. M.C. Lonergan, C.H. Cheng, B.L. Langsdorf, X. Zhou, Electrochemical characterization of polyacetylene ionomers and polyelectrolyte-mediated electrochemistry toward interfaces between dissimilarly doped conjugated polymers, J. Am. Chem. Soc. 124 (2002) 690.

12. M. Pfeiffer, A. Beyer, B. Plonnigs, A. Nollau, T. Fritz, K. Leo, D. Schlettwein, S. Hiller, D. Worhle, Controlled p-doping of pigment layers by cosublimation: basic mechanisms and implications for their use in organic photovoltaic cells, Sol. Energy Mater. Sol. Cells 63 (2000) 83.

13. B.A. Gregg, M.C. Hanna, Comparing organic to inorganic photovoltaic cells: theory, experiment, and simulation, J. Appl. Phys. 93 (2003) 3605.

14. M. Pope, C.E. Swenberg, Electronic Processes in Organic Crystals and Polymers, second ed., Oxford University press, New York, NY, 1999.

15. Z.D. Popovic, A.-M. Hor, R.O. Loutfy, A study of carrier generation mechanism in benzimidazole perylene/tetraphenyldiamine thin film structures, Chem. Phys. 127 (1988) 451.

16. I.H. Campbell, T.W. Hagler, D.L. Smith, J.P. Ferraris, Direct measurement of conjugated polymer electronic excitation energies using metal/polymer/metal structures, Phys. Rev. Lett. 76 (1996) 1900.

17. S. Sze, K.K. Ng, Physics of Semiconductor Devices, third ed., John Wiley & Sons, Hoboken, NJ, 2007.

18. E. Van Kerschaver, G. Beaucarne, Back-contact solar cells: a review, Progress in photovoltaics: research and applications, 14 (2005) 107.

19. A. Barnett, C. Honsberg, D. Kirkpatrick, S. Kurtz, D. Moore, D. Salzman, R. Schwartz, J. Grey, S. Bowden, K. Goossen, M. Haney, D. Aiken, M. Wanlass, K. Emery, 50% Efficient Solar Cell Architectures and Designs, IEEE 4th World Conference on Photovoltaic Energy Conversion, New York. (2006) p. 2560.

20. M. Yamaguchi, Physics and technologies of superhigh-efficiency tandem solar cells, Semiconductors 33 (1999) 961.

Chapter Five

Semiconductor– semiconductor Heterojunction Cells

5.1 INTRODUCTION

Semiconductor–semiconductor heterojunction (HJ) solar cells, composed of semiconductor materials 1 and 2 as seen in Figures 5.1 and 5.2, are the focus of this chapter. One of these materials of an HJ obviously must be an absorber. The other may be an absorber, too, or it may be a window material; i.e., a wider-gap semiconductor that contributes little or nothing to light absorption but is used to create the heterojunction and to support carrier transport. Window materials collect holes or electrons, function as majority-carrier transport layers, and can separate the absorber material from deleterious recombination at contacts. The interface they form with the absorber is also used for exciton dissociation in cells where absorption is by exciton formation. Absorber and window materials may be inorganic semiconductors, organic semiconductors, or mixtures. The physical structure of heterojunctions fits into one of two types: planar or bulk. The physical structure of a planar heterojunction (PHJ) is depicted in a side-view cross-section (perpendicular to the contacts) in Figure 5.1. It is made up of two semiconductor materials that form the active layer (an absorber-absorber or

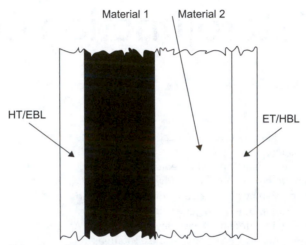

FIGURE 5.1 Side view of a typical planar heterojunction. This cross-section, perpendicular to the contacts, shows material 1 and material 2. The anode is the left-side contact and the cathode is the right-side contact.

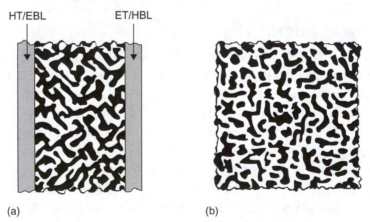

FIGURE 5.2 Bulk heterojunction: (a) Side-view cross-section, perpendicular to the anode (left) and cathode (right) contacts, showing regions containing material 1 or material 2; (b) cross-section parallel to the contacts showing regions containing material 1 or material 2.

absorber-window combination). This particular PHJ also has an ideal hole transport (HT)-electron blocking layer (EBL) and electron transport (ET)-hole blocking layer (HBL). These are shown on the anode and cathode sides, respectively, and have been discussed in detail in Chapter 4.[†]

The physical structure of a bulk heterojunction (BHJ) is seen in a side-view cross-section (perpendicular to the contacts) in Figure 5.2a and in a cross-section parallel to the contacts in Figure 5.2b. A BHJ is also composed of a two-semiconductor active layer but in a mixture that runs from the anode to the cathode as can be discerned from Figure 5.2a.[1] The BHJ structure is often used for organic solar cells.[1-4] In this case, the active-layer starting material usually is initially a mixture of material 1 and material 2 that is then forced to undergo phase separation creating the continuous regions of material 1 and continuous regions of material 2 seen in the figure. In BHJs, the HJ interface between material 1 and material 2 is not a single plane parallel to the cell contacts, as it is in the PHJ, but, in the BHJ this interface has many components, lying at various angles with respect to the cell contacts, and all working electrically in parallel. This myriad of material 1–material 2 HJ interfaces is seen in a cross-section parallel to the contacts in Figure 5.2b. In the case of organic BHJ cells, the cross-sectional dimensions of the material 1 and material 2 regions of Figure 5.2b are in the nano-scale.[1] This size scale is ideal for collecting excitons, the short-diffusion-length excitations produced in organic materials by absorption (see Section 3.4.1.2). The HT-EBL layer at the anode in Figure 5.2a is critical in BHJ devices since examination of the figure shows it blocks electron transport in electron conducting strands coming to the anode. The analogous role is played by the ET-HBL at the cathode.

A one-dimensional band diagram captures the physics of a PHJ; fortunately, this simple band diagram picture often suffices for BHJs also. The latter situation occurs because the strands or "spaghetti-like" paths that carriers follow in BHJ cells can often be conceptually "straightened out" and interpreted in terms of a one-dimensional band diagram.[1-4] Figure 5.3 gives some one-dimensional band diagrams (with the contacts omitted) showing the general features of heterojunctions: (1) the utilization of two different semiconductors that meet at an interface (which may be gradual or abrupt), and (2) the presence of effective force fields (due to changes in electron

---

[†]Window layers differ from HT-EBL and ET-HBL materials in that the former are a part of the principal junction of a device.

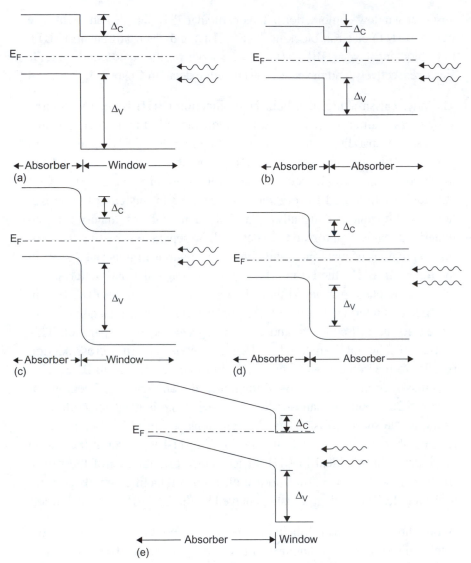

FIGURE 5.3 Some heterojunction solar cell configurations with light entering from the right: (a) p–n absorber-window structure with no built-in electric field; (b) p–n absorber-absorber structure with no built-in electric field; (c) p–n absorber-window structure with a synergistically oriented built-in electric field arising from workfunction differences between the semiconductors; (d) p–n absorber-absorber structure with a synergistically oriented built-in electric field arising from workfunction differences between the semiconductors; and (e) a p–i–n absorber-window structure with a synergistically oriented built-in electric field arising from the window workfunction (doping) and the absorber contact-region doping and contact workfunction. Here Δ_C is the change in the conduction band edge (or the change in the LUMO) that occurs from one material to the other and Δ_V is the change in the valence band edge (or the change in the HOMO) that occurs from one material to the other.

affinity, hole affinity, densities of states, or some combination of these), perhaps along with an electrostatic force field, for charge-carrier separation. As discussed in Section 3.2, meaningful photovoltaic action requires having a structure in which symmetry is broken. In heterojunctions affinity steps and density of states changes and electrostatic fields arising from built-in potentials can all be contributors to this required break in symmetry. We note that heteroface solar cells, both graded and abrupt, are not included here as heterojunctions—but are discussed in Chapter 4—since those devices have a p–n homojunction, located away from the heterostructure region, as the principal source of photovoltaic action.

In Figure 5.3, the quantities Δ_C and Δ_V are the affinity changes. Specifically, Δ_C is the change in the conduction band edge (or, in the terminology of organic materials, the change in the LUMO) that occurs from one material to the other and Δ_V is the change in the valence band edge (or, in the terminology of organic materials, the change in the HOMO) that occurs from one material to the other. The figure shows structures in which there is a built-in electrostatic field oriented so as to aid the affinity-change-caused effective fields in breaking symmetry. These may be developed by choice of the component material workfunctions or by utilizing a p–i–n scheme. As would be expected, a p–n configuration is used when photocarrier collection is accomplished principally by diffusion and a p–i–n configuration is used when photocarrier collection is accomplished principally by drift (see Section 3.4).

The use of two different semiconductor materials for the active (absorber-absorber or window-absorber) layer, which is a characteristic feature of a heterojunction, introduces a new set of problems not encountered in homojunctions, such as chemical compatibility and stability, reproducibility of the chemical and physical interface, and, in the case of crystalline and polycrystalline materials, lattice compatibility at the metallurgical junction. All of these can give rise to defects and therefore gap states that allow a new loss mechanism: HJ interface recombination. On the basis of the new problems inherent to heterojunction cells one might question the interest in these devices. However, there is a strong interest which stems from four features: (a) heterojunctions allow the use of semiconductors that can only be doped either n-type or p-type and yet have attractive properties which may include their absorption length, cost, and environmental impact, (b) heterojunctions allow the exploitation of effective forces, (c) heterojunctions of the window-absorber type can be used to form structures that shield carriers from top-surface or back-surface recombination sinks, and (d) the affinity steps at HJ interfaces can be used to dissociate excitons into free electrons and holes. This latter feature is critical for structures in which light absorption results principally in excitons. In addition, as will be shown in our detailed analysis, heterojunctions can also permit open-circuit voltages that can be larger than the built-in electrostatic potential. This latter point is very interesting.

As just noted, heterojunctions have distinctive advantages and efforts to develop this class of solar cells go back more than half a century, dating from the work on the (n)CdS/(p) Cu_2S materials system reported by Reynolds et al. in 1954.[5] The Reynolds group found that certain types of copper contacts to single-crystal CdS gave open-circuit voltages of 0.45 V and short-circuit currents of 15 mA/cm^2 in direct sunlight. Although it was first believed that a metal-semiconductor barrier (see Chapter 6) at a Cu/CdS interface gave the photovoltaic action in this cell,[6] it was later established that the principal barrier in efficient devices arises at a CdS/Cu_2S heterojunction.[7] In 1956, Carlson et al.[8] made the first thin-film polycrystalline CdS/Cu_2S heterojunction solar cells. Over the ensuing years, this basic structure evolved into many types of thin-film inorganic devices having the configuration seen in Figure 5.4. The

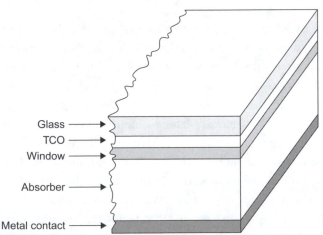

FIGURE 5.4 A thin-film heterojunction of the window-absorber type. Light enters through the glass, the transparent conductive oxide (TCO) contact, and the window (top) semiconductor into the absorber (bottom) semiconductor. Light reflection occurs at the back metal contact.

most efficient are those based on CdS/CuInGaSe$_2$ ($\eta_{AM1.5}$ ~19.5%) and CdS/CdTe ($\eta_{AM1.5}$ ~16.5%) PHJs.[9] These exploit the direct band gap, strong absorption properties of CuInGaSe$_2$ and CdTe (see Chapter 3) while using the n-type window material CdS for junction formation. Thin-film organic devices having the BHJ or PHJ configuration have also been explored using polymeric absorber-hole conductor materials and fullerene-based electron conductor materials, and efficiencies of $\eta_{AM1.5}$ ~ 6% have been achieved.[10] To date, the most efficient, simple heterojunction has been a structure based on crystalline silicon. This device has attained efficiencies above 22%.[11]

5.2 OVERVIEW OF HETEROJUNCTION SOLAR CELL DEVICE PHYSICS

5.2.1 Transport

Figure 5.5 may be used to begin to explore what happens inside a heterojunction when light impinges and the absorption process, either directly or indirectly through exciton dissociation, produces free electrons and holes. We assume for the sake of a very general discussion that pho-togenerated free electrons and holes are produced in the both the top and bottom layers. These free carriers are then subject to the carrier-loss

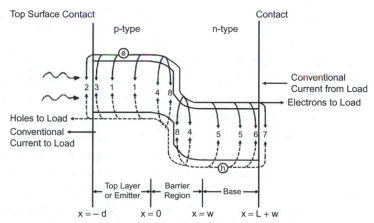

FIGURE 5.5 A p-on-n (p–n) heterojunction of the form absorber-absorber. The top layer is material 1 and the bottom layer is material 2. Light is seen to impinge at x = −d. Paths (processes) 1–8 are all loss mechanisms: Paths 1, 4, and 5 are bulk recombination; Paths 2, 3, 6, and 7 are losses at the top-surface contact and base-layer contact; and Path 8, the unique feature of heterojunctions, is the interface-state recombination path. Particle current flow is shown inside the cell. The flow arrows are at the band edges to convey thermalization of photogenerated carriers.

mechanisms of (a) bulk recombination (loss mechanisms 1 and 5 of the figure), (b) bulk recombination in the barrier region (mechanism 4), (c) recombination at the top contact (mechanisms 2 and 3), (d) recombination at the back contact (mechanisms 6 and 7) and (e) recombination through localized states situated at the HJ interface (mechanism 8). The distinction between mechanisms 2 and 3 and between 6 and 7 has been discussed in depth in Section 2.3.2.9.

From Chapter 4 we know that the net result of all the activity taking place under illumination in Figure 5.5 can be expressed mathematically by picking any plane and determining the conventional current density J crossing that plane when the cell is developing a voltage V. Taking the plane x = L + W (actually we use a plane just to the right of processes 6 and 7 as discussed in Section 2.3.2.9) to do our "bean-counting" using electrons and the continuity concept as we did in Chapter 4, it can be seen that the net number of electrons leaving the device at x = L + W per time per area causes the conventional current density J flowing into the device at x = L + W given by

$$
J = - \left[e \int_{-d}^{L+W} \int_{\lambda} G_{ph}(\lambda,x)d\lambda dx \right.
$$
$$
\left. - e \int_{-d}^{L+W} \mathcal{R}(x)dx - J_{ST}(-d) - J_{SB}(L+W) - J_{IR} \right] \qquad (5.1)
$$

Here, $G_{ph}(\lambda, x)$ is the free carrier photogeneration function introduced in Section 2.2.6 and its λ integration is over the impinging spectrum. The integral $\int_{-d}^{L+W} \mathcal{R}(x)dx$ accounts for loss mechanisms 1, 4, and 5, $J_{ST}(-d)$ accounts for the electrons (and consequently holes) lost at the top (front, light entering) surface through loss mechanisms 2 and 3, and $J_{SB}(L + W)$ accounts for the electrons (and consequently holes) lost at the back surface through mechanisms 6 and 7. We did not see the term J_{IR} in Chapter 4 in our discussion of homojunctions. We explicitly call it out here for heterojunctions to stress this possible HJ interface recombination loss path (mechanism 8). The minus sign in front of all the terms on the right-hand side in the equation is needed to conform to our x-direction and power quadrant sign conventions; i.e., when developing power, the conventional current density is negative and in the power quadrant and is flowing in the negative x-direction in Figure 5.5. When the photogenerated free electrons and holes are produced directly by absorption,

$G_{ph}(\lambda, x)$ captures the resulting carrier generation distribution. When the photogenerated free electrons and holes are produced by exciton dissociation, at the HJ interface, we will take $G_{ph}(\lambda, x)$ to be almost delta-function-like carrier generation confined to the interface region.[12] This delta-function-like x-dependence arises from the exciton dissociation enabled by the electron and hole affinity steps at the interface.[10] In form, Eq. 5.1 is exactly what we established for homojunctions, with the addition of the explicit interface recombination term. If we knew all the HJ material and absorption properties as well as the voltage and illumination intensity dependence of all these different terms on the right-hand side of Eq. 5.1, we would have the light J-V behavior of the cell.

Loss mechanisms 1–7 of Figure 5.5, together with loss mechanism 8, determine the shape of the light J-V characteristic. They determine the quantum efficiency, which is discussed in detail in Chapter 4, and the short-circuit current density J_{sc}. They determine the open-circuit voltage V_{oc} necessary to drive the contact Fermi levels sufficiently apart to ensure that all the photogenerated free electrons or holes are internally annihilated through Paths 1–8. The stronger the kinetics of Paths 1–8 are, the poorer is the cell performance and the more reduced is the open-circuit voltage. Whatever the results of this struggle between generation and recombination, the open-circuit voltage will be such that the right-side electrode of the structure in Figure 5.5 will be negative with respect to the left-side electrode. That is, at open circuit, the Fermi level in the contact of the right side will be raised up in energy by V_{oc} with respect to the Fermi level in the contact of the left side.

Just as in the case of homojunctions, a variety of physical phenomena are encompassed by Eq. 5.1. The variety of the phenomena present allows for the possibility that free electron and hole transport in the semiconductor regions of a heterojunction can be controlled, for example, by minority-carrier diffusion, ambipolar diffusion, drift, space-charge-limited currents, etc. Of course, transport in the bulk regions is in series with transport at the HJ interface, which can be drift-diffusion controlled or even thermionic emission–controlled (see Section 2.3.2). In addition, free carrier photogeneration can be due to exciton dissociation or direct photocarrier production, as we have noted. All of these phenomena give heterojunctions a rich range of possible behavior.

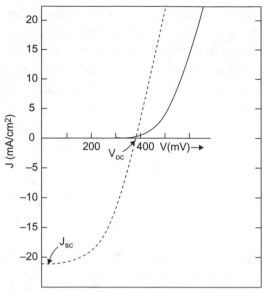

FIGURE 5.6 Experimental light (---) and dark (—) J-V's for a Cu$_2$ S/CdS heterojunction solar cell structure. It is evident that superposition is not valid for this structure. (After Ref. [13].)

When we delve into a detailed evaluation of Eq. 5.1 for heterojunctions, we will use numerical techniques to keep track of all the interplay among electrostatic-field forces, effective forces, drift, diffusion, trapping, and recombination in a heterojunction. This is done in Section 5.3. We will also use an analytical approach in Section 5.4 to evaluate Eq. 5.1 in terms of a tractable, albeit involved, analytical expression. In the analytical approach, to get a manageable mathematical system, we will be forced to make a series of assumptions, just as we did for homojunctions in Chapter 4. The assumptions will lead to a fully linearized equation set (differential equations plus boundary conditions). We will show that this linearization, as it did in Chapter 4, leads to superposition being valid; i.e.,

$$J = -[J_{sc} - J_{DK}(V)] \qquad (5.2)$$

is rigorously true in that case. We examine, in detail in Sections 5.3 and 5.4, all the approximations needed to have Eq. 5.1 turn into Eq. 5.2 for heterojunctions. In Section 5.3 we examine situations where Eq. 5.2 obviously does not work and situations where, surprisingly, it does. Figure 5.6 presents an experimental example of the former: the figure gives experimental dark and light J-V characteristics for a CdS/Cu$_2$S cell that display a crossover behavior that is clearly inconsistent with Eq. 5.2. Numerical modeling can capture situations where superposition does and does not work for heterojunctions since it solves the full non-linear, coupled set of transport, trapping, recombination, and Poisson equations. We worry about superposition because it provides a useful "test". Its presence or absence gives us insight into the importance of the non-linear phenomena present in the full mathematical set (Section 2.4) describing solar cell behavior.

5.2.2 The heterojunction barrier region

A heterojunction has at least an electron or hole effective field present, since at least one of the affinities changes at the interface. In addition, it is usually designed to have an electrostatic field present at and about the interface as well. In the example of Figure 5.5, the symmetry breaking is seen to be accomplished by both. The job of these forces in this figure is (a) to draw photogenerated electrons away from loss mechanisms in the top layer and to draw photogenerated holes away from loss mechanisms in the base (back) layer and (b) to sweep photogenerated carriers past the loss mechanisms in the space-charge region and at the HJ junction. From Section 2.3 we know that the sum of the electrostatic and effective forces doing all this for electrons is F$_e$, which is given by

$$F_e = -e\left[\xi - \frac{d\chi}{dx} - kT\frac{d\ln N_c}{dx}\right] \tag{5.3}$$

where ξ is the electrostatic field and the remaining terms $\xi'_n = -[d\chi/dx - kT(d\ln N_c/dx)]$ constitute the electron effective force field ξ'_n acting on an electron. The electron effective force field is clearly something we did not have to deal with in a homojunctions except at selective contact structures.

From Section 2.3, we also know that the sum of the electrostatic force and hole effective force acting on a hole in the heterojunction of Figure 5.5 is F$_h$, given by

$$F_h = e\left[\xi - \frac{d}{dx}(\chi + E_G) + kT\frac{d\ln N_v}{dx}\right] \tag{5.4}$$

where, as noted, ξ is the electrostatic field and the remaining terms $\xi'_p = [-(d/dx)(\chi + E_G) + kT(d\ln N_v/dx)]$ give the hole effective force field ξ'_p acting on the hole. The hole effective force field is not present in a homojunction, except possibly at selective contacts that may be employed.

Inspection of Eq. 5.3 shows that it may be rewritten as

$$F_e = -e\left[\frac{dE_C}{dx} - kT\frac{d\ln N_c}{dx}\right] \tag{5.5}$$

This formulation stresses that the electric field ξ and the electron affinity variation $d\chi/dx$ from one material to the other, taken together, cause the conduction band edge E_C to vary with position from one material to the other. These act together to create an effective "total electron barrier" $V_{TEB} \equiv \left|\int_{\substack{Barrier \\ Region}} (dE_C/dx)dx\right|$ for the conduction band at a heterojunction that is given by

$$\left|\int_{\substack{Barrier \\ Region}} \frac{dE_C}{dx}dx\right| = \left|\int_{\substack{Barrier \\ Region}} \left(\xi - \frac{d\chi}{dx}\right)dx\right| \tag{5.6}$$

This integrates to

$$V_{TEB} = V_{Bi1} + V_{Bi2} + \Delta_C \tag{5.7}$$

where V_{Bi1} is the built-in electrostatic potential energy developed in material 1, V_{Bi2} is the built-in electrostatic potential energy developed in material 2, and Δ_C is the electron affinity change from material 1 to material 2. The quantities V_{Bi1} and V_{Bi2} include the total electron potential energy built-in across the device from contact to contact.

Inspection of Eq. 5.4 shows it may be rewritten in the form

$$F_h = -e\left[\frac{dE_V}{dx} - kT\frac{d\ln N_V}{dx}\right] \tag{5.8}$$

This formulation stresses that both the electric field ξ and hole affinity variations $(d(\chi + E_g)/dx)$ cause the valence band edge E_V to vary with position. Comparison of Eqs. 5.8 and 5.4 shows that the valence band edge position change due to the electrostatic force and the effective force arising from the hole affinity changes may be thought of as creating an effective "total hole barrier" $V_{THB} \equiv \left|\int_{Barrier\ Region} (dE_V/dx)dx\right|$ in the valence band of a heterojunction given by

$$\left|\int_{\substack{Barrier \\ Region}} \frac{dE_V}{dx}dx\right| = \left|\int_{\substack{Barrier \\ Region}} \left[\xi - \frac{d(\chi + E_g)}{dx}\right]dx\right| \tag{5.9}$$

which integrates to

$$V_{THB} = V_{Bi1} + V_{Bi2} + \Delta_V \tag{5.10}$$

Here Δ_V is the hole affinity change from material 1 to material 2. In both Eq. 5.7 and Eq. 5.10, the total built-in electrostatic potential energy $V_{Bi} = V_{Bi1} + V_{Bi2}$ must be the same and arises from the charge transfer driven by the need for the entire materials system, including contacts, to have one Fermi level in TE. In the case where the contacts are materials (metals or TCOs) which may be considered to be infinite reservoirs of electrons (i.e., their Fermi levels do not move with respect to the local vacuum level), then the total built-in potential must obey

$$V_{Bi} = V_{Bi1} + V_{Bi2} = \phi_1 - \phi_2 \tag{5.11}$$

where ϕ_1 is the workfunction of contact 1 and ϕ_2 is the workfunction of contact 2 (and $\phi_1 > \phi_2$ has been assumed). This last statement holds, of course, for any cell. Interestingly, a comparison of Eqs. 5.7 and 5.10 shows that the total electron and hole effective barriers may not be equal. Reconsideration of Eqs. 5.5 and 5.8 allows us to note that the total barriers are not all that breaks symmetry. The density of states changes can also make an impact.

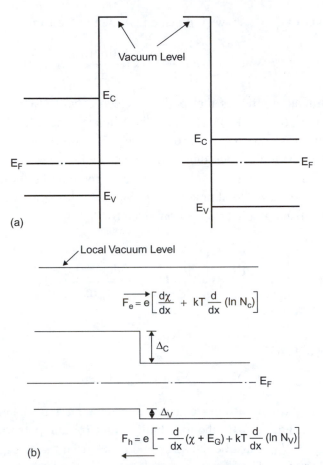

FIGURE 5.7 (a) Heterojunction composed of two semiconductors, which are such that when (b) combined in a heterojunction, the resulting electron and hole barriers have no electrostatic-field component. In this case, there are only effective forces to break the symmetry, collect photogenerated carriers, and give rise to photovoltaic action.

This discussion has established that effective forces are always present in heterojunctions and that, in principle, there can be HJ structures where the fields in the barrier region arise solely from the effective-force-field terms of Eqs. 5.3 and 5.4. An example is seen in Figure 5.7, which depicts a pair of semiconductors that, when placed in a heterojunction, have a barrier region consisting only of effective forces. As seen in Figure 5.7a, before contact, both materials 1 and 2 have been chosen to have the same electrochemical potential when measured with respect to a common reference energy (the vacuum level); i.e., both

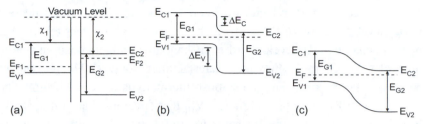

FIGURE 5.8 The same two semiconductors used to form two different HJ barriers. Here (a) is the energy band diagram for the two semiconductors before contact, depicted with respect to vacuum-level reference energy while (b) and (c) are two energy band pictures out of an infinity that are possible after contact. Case (b) is the abrupt HJ and case (c) is a junction that occurs for a particular material-grading profile from material 1 to material 2. The figure demonstrates that an endless variation in effective barrier shapes [$E_C = E_C(x)$ and $E_V = E_V(x)$] can exist for what is nominally the same HJ.

semiconductors have the same workfunction. This is also assumed to be the case for their contacts. Consequently, when a junction is formed between the materials and allowed to come to TE, as seen in Figure 5.7b, there is no charge exchange. Since there is no charge exchange, there can be no electrostatic field and no V_{Bi1} or V_{Bi2} in TE. There are, however, effective fields at the interface. The effective force on electrons acts to sweep (drift) them to the right in the figure; the effective force on holes acts to sweep (drift) them to the left in the figure. In TE, the former is balanced by electron diffusion to the left at the HJ interface plane and the latter is balanced in TE by hole diffusion to the right at the interface plane.

Because of the interplay between ξ and $(d\chi/dx)$ as well as between ξ and $d(\chi + Eg)/dx$, the total electron and hole effective barriers in a heterojunction can take on various positional dependencies (i.e., various shapes) for the same electrostatic contribution $V_{Bi} = V_{Bi1} + V_{Bi2}$ and the same affinity step magnitudes Δ_C and Δ_V. The spatial dependence of the doping and of the band edge variations leading to Δ_C and Δ_V (i.e., whether the junction is abrupt or graded) dictate this shape. Some results of this interplay are shown in Figure 5.8. The two semiconductors, which are going to be used to form heterojunctions with varying degrees of grading, are seen before contact in Figure 5.8a. The electrode contact (not shown) to each semiconductor is taken to have the same workfunction as its adjacent semiconductor to keep things simpler. The respective Fermi-level positions of these semiconductors before contact tell us that negative charge must develop in the left-hand material and positive charge must develop in the right-hand material after contact. This occurs to equate the Fermi levels and to establish thermodynamic equilibrium. The built-in electrostatic potential energy V_{Bi} that develops due to this work-function difference must be the same for all junctions created between the two materials. The band variations as a function of position—and therefore the effective barriers—that result for an abrupt (step-function) junction between the semiconductors (i.e., when no material grading is used) are seen in Figure 5.8b. When grading is present, one of the possible band-variation—and therefore one of the effective barrier—profiles that can result is seen in Figure 5.8c. Obviously, there is an infinity of such possible profiles, the details of which depend on the details of the material grading. And, in all these cases, doping could also be graded.

The electrostatics can get more complicated for HJs than what appears in Eq. 5.11, and we need to face that possibility at this point. The problem is that our use of the condition of Eq. 5.11 neglects the presence of what are termed HJ permanent interface dipoles. Neglecting the possibility of permanent interface dipoles at HJs is termed the Anderson heterojunction model.[14] Now that we are aware that these permanent dipoles can exist at HJ interfaces, there are some basic questions to be asked: what are they, what causes them, and how does their presence impact Eqs. 5.7, 5.10, and 5.11? First of all, interface permanent dipoles are equal amounts of fixed (not varying with voltage or illumination) positive and negative charge straddling the HJ interface. Permanent interface dipoles can arise, for example, from interface chemical reactions or interdiffusion and, for crystalline materials, from bonding issues due to lattice mismatch, such as that depicted in Figure 5.9.

We will discuss permanent interface dipoles and the modifications they cause in our electrostatics considerations with the help of Figure 5.10. Figure 5.10a shows two semiconductors before junction formation, while Figure 5.10b shows them after HJ formation in the case where an interface dipole is not present. We note that the particular semiconductors chosen for this discussion have the same hole affinity to keep things

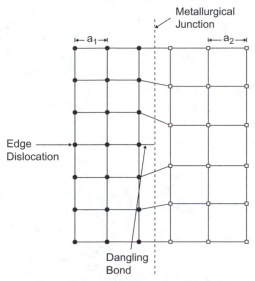

FIGURE 5.9 A cross-section perpendicular to the idealized junction plane of an abrupt heterojunction between two simple cubic crystals with lattice constants a_1 and a_2.

simpler. Also they are not ideal from a solar cell point of view since they are seen in Figure 5.10b to give rise to what is termed a "spike" in the conduction band in an HJ; i.e., the electrostatic and electron effective force fields are in opposition. This spike case is being used just to make the possible impact of a dipole more dramatic. Under cell operation, photogenerated electrons, as particles, will be moving left to right in this figure and the spike presents a substantial impediment to this flow. The spike is interesting, but the main point of the figure is to show what will happen if a dipole with an electron potential energy contribution of magnitude Δ is introduced. Figure 5.10c specifically shows the case of the permanent dipole of electron potential energy magnitude Δ oriented opposite to the spike. Its impact is seen to be quite significant.

The presence of the dipole in Figure 5.10c does not affect the total electrostatic potential energy V_{Bi} developed across the barrier region because that must always equal the workfunction difference $\phi_1 - \phi_2$, but it does strongly affect how the electrostatic potential energy is developed spatially, as can be seen by looking at the local vacuum-level profile and recalling that its derivative is always the electrostatic field. The electrostatic potential

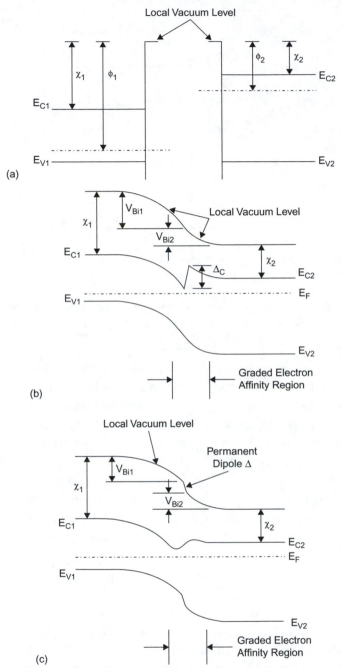

FIGURE 5.10 (a) A pair of hypothetical semiconductors used to form HJs before contact. Both materials have the same hole affinity for simplicity. The semiconductors are used to form graded junctions in TE. Case (b) has no permanent dipole (the Anderson model) whereas case (c) has a permanent dipole present (as manifested in the local vacuum-level behavior). In (c), the permanent dipole Δ and electron affinity step Δ_C are seen to counter one another.

energy developed in material 1, V_{Bi1}, and in material 2, V_{Bi2}, now must adjust their vales to fulfill

$$V_{Bi} = V_{Bi1} + V_{Bi2} + \Delta = \phi_1 - \phi_2 \qquad (5.12)$$

instead of Eq. 5.11. For the dipole orientation we have chosen in Figure 5.10c, the Δ in Eq. 5.12 is a positive quantity. Its orientation (seen in the local vacuum-level variation) makes the impact of the spike less severe. To sum up the effects of an interface permanent dipole: the effective total electron and hole barriers must obey modified versions of Eqs. 5.7 and 5.10 when a permanent interface dipole is present. These modified versions are given, respectively, by

$$V_{TEB} = V_{Bi1} + V_{Bi2} + \Delta_C + \Delta \qquad (5.13)$$

and

$$V_{THB} = V_{Bi1} + V_{Bi2} + \Delta_V + \Delta \qquad (5.14)$$

where it is assumed that the built-in potentials, the affinity steps, and the dipole electron potential energy contribution are all in the same sense.

To actually determine V_{Bi1} and V_{Bi2} in either the Anderson or non-Anderson situation, one has to solve Poisson's equation with the accompanying boundary conditions for the material-1 and material-2 sides of the barrier remembering that their sum is subject to the conditions of Eq. 5.11 or Eq. 5.12, as is appropriate. The charge density that appears in the Poisson equations has the usual components (see Chapter 2); i.e., the free electron, free hole, trap, and doping contributions. In addition, there is a new component that can occur in HJ structures: interface gap states. These states cause mechanism 8 in Figure 5.5, can store charge, and their occupancy can vary with illumination and voltage. The interface gap states, which owe their origins to things like chemical reactions, interdiffusion, and, in the case of crystalline materials, lattice mismatch (Fig. 5.9), have much the same sources as interface dipoles. However, there is a very crucial distinction between interface dipoles and interface states: the charge state does not change in an interface dipole. When the terms "interface states" or "interface gap states" are used, it is implicit that these defects have populations that can change with illumination and voltage. The presence of these variable-occupancy localized states at the interface means that the electric-field impact and the concomitant electron potential energy distribution impact caused by the charge in these states, as well as their recombination role, are variable and depend on the extent to which the states have been emptied or filled during cell operation. While the interface states supporting recombination probably change occupancy rapidly, those affecting fields and potentials could actually change occupancy with varying time constants. If cell operation were to be switched from one J-V point to another, the interface states may empty and fill to varying degrees, depending on the time allowed for switching.

5.3 ANALYSIS OF HETEROJUNCTION DEVICE PHYSICS: NUMERICAL APPROACH

As was done for homojunctions, we will now undertake a more in depth study of heterojunction operation. This will be done by first using the very general approach of numerical analysis. With this approach, we can develop a good quantitative feel for how the devices are operating without getting into many equations. (The analytical analysis approach is given in Section 5.4 for those who wish to pursue that perspective.) In the computer modeling utilized in this section, the full set of non-linear device physics equations described in Chapter 2 is solved in one dimension. Although we saw opportunities in two-dimensional and even three-dimensional cell structures in Section 3.4.1.4, we stay with one dimension here once again to focus on the key aspects of device performance. Using a numerical approach and solving the full set of equations in exploring HJ behavior allows us to avoid the assumptions that have to be made in an analytical analysis—in other words, we avoid missing a lot of device physics by using numerical computation. The numerical analysis approach is particularly useful for HJs because it explicitly includes the distinguishing features of this class of cells: the effects of electron and hole affinity changes with position as well as the effects of interface recombination and charge storage. In the discussion of HJs, we focus on p–n type structures. The use of drift collection in p–i–n configurations is discussed in the context of homojunctions in Section 4.3.5.

5.3.1 Absorption by free electron–hole pair excitations

In this section the absorption process is taken to directly produce free electrons and free holes and is modeled using the Beer-Lambert expression discussed in Chapter 2. Recombination is taken to be the bimolecular mechanism of Shockley-Read-Hall trap-assisted recombination.

5.3.1.1 EFFECTIVE-FORCE-FIELD BARRIER ONLY

The first heterojunction to be considered is the simplest: it is seen in Figure 5.11 to have only electron and hole affinity steps at the HJ interface. The semiconductors and their contacts all have the same workfunction. Consequently, there is no built-in potential V_{Bi} and therefore no built-in electrostatic field anywhere in the structure at TE. Both of these points about V_{Bi} and the electrostatic field may be read from the vacuum level, as first discussed in Chapter 2 and Appendix D. We call this simple cell, which only has effective forces present, our baseline device and start our numerical analysis study of heterojunctions with it. As noted, this device

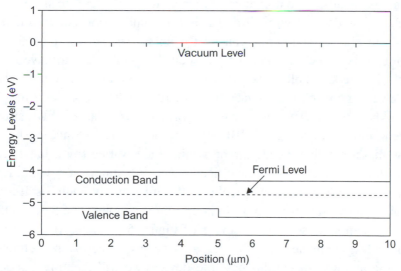

FIGURE 5.11 The band diagram of a very simple heterojunction structure in TE. Hypothetical materials 1 and 2 are both absorbers and happen to have the same energy band gap but different electron and hole affinities. There is no built-in electrostatic field anywhere in the device, as can be seen from the local vacuum level, so only effective forces are present at the interface. The lack of a built-in electric field and therefore of a built-in electrostatic potential V_{Bi} is achieved by the lack of workfunction differences (1) between the contacts themselves and (2) between the contacts and materials 1 and 2.

has no built-in potential so V_{Bi} cannot limit V_{oc}. In the case of the simple p–n homojunction of Figure 4.3, V_{Bi} did limit V_{oc}. On the other hand, we have already seen a situation (Fig. 3.9) where there is no relationship between V_{Bi} and V_{oc}. In contemplating these findings, we note that the difference lies in the presence or absence of an effective-force-field barrier component. We explore this further in several parts of Section 5.3.

The device of Figure 5.11 is composed of two hypothetical semiconductors that are described by the parameters in Table 5.1. The affinity steps at the HJ interface lead to a material-2 conduction-band minus material-1 valence-band difference of 0.87 eV. If the materials were organics, this difference would be called the HOMO1 − LUMO2 difference (measuring positively down from the vacuum level). This quantity is sometimes referred to as the "effective band gap." (For a discussion of the origin of this name, see Section 5.3.1.4.) The device of Figure 5.11 is an idealization in the sense that there are no HJ interface defects and therefore no interface recombination or charge trapping. The surface recombination speeds at the front contact are seen from Table 5.1 to have been picked to reduce electron front-surface recombination but simultaneously to give a good ohmic contact for holes. The surface recombination speeds at the back contact are seen from Table 5.1 to have been picked to reduce hole back-surface recombination but simultaneously to give a good ohmic contact for electrons. Chapter 4 describes in detail how to achieve such contacts in practice using an HT-EBL structure at the front and an ET-HBL structure at the back. Since we know how these effective speeds can be attained and varied (by using selective ohmic contacts of varying efficacy), we will simply stay with these values throughout Section 5.3. We will omit actually adding selective ohmic contact structures to try to keep things simpler. The S-R-H bulk recombination present in the device of Figure 5.11 is described by the bulk defect parameters of Table 5.1. After using numerical modeling to explore and understand this baseline device, we will then make modifications to it and explore the device physics and performance changes caused by these modifications. The changes are not pursued here with the goal of designing the most efficient heterojunction. As is our custom, that is left to the reader. The changes are pursued with the goal of learning as much as we can about heterojunctions from a few examples of device numerical analysis. As we embark on this path, it can be noted

Table 5.1 Baseline Heterojunction Device

Parameter	Material 1	Material 2
Length	5000 nm	5000 nm
Band gap	$E_G = 1.12$ eV	$E_G = 1.12$ eV
Electron affinity	$\chi = 4.05$ eV	$\chi = 4.30$ eV
Absorption properties	Absorption data for Si used	Absorption data for Si used
Doping density	$N_A = 8.0 \times 10^{12}$ cm^{-3}	$N_D = 8.0 \times 10^{12}$ cm^{-3}
Front-contact workfunction, TE Fermi level position, and surface recombination speed	$\phi_W = 4.75$ eV, $E_F - E_V = 0.42$ eV $S_n = 1 \times 10^3$ cm/s $S_p = 1 \times 10^7$ cm/s	N.A.
Back-contact workfunction, TE Fermi level position, and surface recombination speed	N.A.	$\phi_W = 4.75$ eV, $E_C - E_F = 0.45$ eV $S_n = 1 \times 10^7$ cm/s $S_p = 1 \times 10^3$ cm/s
Electron and hole mobilities	$\mu_n = 1350$ cm^2/vs $\mu_p = 450$ cm^2/vs	$\mu_n = 1350$ cm^2/vs $\mu_p = 450$ cm^2/vs
Band effective densities of states	$N_C = 2.8 \times 10^{19}$ cm^{-3} $N_V = 1.04 \times 10^{19}$ cm^{-3}	$N_C = 2.8 \times 10^{19}$ cm^{-3} $N_V = 1.04 \times 10^{19}$ cm^{-3}
Bulk defect properties	Donor-like gap states from E_V to mid-gap $N_{TD} = 5 \times 10^{13}$ cm^{-3}eV^{-1} $\sigma_n = 5 \times 10^{-13}$ cm^2 $\sigma_p = 5 \times 10^{-15}$ cm^2 Acceptor-like gap states from mid-gap to E_C $N_{TA} = 5 \times 10^{13}$ cm^{-3}eV^{-1} $\sigma_n = 5 \times 10^{-15}$ cm^2 $\sigma_p = 5 \times 10^{-13}$ cm^2	Donor-like gap states from EV to mid-gap $N_{TD} = 5 \times 10^{13}$ cm^{-3}eV^{-1} $\sigma_n = 5 \times 10^{-13}$ cm^2 $\sigma_p = 5 \times 10^{-15}$ cm^2 Acceptor-like gap states from mid-gap to E_C $N_{TA} = 5 \times 10^{12}$ cm^{-3}eV^{-1} $\sigma_n = 5 \times 10^{-15}$ cm^2 $\sigma_p = 5 \times 10^{-13}$ cm^2
HJ interface light reflection	Neglected	
Back light reflection		Total reflection

from Table 5.1 that any reflection or absorption of the AM1.5G spectrum as it enters from the left of the contact is neglected; however, reflection at the back contact is included. Light reflection at the internal interface and interference effects are not addressed.

Figure 5.12a gives the semi-log and Figure 5.12b gives the linear light and dark J-V that come from a numerical analysis of the baseline cell of Figure 5.11. As is our practice, the linear J-V characteristics are plotted with the fourth quadrant serving as the power quadrant. The simulation results show that the symmetry breaking caused by the affinity steps (electron and hole effective forces) is sufficient to give rise to significant photovoltaic action and consequently that this simple baseline structure works as a solar cell, in spite of there being no built-in electric field. This is no surprise, since the phenomenon has been discussed in Section 3.2.4. The simulation results for the device of Figure 5.11 give a dark J-V that is seen from Figure 5.12a to be of the form $J \sim e^{V/nKT}$ with a diode n-factor that is slightly >1. These results give a light J-V with FF = 0.58, $V_{oc} = 0.37 \text{V}$, $J_{sc} = 28.4 \text{mA/cm}^2$, and $\eta = 6.1\%$. This performance is much better than the no built-in-potential example of Section 3.2.4 primarily because of the presence of an effective force field for holes as well as for electrons in Figure 5.11. Again, we stress that the rule of thumb for homojunctions that V_{oc} is limited by the built-in potential is obviously not applicable to heterojunctions, since there is no built-in potential V_{Bi} in this example (nor in the one of Section 3.2.4). The idea of superposition of a photocurrent, which only depends on illumination, with a dark (bucking) current, which only depends on voltage—the idea expressed by Eq. 5.2—is seen to fail for the cell of Figure 5.11, with $V_{oc} \sim 0.05 \text{ V}$ lower than what superposition predicts. It is noteworthy that this is the same behavior seen in the experimental data of Figure 5.6. The J_{sc} of this cell structure is rather good when one notes from Figure 3.20 that 20 μm (10 μm in, plus 10 μm back out, due to total reflection at the back contact) of single-crystal silicon, the material whose absorption data we are using (Table 5.1), can only yield a maximum photocurrent about 30 mA/cm^2, if the quantum efficiency were 100%.

Numerical analysis allows us to easily "open up" a solar cell and peer inside to see its full workings, so let us do that with the device of Figure 5.11. To begin to delve into this cell's detailed operation, we note that TE electrons are diffusing from right to left at the heterojunction

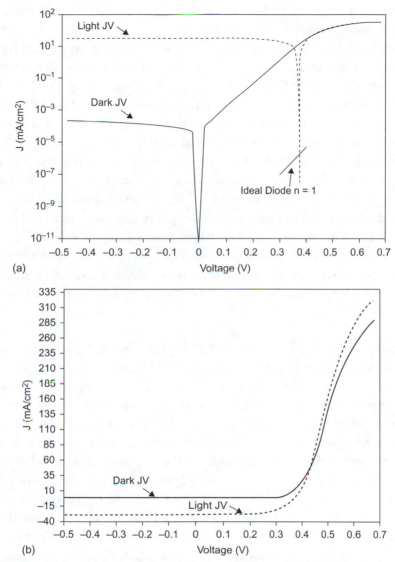

FIGURE 5.12 Computer simulation results for the light and dark J-V behavior for the cell of Table 5.1 and Figure 5.11 depicted using (a) a semi-log and (b) a linear plot. The sign convention for the linear plot has the fourth quadrant as the power quadrant. Superposition is seen to be slightly off for this cell; i.e., the V_{oc} is about 0.05 V lower than is predicted from Eq. 5.2. The n-factor of an ideal diode is also shown in (a).

interface plane in Figure 5.11 due to the electron population differences, which may be read from the conduction band edge position versus the Fermi level. Simultaneously, electrons must be drifting from left to right due to the electron effective force field (electron affinity step) at the HJ interface plane in Figure 5.11. These flows are exactly equal in TE, since it is necessary that $J_n = 0$, due to the principle of detailed balance. There is a similar detailed balance between the hole diffusion and hole effective force drift components at this interface plane, since $J_p = 0$ is also required at TE. Under light, the materials system is driven out of TE, detailed balance does not apply for either carrier, and, from the way the symmetry is broken in Figure 5.11, a conventional photocurrent must be set up that is flowing in the negative direction (right to left in Fig. 5.11). The short-circuit simulation results show all this in Figures 5.13a and b. A quick glance at these plots shows the transport causing the short circuit current is a rich mixture of electron and hole drift and diffusion components. A rather interesting symmetry is seen between Figures 5.13a and 5.13b.

Exploring this all more deeply, we see that Figure 5.13a shows that those electron current components which are negative must be dominating the electron contribution to J_{sc}. This must be true because the total electron current density contribution to the short-circuit current density must be negative in the context of the positive x-coordinate direction of Figure 5.11. With this in mind, it can be seen that electrons, as particles, are moving toward the HJ interface in the top absorber material due to diffusion's dominating up to about ~2 μm. Diffusion is occurring here because electron photogeneration is intense in this region due to light entry at the front contact. Diffusion never brings electrons to the front surface. This deleterious flow pattern is avoided in this example because there is an inefficient sink there; i.e., S_n is small. After ~2 μm, electric field drift is seen to aid diffusion in bring the photogenerated electrons to the effective-force-field barrier. This electric field does not exist at TE but develops under light and it is caused by the free photocarriers themselves. This is certainly one of the non-linear phenomena in the equation set of Section 2.4; i.e., the carriers are generating the electric field that is aiding in collecting the minority carriers. At that HJ interface, the residual diffusion current and effective field drift currents of TE can be seen, but there is no longer an exact balance. In fact, effective field drift is dominating

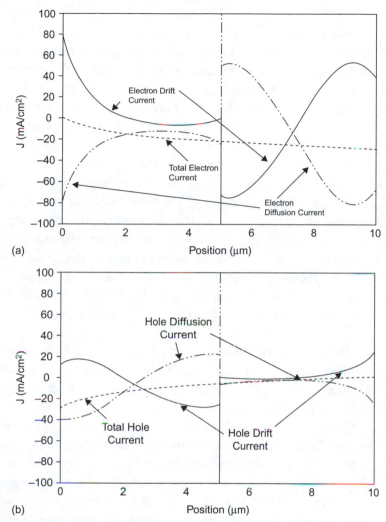

FIGURE 5.13 The computed current density components for the device of Figure 5.11 in short circuit. The sign convention is that the positive direction for conventional current densities and the electric field is left to right. (a) The electron diffusion and drift conventional current density components and the total electron conventional current density. (b) The hole diffusion and drift conventional current density components and the total hole conventional current density.

in carrying the electrons (as particles) across the HJ interface into the bottom absorber. Once the electrons are in the bottom absorber, transport is dominated by electric field drift until about ~6.8 μm. This electrostatic field also does not exist in TE. From that point until the back contact, electron transport is dominated by diffusion—driven by the attractiveness of the back contact ($S_n = 1 \times 10^7$ cm/s, from Table 5.1). This plot shows that there is a great deal of interplay between diffusion and drift taking place across the structure. Figure 5.13b shows that the hole current components which are negative, must be dominating the hole contribution to $J_{sc,}$ since the total hole current density contribution to the short-circuit current density is negative. Examination of the figure shows that the hole transport behavior at the short-circuit condition is the mirror image of that seen for electrons. In the very back of the cell, diffusion dominates in carrying the hole current, but barely. Then diffusion works with electric field drift but effective force drift takes over at the HJ interface plane forcing the holes into the top absorber. On entering the top absorber hole electric field drift dominates initially but then finally diffusion dominates in bringing the holes to the top contact due to its serving as a hole sink. Of course, the total electron current density in Figure 5.13a plus the total hole current density in Figure 5.13b is a constant (28.4 mA/cm^2) across the whole device, as can be readily verified from Figure 5.13.

The band bending and recombination at open circuit are shown for this device in Figure 5.14. As can be seen from the two distinct curvature regions of the bands, or perhaps more clearly from the curvature regions of the local vacuum level (Fig. 5.14a), negative charge is being developed on the right side of the HJ interface and positive charge is being developed on the left side at open circuit. This charge pile-up possibility is a unique feature of heterojunctions due to the presence of the affinity steps that allow it to occur. We never saw this in homojunctions. The piled-up charge creates an electric field ξ oriented in the positive x-direction and located at and around the HJ interface, as seen from the spatial variation of the vacuum level. From Eq. 2.47, we know that the integral of the electric field present at open circuit (the electric field at and around the HJ as well as in the rest of the structure) taken across the device $\int \xi dx$ must now equal Voc (the ξ_0 term in Eq. 2.47 is zero here since there is no built-in electrostatic field in this example). In other words, this integral must equal the contact Fermi-level splitting that has made the Fermi level of the electrode on the right higher in energy than the Fermi level of the electrode on the left by V_{oc}. The feedback built into the equations of Section 2.4 insures that this V_{oc} gives the Fermi-level splitting needed to get recombination to equal generation.

This recombination is due to electron loss at the front contact (mechanisms 1 and 2), which we are modeling with Eq. 2.40 (i.e., in this case by $S_n(n-n_0)$), hole loss at the back contact (mechanisms 6 and 7), which we are also modeling with Eq. 2.40 (i.e., in this case by $S_p(p-p_0)$), and bulk recombination $\int_{-d}^{L+W}\mathscr{R}(x)\,dx$, which accounts for mechanisms 1, 4, and 5 of Figure 5.5. HJ interface recombination

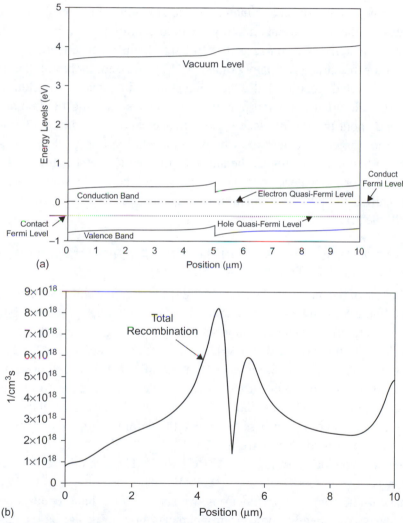

FIGURE 5.14 Computer simulation results giving: (a) Plot of the band bending for the device of Figure 5.11 at open circuit and (b) plot of the bulk recombination integrand $\mathscr{R}(x)$ at open circuit. The electric field is the derivative of the vacuum level as a function of position and can therefore be discerned from (a).

(mechanism 8 of Figure 5.5) is not included since a perfect interface has been assumed in this example. Figure 5.14b shows a plot of the integrand $\mathscr{R}(x)$ at open circuit. Interestingly, the presence of all the electrons just to the right of the HJ interface and all the holes just to the left of this interface at open circuit causes the two spikes seen in Figure 5.14b. From Eq. 2.9, we can see that these spikes occur where the np product is largest. This situation would be totally missed in a linearized (lifetime) model for recombination, since only n would control recombination in the top material and only p would control recombination in the bottom material. Of course, if the full set of linearized equations were used for this structure, many points would be missed, including the carrier generated electric filed which plays a role in carrier collection, the charge pile up at the heterojunction, and the electrostatic field creation in and about the HJ interface region. If one surveys the numerical output (not shown) to determine which loss mechanism or mechanisms are controlling in this example, the answer turns out to be mechanisms 1, 2, 6, and 7 for the material parameters chosen in Table 5.1. Control is easily shifted to bulk recombination by adjusting the gap-state parameters in Table 5.1.

5.3.1.2 ADDITION OF A BUILT-IN ELECTROSTATIC-POTENTIAL BARRIER TO THE EFFECTIVE-FORCE-FIELD BARRIER

The first of the modifications to be done to the baseline cell of Figure 5.11 and Table 5.1 can be found in Table 5.2. These changes produce a built-in potential V_{Bi} having components V_{Bi1} and V_{Bi2} developed across a built-in electrostatic field at the HJ region of the device as seen for TE in Figure 5.15. The structure looks like a "normal" HJ solar cell now, with affinity steps and a built-in electrostatic potential. As seen from Table 5.2 and Figure 5.15, this "normal" device was achieved by increasing the doping in both absorbers to change the before-contact workfunction of the top semiconductor from 4.75 eV to 5.09 eV and that of the bottom semiconductor from 4.75 eV to 4.40 eV. These modifications lead to a built-in potential of $V_{Bi} = 0.69$ eV. Since the same changes were made to the contact workfunctions, we stay with flat band conditions at contacts. We already know from Chapter 4 how to design an advantageous high-low junction or other selective ohmic contact structures at device electrodes, so we will not repeat that here. Instead, we just keep the contacts simple and described by surface recombination speeds. The

Table 5.2 Adding an Electrostatic Built-in Potential to the Baseline Heterojunction Device

Changes	Cell of Figure 5.15
Front-contact workfunction changed to	5.09 eV
Back-contact workfunction changed to	4.40 eV
Top-layer doping changed to	$5 \times 10^{17} \, cm^{-3}$
Top-layer electron affinity changed to	___
Top-layer band gap changed to	___
Top-layer thickness changed to	___
Bottom-layer doping changed to	$5 \times 10^{17} \, cm^{-3}$
Bottom-layer thickness changed to	___
Heterojunction interface property changes	___

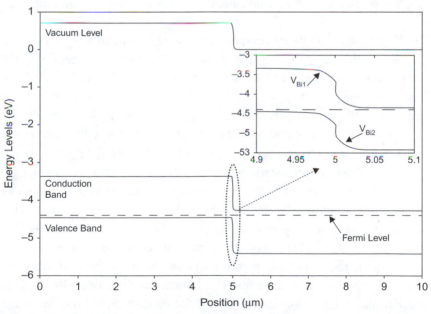

FIGURE 5.15 The computed HJ structure of Figure 5.11 but with the changes of Table 5.2. An electrostatic built-in potential $V_{Bi} = V_{Bi1} + V_{Bi2} = 0.69 \, eV$ is now present at the HJ region in addition to the 0.25-eV affinity steps. Both types of contributions to the total electron and total hole effective barriers are better seen in the inset. The structure is shown in thermodynamic equilibrium.

semi-log and linear J-V characteristics for this cell, as determined by numerical simulation, are presented in Figure 5.16. The addition of the built-in electrostatic-field barrier is seen to improve performance over that of the device of Figure 5.11. The structure of Figure 5.15 gives an open-circuit voltage of $V_{oc} = 0.65\,V$, $J_{sc} = 26.4\,mA/cm^2$, FF = 0.84, and an efficiency of $\eta = 14.3\%$. The figure also shows that the dark current diode n-factor is a function of voltage in this case, but becomes close to the ideal unity value as the voltage approaches that of open circuit. It also shows that superposition works quite well. This is because the role of drift in helping diffusion collect minority carriers is no longer important and the piling-up phenomenon present for the baseline device has been eliminated, as we will see shortly when we look into cell operation details. Interestingly, the short-circuit current is somewhat smaller for the device of Figure 5.15 than for that of Figure 5.11. This is due to the doping change and a resulting shift in which gap states are supporting the recombination traffic. The doping change has caused a shift to gap states with higher recombination cross-sections for the minority carriers of each absorber. That causes an increase in recombination.

Turning to the simulation output to explore the details of the device physics of this "normal" heterojunction cell, we start by examining the short-circuit condition and plot some of the results in Figure 5.17. The band diagram in the figure shows that there is a "bubbling up" of the (originally minority-carrier) electron quasi-Fermi level in the top material and a corresponding bubbling up of the (originally minority-carrier) hole quasi-Fermi level in the bottom material. This is a manifestation of the minority-carrier population changes caused by the light absorption. The slopes of the minority-carrier quasi-Fermi levels show the direction in which the electron and hole current densities are flowing in the top and bottom materials, respectively, since $J_n = en(dE_{Fn}/dx)$ and $J_p = -ep(dE_{Fp}/dx)$. These expressions continue to use our convention that E_{Fn} is measured positively up from the Fermi-level position in the p-material contact and E_{Fp} is measured positively down from the Fermi-level position in the n-material contact. Since the current is the product of population and quasi-Fermi level gradient, once the electron quasi-Fermi level enters the n-type material, the population n becomes large, and E_{Fn} becomes essentially flat. Of course, the same thing happens to the hole quasi-Fermi level once it enters the p-material.

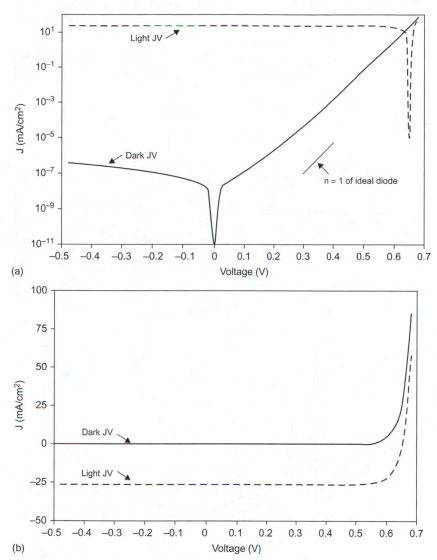

FIGURE 5.16 Computer simulation results for the light and dark J-V behavior for the cell of Figure 5.15 depicted using (a) semi-log and (b) linear plots. The introduction of the built-in electrostatic field improves the open-circuit voltage and fill factor, and therefore efficiency. Superposition works well for this cell.

Interestingly, the quasi-Fermi levels are not flat across the total barrier (electrostatic barrier and affinity step barrier) region in short circuit, as seen in the inset of Figure 5.17a. Instead, they begin the "bubbling up" described above at the HJ interface.

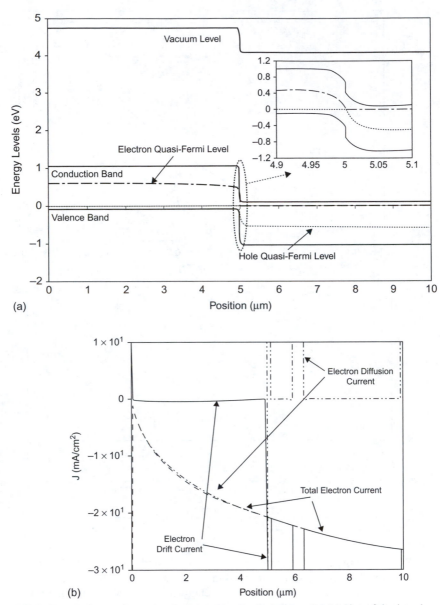

FIGURE 5.17 Some short-circuit computer simulation results. (a) Plot of the band bending at short circuit for the cell of Figure 5.15. The electric field is the derivative of the vacuum level in this band diagram. The inset shows details at the HJ. (b) The electron conventional current components and total current.

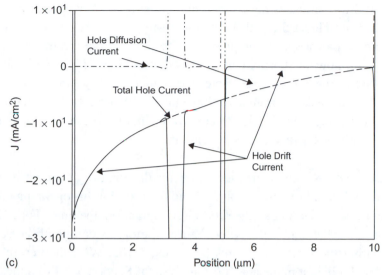

(c)

FIGURE 5.17 (c) hole conventional current components and total current at short circuit.

While the slopes of E_{Fn} and E_{Fp} show that electrons are flowing out of the top material to the HJ interface and that holes are flowing out of the bottom material to the HJ interface at short circuit, the plots of Figure 5.17b and c give the details. These plots make it clear that in the cell of Figure 5.15 diffusion is much bigger than drift in the minority carrier collection region. Drift is seen to be important at the HJ and where a carrier is the majority species. This is the behavior we saw in Chapter 4 for p–n homojunctions. We will utilize this is Section 5.4 when we undertake an analytical analysis of this cell. It is interesting to note that the roles of diffusion and drift are much more intertwined in the case of Figure 5.13. These two current components are very similar in magnitude for the minority carriers in that cell structure, as seen in Figure 5.13. It is also interesting to look at the vestiges, still noticeable at short circuit, of the diffusion-drift balance in place for each carrier at TE. Taking a look at this for electrons first, we see from Figure 5.17b that a vestige of the balancing of electron diffusion and electrostatic field drift currents at TE is very noticeable in sections of the electrostatic-field region. At short circuit, electrostatic field drift is dominating. Figure 5.17b also shows that, at the HJ interface plane, there was another precarious balance between a large electron diffusion current and a large electron drift current. The latter being due, in this case, to the presence of an electron effective force field (electron affinity change). At short circuit, drift dominates here, too. Making a similar assessment for holes, we see from Figure 5.17c that a vestige of the hole diffusion and electrostatic drift balance at TE is present in sections of the electrostatic-field region but now, at short circuit, drift is dominating. Figure 5.17c shows also that at the HJ interface there was a precarious balance between a large hole diffusion current and a large hole drift current. The latter being due, in this case, to a hole effective force field (hole affinity change). Drift dominates here, too, in short circuit.

Figure 5.18 presents some simulation results for the maximum power point condition. Looking first at the band diagram for this operating point, it can be seen that, unlike the short-circuit situation, the quasi-Fermi levels are now flat across the whole barrier region, as can be observed by comparing the insets of Figures 5.17a and 5.18a. We will keep this in mind and utilize it in the analytical analysis of Section 5.4. From the definition of short circuit, the Fermi levels in the contacts must be at the same energy. As we know, this condition is removed at any other operating point and the contact Fermi levels split apart developing the voltage between the device electrodes. The majority-carrier quasi-Fermi levels will follow this split since the gradient in the majority carrier quasi-Fermi levels must be small. This point follows from $J_n = en(dE_{Fn}/dx)$ and $J_p = -ep(dE_{Fp}/dx)$ and the fact that n and p are large where they are majority carriers. The splitting of the majority-carrier quasi-Fermi levels, which are relatively stuck in their position in the band gap by doping, must force a reduction of the band bending for this cell design of Figure 5.15. This is seen in the numerical analysis results of the inset of Figure 5.18a. Indeed, examination shows that the band bending in material 1 has decreased from the TE value V_{Bi1} and the band bending in material 2 has decreased for its TE value V_{Bi2} seen in Figure 5.18a. Thinking in terms of Eq. 5.2 and superposition, which we know from Figure 5.16a works for this particular cell, the reduction of the electrostatic barrier may be thought of as increasing (i.e., forward biasing) the dark diode current that is opposing the photocurrent J_{sc}.

Now let us change our perspective and look at the light $J = J(V)$ for this cell in the context of its being the net result of generation minus recombination, the idea expressed in Eq. 5.1. Unlike superposition, this idea is always correct and it must therefore be true for any point on the J-V curve, including the maximum power point. We use this generation

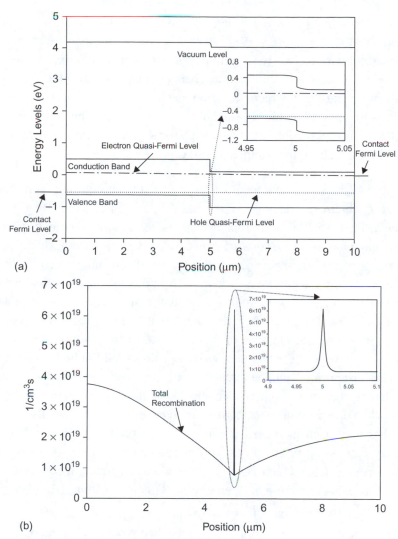

FIGURE 5.18 Some simulation results for the maximum power point for the cell of Figure 5.15. (a) Band bending with details at the HJ shown in the inset. The electric field is the derivative of the vacuum level. (b) The bulk recombination quantity $\mathcal{R}(x)$ at the maximum power point.

minus recombination view at the maximum power point and consider Figure 5.18b, which gives the integrand in the expression $\int_{-d}^{L+W} \mathcal{R}(x)\,dx$ (loss mechanisms 1, 4, and 5 from Fig. 5.5). The area under the plot in Figure 5.18b is the total bulk recombination $\int_{-d\lambda}^{L+W} \mathcal{R}(x)\,dx$ for the maximum power point. The numerical simulation for this case gives

$\left|J_{ST}(-d)/e\right| = S_n n(-d) \approx 10^{15}/cm^2$-s (loss mechanisms 2 and 3) and $\left|J_{SB}(L+W)/e\right| = S_p p(L+W) \approx 10^{15}/cm^2$-s (loss mechanisms 5 and 6) for contact losses for this maximum power point situation. From these values for the surface recombination mechanisms and from the plot of Figure 5.18b, it is clear that the recombination mechanism in control at the maximum power point for this case is bulk recombination. Generation minus total bulk recombination establishes the J and V at the maximum power point for this cell for the parameters used.

Figure 5.19 presents some simulation results for the cell of Figure 5.15 for the open-circuit condition. The band diagram for this operating point given in Figure 5.19a shows that, as was the case at the maximum power point, the majority carrier quasi-Fermi levels extent in a flat fashion from the majority carrier region across the whole electrostatic and effective-force-field barrier region. The additional splitting of the majority-carrier Fermi levels seen at open circuit versus that seen at the maximum power point condition is needed to further reduce the electrostatic band bending, as is apparent in the inset of Figure 5.19a. In fact, the local vacuum-level plot shows that for this particular cell, almost all the built-in electric field has disappeared at open circuit. Clearly, this cell is not far away from developing the charge pile-up situation encountered in Section 5.3.1.1. From the superposition perspective, the dark diode forward-biased current has now been increased at this operating point to a level where it exactly equals the photocurrent. From the generation versus recombination perspective, which is, of course, always correct, recombination exactly equals generation at open circuit. This means that, from Eq. 5.1

$$\int_{-d\lambda}^{L+W}\int G_L(\lambda, x)d\lambda dx = \int_{-d}^{L+W}\mathscr{R}(x)dx + J_{ST}(-d)/e + J_{SB}(L+W)/e$$

must be true at open circuit (We have been neglecting recombination path 8, so far). Figure 5.19b gives the integrand in the expression $\int_{-d}^{L+W}\mathscr{R}(x)dx$ as a function of x while numerical data from this case shows (not given) that $\left|J_{ST}(-d)/e\right| = S_n n(-d) \approx 10^{16}/cm^2$-s and $\left|J_{SB}(L+W)/e\right| = S_p p(L+W) \approx 10^{16}/cm^2$-s. It can be seen from doing an "eye-ball" integration of the bulk recombination in Figure 5.19b and comparing that result with the contact losses at open circuit that bulk recombination predominately determines V_{oc} for this particular cell structure.

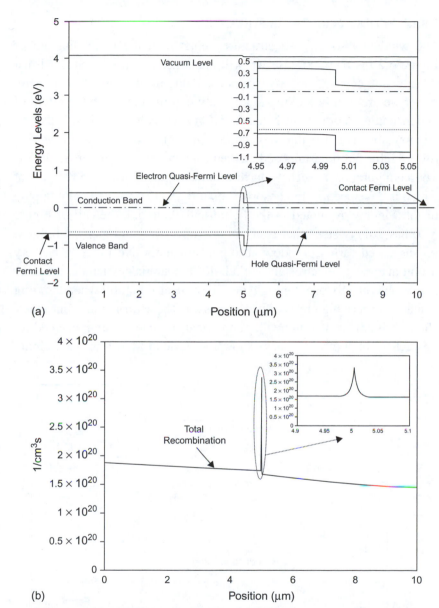

FIGURE 5.19 Some simulation results for open circuit for the cell of Figure 5.15. (a) Band bending with details at the HJ shown in the inset. The electric field is the derivative of the vacuum level as a function of position. (b) A plot of the bulk recombination $\mathcal{R}(x)$ occurring at every interior point in the device at open circuit.

5.3.1.3 WINDOW-ABSORBER STRUCTURE

The window-absorber cell structure depicted in the band diagram of Figure 5.20 is the next case we will examine. Again it is stressed that we are not engaged in any sort of an optimization procedure here; rather we are simply looking at a variety of structures with the intent of understanding more about heterojunction solar cell operation. This first window-absorber structure has the modified properties listed in the first row of Table 5.3 superimposed on the basic properties listed in Table 5.1. The configuration is seen to have a built-in potential of 0.54 eV, all of which is developed in the absorber, material 2, due to the heavy doping of the window. As noted in the figure and Table 5.3, the cell of Figure 5.20 has affinity steps that result in an "effective band gap" of 0.87 eV. As discussed earlier, window-absorber structures like this can be very useful in creating electrostatic-field built-in potentials when the absorber is available in only one doping type. They are also helpful when the front contact has very high recombination losses. The semi-log and linear J-V characteristics, as determined by numerical simulation, are presented for this cell in Figure 5.21. These show the device to have an open-circuit

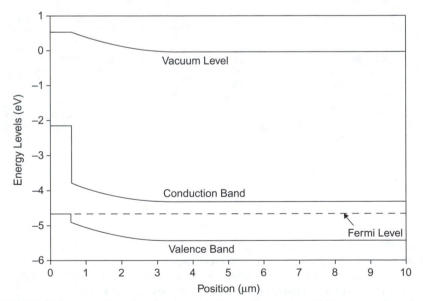

FIGURE 5.20 A window-absorber heterojunction structure in TE. The properties of this structure are listed in Table 5.3, together with Table 5.1. The window material is heavily doped p-type. A 0.54-eV built-in potential and its built-in electric field are present at the HJ region in addition to the affinity steps. Under illumination, light impinges from the left.

Table 5.3 Changes Made to the Baseline Cell for Some Window-Absorber Examples

Cell	Front-Contact Workfunction changed to	Back-Contact Workfunction changed to	Top-Layer Doping changed to	Top-Layer Electron Affinity changed to	Top-Layer Band Gap changed to	Top-Layer Thickness changed to	Bottom-Layer Doping changed to	Bottom-Layer Thickness changed to	Heterojunction Interface Property Changes
Fig. 5.20	5.17 eV	4.63 eV	1×10^{19} cm^{-3}	2.67 eV	2.50 eV (same absorption data used but now cut off at 2.50 eV)	600 nm	5×10^{14} cm^{-3}	1×10^4 nm	——
Fig. 5.24	5.17 eV	4.63 eV	1×10^{19} cm^{-3}	2.67 eV	2.50 eV (same absorption data used but now cut off at 2.50 eV)	600 nm	5×10^{14} cm^{-3}	1×10^4 nm	Acceptor-like interface states at $E_V + 0.56$ eV in first 10 nm of the absorber $N_{TA} = 1 \times 10^{19}$ cm^{-3} $\sigma_n = 1 \times 10^{-14}$ cm^2 $\sigma_p = 1 \times 10^{-13}$ cm^2
Fig. 5.27	5.17 eV	4.63 eV	1×10^{19} cm^{-3}	2.67 eV	2.50 eV (same absorption data used but now cut off at 2.50 eV)	600 nm	5×10^{14} cm^{-3}	1×10^4 nm	Donor-like interface states at $E_V + 0.27$ eV in first 10 nm of the absorber $N_{TD} = 1 \times 10^{19}$ cm^{-3} $\sigma_n = 1 \times 10^{-12}$ cm^2 $\sigma_p = 1 \times 10^{-18}$ cm^2
Fig. 5.30	5.17 eV	4.63 eV	1×10^{19} cm^{-3}	2.67 eV	2.50 eV (same absorption data used but now cut off at 2.50 eV)	600 nm	5×10^4 cm^{-3}	1×10^4 nm	Donor-like interface states at $E_V + 0.45$ eV in first 10 nm of the window at the HJ $N_{TD} = 1 \times 10^{19}$ cm^{-3} $\sigma_n = 1 \times 10^{-8}$ cm^2 $\sigma_p = 1 \times 10^{-30}$ cm^2 and Acceptor-like interface states at $E_V + 0.56$ eV in first 10 nm of the absorber at the HJ $N_{TA} = 1 \times 10^{19}$ cm^{-3} $\sigma_n = 1 \times 10^{-14}$ cm^2 $\sigma_p = 1 \times 10^{-13}$ cm^2

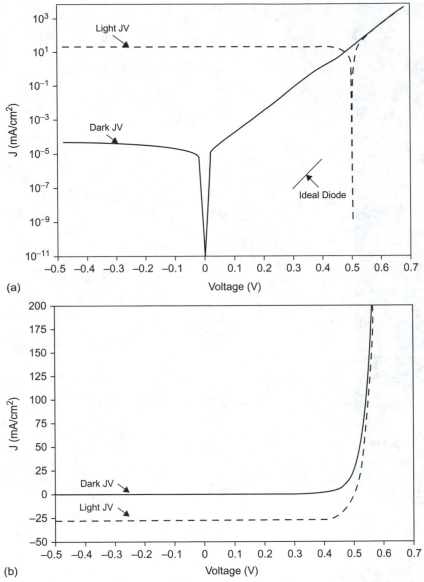

FIGURE 5.21 Computer simulation results for the light and dark J-V behavior for the cell of Figure 5.20 depicted using (a) semi-log and (b) linear plots. Numerical analysis says superposition works well for this cell and gives a dark diode n-factor close to unity.

voltage of 0.50 V, J_{sc} = 27.9 mA/cm², FF = 0.76, and a conversion efficiency of 10.6%. The length of the absorber is 10 μm in this device (see Table 5.3) so that the photocurrent produced is in the same range as that of the devices of Figures 5.11 and 5.15. Figure 5.21 also shows that the dark current diode n-factor is close to, but somewhat larger than, unity, and that superposition works well for this cell.

Figure 5.22 allows us to pry open this cell and to peer into its operation at short circuit. The current components in Figures 5.22a and b exhibit the same behavior trends we saw in the previous heterojunction example, which, as we recall, also had both an electrostatic-field and effective-field barrier region as well. For example, from Figure 5.22a and b we see electrons move in the absorber due to the domination of drift transport (except near the back contact, where the sink of the back-contact recombination modifies the pattern) and holes move in the absorber due to the domination of diffusion (except as the holes enter into the electrostatic-field barrier region).

Some detailed simulation results are given in Figure 5.23 for the maximum power point. The band diagram shows the behavior we have come to expect, including the flat quasi-Fermi level behavior for both carriers across the total barrier region. The local vacuum level shows that the built-in electrostatic potential and its electrostatic field have almost been annihilated in this cell at the maximum power point. However, things have not reached the situation where there is carrier pile up at the HJ such as we saw in the baseline device. This can be deduced from the lack of electrostatic field reversal in the region. Figure 5.23b is the integrand of $\int \mathscr{R}(x)\,dx$. It shows a peaking at the HJ interface due to the carrier population increases at the maximum power point condition with the collapse of the electrostatic barrier. The integral of the plot in Figure 5.23b gives a total bulk recombination of the order of $\sim 10^{16}/cm^2$-s, while the simulation results (not shown) give hole recombination losses at the back contact, which are of the order of $\sim 10^{16}/cm^2$-s also. Consequently, due to our material parameter selections in Tables 5.1, 5.2, and 5.3, both bulk and back-surface recombination are responsible for the carrier losses at the maximum power operating point for this cell.

5.3.1.4 WINDOW-ABSORBER STRUCTURE WITH ABSORBER INTERFACE RECOMBINATION

Earlier in this chapter we expressed concern about the possible impact of localized states at and around a HJ's metallurgical junction on device performance (mechanism 8 in Fig. 5.5). Having developed a good feel

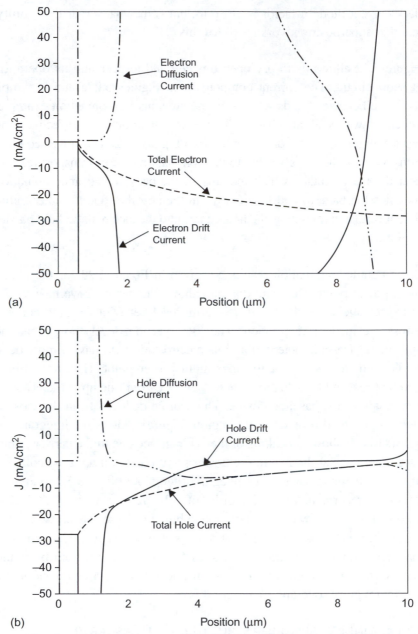

FIGURE 5.22 Some short-circuit simulation results. (a) Electron conventional current density components and total conventional current density and (b) the hole conventional current density components and total conventional current density.

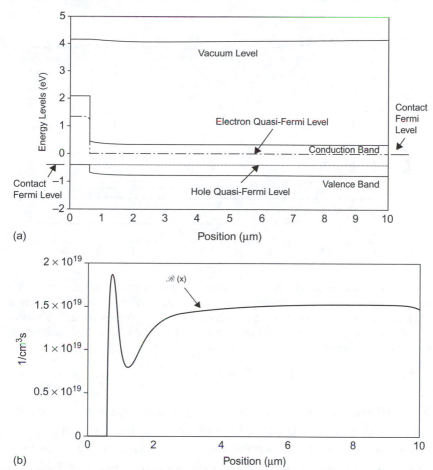

(a)

(b)

FIGURE 5.23 Some simulation results for the maximum power point for the cell of Figure 5.20. (a) Band bending at the maximum power point with the electric field given by the derivative of the local vacuum level; (b) Plot of the bulk recombination $\mathcal{R}(x)$ as a function of position at the maximum power point.

for heterojunction behavior without such defects, we are ready now to add them and examine their impact using numerical analysis. We do this using the cell shown in TE in Figure 5.24. It is the same device as that of Figure 5.20, except now it has defects at the HJ interface. As seen from the second row of Table 5.3, the defects have been added to a 10-nm interface layer in the absorber at the HJ. Taking the number density of atoms in the absorber to be about $10^{21}/cm^3$, it can be seen from Table 5.3 that the defect density used is such that 1 in every 100 atoms

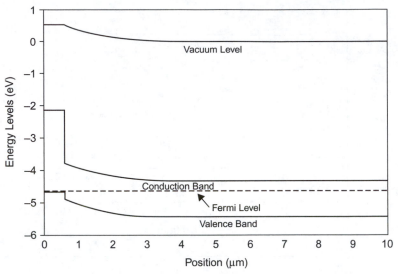

FIGURE 5.24 A window-absorber type heterojunction TE band diagram for a structure with mid-gap, acceptor-like interface recombination defects in the first 10 nm of the absorber material. The defects are described in the second row of Table 5.3. Since these particular interface states are uncharged at TE, the band diagram is exactly the same as that of Figure 5.20.

in the 10-nm interface layer is a defect (impurity, interdiffused species, vacancy, etc) in this HJ example. The defect states have been chosen to be effective, from a loss point of view, S-R-H interface recombination sites. This can be noted from their properties listed in Table 5.3, which exhibit the telltale signs of efficient S-R-H recombination paths; i.e., relatively large capture cross-sections with both cross-sections relatively close in magnitude (making them attractive to both carriers) and an energy location for these states away from the band edges. The defects have been taken to be acceptor-like, so the capture cross-section for holes is higher due to coulombic attraction, as discussed in Chapter 2. In the numerical analysis undertaken for this example, it has also been assumed that the interface recombination path has electrons at the absorber conduction band edge (absorber E_C or LUMO) recombining with holes at the absorber valence band edge (absorber E_V or HOMO); hence, the band gap involved is that of the absorber. However, there are many more holes at the valence band edge (window E_V or HOMO) of the window. Consequently, due to intermixing of the materials at the interface, tunneling, etc., it is very possible that the recombination path could be from the absorber LUMO to the window HOMO. This would result in an effective band gap, a quantity mentioned earlier, for this recombination process that would equal the window HOMO − absorber LUMO difference.

The J-V characteristics that result from numerical analysis of the cell of Figure 5.24 are presented in Figure 5.25. They show that the presence of the interface states does affect cell performance, with the open-circuit voltage dropping by 0.11 V from that of the Figure 5.20 device. Overall, the cell now has an open-circuit voltage of 0.39 V, $J_{sc} = 27.6\,mA/cm^2$, FF = 0.74, and a conversion efficiency of 8.1%. Interestingly, the J_{sc} and FF are only very slightly decreased from those seen for the cell of Figure 5.20. Even with this interface recombination present, the dark current diode n-factor is seen to be close to unity and superposition still works quite well in Figure 5.25a. This example indicates what recombination through interface defects can do to a heterojunction. Obviously, the impact of HJ interface recombination can be increased or decreased by modifying the interface defect properties, number, and spatial extent.

The TE band diagram of the cell of Figure 5.24 is seen to be exactly the same as that of the cell of Figure 5.20. This is done to try to keep things simpler and has been accomplished by taking the HJ interface recombination sites of the example to be acceptor-like states lying above the Fermi level at TE. The current-density-component plots at short circuit for this device are similar to those of the cell of Figure 5.20 and are not shown here. The key difference in the behavior of the cell of Figure 5.24 shows up in the recombination plots for short and open circuit given in Figure 5.26. They give the interior (bulk and HJ interface state supported) recombination component contributions from Eq. 5.1. These plots show that the presence of the interface recombination centers causes huge recombination at the HJ. Looking specifically at the open-circuit situation in Figure 5.26, the bulk $\mathscr{R}(x)$ and HJ interface state supported recombination are seen from the figure to contribute the integrated recombination rates of $\sim 10^{16}/cm^2$-s and $\sim 10^{17}/cm^2$-s, respectively. The simulations also give a back-contact recombination rate J_{SB}/e of $\sim 10^{16}/cm^2$-s at open circuit for this case. There are no

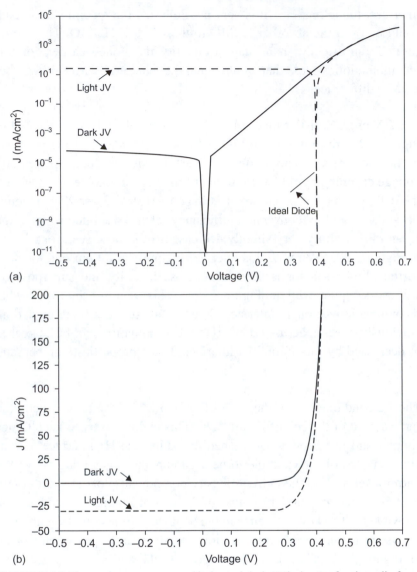

FIGURE 5.25 The computer generated light and dark J-V behavior for the cell of Figure 5.24 depicted using (a) semi-log and (b) linear plots. Superposition works well for this cell.

significant losses, of course, at the contact to the window cell. From this information, it follows that the HJ interface recombination through the HJ defect states in this cell is responsible for the open-circuit voltage deterioration.

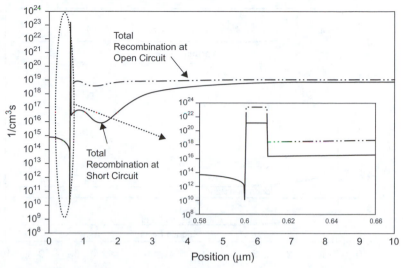

FIGURE 5.26 Numerical simulation results for the total interior recombination as a function of position. Includes the HJ interface recombination component (seen at $x = 0.6\,\mu m$) and $\mathscr{R}(x)$. Shown for short-circuit and open-circuit conditions.

5.3.1.5 WINDOW-ABSORBER STRUCTURE WITH ABSORBER INTERFACE TRAPPING

The cell of this example, shown in TE in Figure 5.27, is again the same as that of Figure 5.20, except that it, too, has defects present in a 10 nm interface layer in the absorber. They are, however, of a different type than those present in the example of Figure 5.24. The properties of these interface states, seen in the third row of Table 5.3, have been chosen to make them function as traps, as opposed to recombination sites. The telltale sign of trapping capabilities is the relatively large capture cross-section for one carrier combined with a much smaller capture cross-section for the other. The capture cross-section for electrons is the higher one for the donor-like states chosen for this example due to coulombic attraction. This high capture cross-section for electrons and very low capture cross-section for holes means the electron population of these states will follow the electron quasi-Fermi level during cell operation. Given the spatial and energy gap positions of these states, they will be positively charged in thermodynamic equilibrium, resulting in an intense electrostatic field in the interface layer. This can be seen from the step in the vacuum level (in the electron potential energy) across the 10 nm absorber interface layer seen in Figure 5.27. These defect states will increasingly

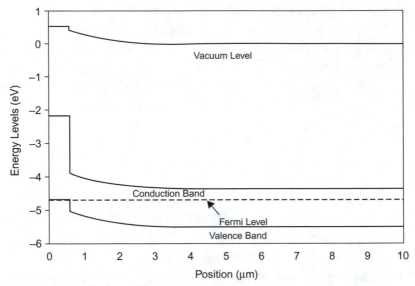

FIGURE 5.27 A window-absorber HJ structure with absorber interface trapping in TE. The differences between this structure and the previous examples are listed in the third row of Table 5.3. The presence of the charge in the HJ interface states at TE can be discerned from the vacuum-level "step-like" behavior at the interface. This shows that a change has taken place in the spatial distribution of the built-in potential from that seen for the cell of Figure 5.20. The overall built-in potential remains the same as that of Figure 5.20, since the window (and its contact) and absorber (and its contact) workfunctions were not changed.

become neutral as the cell develops a voltage under illumination, the electron quasi-Fermi level for electrons rises, and they capture electrons. When the population of these interface states changes with biasing, it then causes the barrier in the absorber to change appropriately; i.e., the step seen in the vacuum level in Figure 5.27, due to the charge in these states at TE, will be reduced with biasing in the dark—or under light. Interface trapping, such that occurring in this example we are about to explore in more detail, can be troublesome in actual HJ cells, as will soon to become apparent. HJ interface states like these can arise for all the reasons discussed earlier.

The J-V characteristics that result for this device, presented in Figure 5.28, show that the HJ trapping states can have a significant impact on cell performance. The device open-circuit voltage (now 0.48 V) is somewhat less than that for the cell of Figure 5.20 but it is not as low as it is for the cell (Fig. 5.24) that has the HJ interface recombination path. The short-circuit

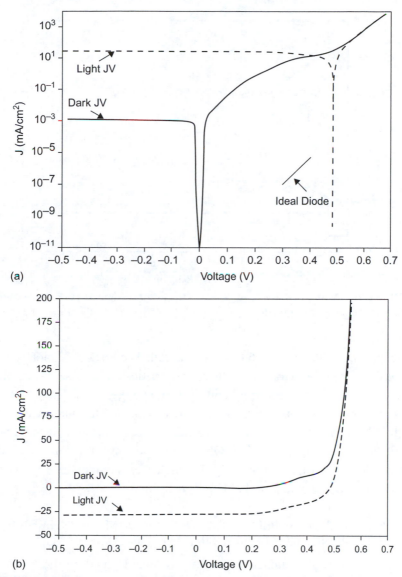

FIGURE 5.28 Computer generated light and dark J-V behavior for the cell of Figure 5.27 depicted using (a) semi-log and (b) linear plots. These characteristics display "kinks" in the light and dark J-V plots and a dark diode n-factor that varies with voltage.

current is seen to be lower than that of the Figure 5.20 and Figure 5.24 cells, but only slightly ($J_{sc} = 26.9 \, \text{mA/cm}^2$). However, the fill factor ($FF = 0.51$) has deteriorated considerably compared to that for the cells of Figures 5.20 and Figure 5.24. As a result, the power-conversion

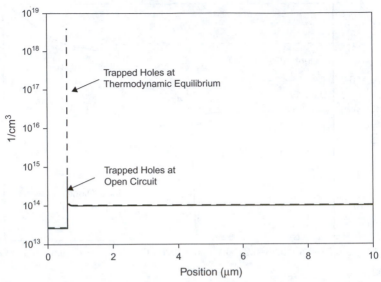

FIGURE 5.29 The trapped hole population in the interface defects at TE and at open circuit for the device of Figure 5.27.

efficiency has fallen to $\eta = 6.6\,\%$ for this cell. The linear J-V plots show "kinks" in both the dark and light characteristics and the semi-log plot has an n-factor that is not unity and, in fact, changes with voltage. Both traits are manifestations of the varying occupancy of the interface traps with voltage which is seen in Figure 5.29. Superposition is seen to not work that well for this device. Obviously, the impact of HJ interface trapping can be varied by changing the properties, number, energy distribution, and spatial extent of the defects causing it.

5.3.1.6 WINDOW-ABSORBER STRUCTURE WITH WINDOW INTERFACE TRAPPING AND ABSORBER INTERFACE RECOMBINATION

We now explore the impact of interface defects located on the window side of the HJ interface. The structure of the HJ solar cell shown in TE in Figure 5.30 is the same as that of the cell of Figure 5.24, except that the device in Figure 5.30 also has traps in the window in a 10 nm window interface layer as well as recombination defects in the first 10 nm of the absorber. Numerical analysis of the device shows that this collection of defects has a significant impact on the applicability of superposition but a smaller impact on overall performance. The properties of the two types of defects present in the interfacial region of this HJ are

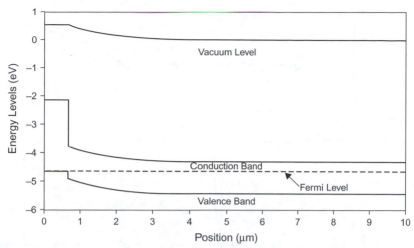

FIGURE 5.30 A window-absorber HJ structure band diagram for TE. Structure has window interface trapping as well as absorber interface recombination. The differences between this structure and the previous examples are listed in the fourth row of Table 5.3. Since the window and absorber interface states are uncharged at TE, the band diagram is exactly the same as that of Figure 5.20.

presented in the fourth row of Table 5.3. As may be noted from the capture cross-section choices in Table 5.3, electrons that get stuck in the trapping defects in the window material interface layer essentially have no communication with the material's valence band. The numerical modeling done here assumes the electrons find themselves in these defects when the cell is under illumination through production by absorption in the window (see window absorption properties in Table 5.3) and then trapping. Of course, they could also populate the traps directly through sub-band-gap absorption. One can envision the traps emptying out in the dark by tunneling with an appropriate time constant.

Based on what we have learned from the previous numerical analyses, we expect the acceptor defect states in the window interface layer to charge negatively under light (due to "bubbling up" of the electron quasi-Fermi level in the window material—see Fig. 5.22a) but not to charge under dark biasing. The latter is the case since there is no window electron quasi-Fermi level "bubbling up" without illumination present. This difference in charging between light and dark is not what was seen in the trapping of the device of Figure 5.27. It should lead to loss of superposition, since the charging a some voltage V is different with and without

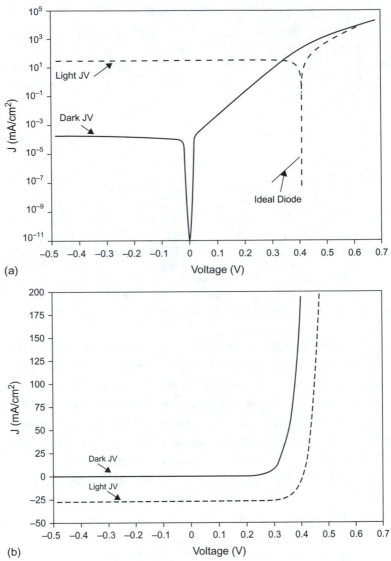

FIGURE 5.31 Computer simulation results for the light and dark J-V behavior for the cell of Figure 5.30 depicted using (a) semi-log and (b) linear plots. Superposition fails badly for this cell, although the cell has a dark diode n-factor of unity.

illumination. Examining the actual results of the numerical analysis of this cell's operation, shows that the electron and hole current components (not given) have no significant behavioral features that distinguish them from those of the cell of Figure 5.24. Specifically, total current density-voltage (J-V) light and dark simulation results in Figure 5.31 show that the performance characteristics are $V_{oc} = 0.41$ V, $J_{sc} = 27.16$ mA/cm^2, FF $= 0.75$, and $\eta = 8.3\%$. Interestingly, the V_{oc} and η are somewhat increased from those seen for the cell of Figure 5.20, while the FF is unchanged and the J_{sc} is very slightly decreased. Even with all the window interface region trapping and absorber interface region recombination present, the dark current diode n-factor is still essentially unity for this particular cell, but, because the trapping is different under light than in the dark, superposition is seen to totally fail, as expected. The result is that device performance is better under illumination than would be predicted from the dark J-V behavior. When superposition fails in heterojunctions, it can be for the reasons we discussed in the context of homojunctions. In addition, it can be due to the HJ interface trapping effects such as the type we have just seen. Earlier, we learned that it can also happen in heterojunctions due to a significant role for drift in minority carrier collection and to charge pile-up facilitated by the affinity steps.

Obviously the impact of HJ interface trapping and recombination can be varied by changing the properties, number, energy position, spatial position, and spatial extent of the defects causing it. As an example, note that the device we just discussed would give very different behavior with some small changes. For example, if the defect properties remained the same in the absorber but the defects were donor-like in the window and were depopulated by sub-band-gap absorption, then such a cell would show behavior similar to that seen in the experimental data of Figure 5.6.

Diode n-factor behavior and superposition, or lack thereof, are always features to assess in a cell's light and dark J-V behavior due to the insight they may provide into cell operation. Numerical modeling of experimental light and dark J-V behavior is a useful tool in tracking down the cause of n-factor behavior or the cause of the lack of superposition in a given cell structure. It therefore can be a very useful tool for gaining direction for design improvement and performance optimization.

5.3.2 Absorption by exciton generation

This section examines excitonic HJ solar cell structures: heterojunctions in which the absorption process produces mobile excitons. In the case of such exciton-producing absorbers, the HJ interface plays the additional role of dissociating the excitons and thereby being the generator of the free electrons and free holes needed for photovoltaic action. It does this by providing the energy, as discussed in conjunction with Figure 3.29, needed to dissociate the neutral excitons that have diffused through the absorber to the HJ. As seen in that figure, dissociation is driven by electrons falling in energy from the absorber into a material at the HJ with a larger electron affinity, The energy given up in this transition drives the dissociation providing the electrons. These electrons populate the LUMO(A) (conduction band edge) of the larger electron affinity material at the HJ interface. The material taking these electrons is termed the acceptor and the absorber material providing the electrons is said to be the donor. The corresponding free holes produced by the absorption, diffusion, and HJ exciton dissociation populate the HOMO(D) (valence band edge) of the donor side of the HJ. Subsequently, the newly generated free electrons move away from the interface and travel through the acceptor to the solar cell cathode, while the newly generated free holes travel through the donor to the anode. This picture applies to planar and bulk heterojunctions that utilize exciton-producing absorption and HJ dissociation. We stress that, because of the need to dissociate the excitons at the HJ, the absorber must be the donor material. As was the case in the previous numerical analysis section, the computer-simulation study we are about to embark on is not undertaken to optimize performance but is done in the spirit of exploring the richness of the device physics possibilities.

In the numerical analysis used in this section, the appearance of the electrons in the acceptor material at the HJ interface and the corresponding appearance of the holes in the donor material at the HJ interface due to exciton dissociation are modeled as in Figure 5.32, with a factious 1-nm-wide "generation layer."[12] The generation rate for the interface layer equals the exciton dissociation rate, which is taken as the same as the exciton arrival rate by diffusion. The exciton arrival rate at the HJ is dictated by the absorption in the absorber material and the exciton diffusion length. As a result of the use of the generation layer, free electrons appear at the LUMO(A) energy level and free holes appear at the HOMO(D) level at the interface in a delta-function-like generation pattern 1 nm wide. Since most exciton-producing absorbers are organics, we have changed to using the HOMO and LUMO designations instead of the corresponding valence and conduction band edge terminology. In addition a dielectric constant of 3 is used in the computer simulations of

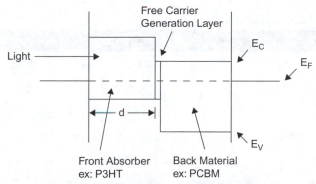

FIGURE 5.32 The factious generation layer used in the numerical analysis of this section to model exciton dissociation and free-carrier generation at the HJ interface. Exciton diffusion to, and dissociation at, the HJ results in the generation of electrons in the acceptor conduction band (LUMO(A)) and of holes in the donor valence band (HOMO(D)) of a 1-nm layer at the interface.

this section for the same reason; i.e., many of these cells use organics and organics do not readily polarize.

We will now use numerical analysis to examine three examples of heterojunctions based on exciton-producing absorbers. This brief study will underscore some of the unique features of this type of heterojunction.

5.3.2.1 STRUCTURE WITH EFFECTIVE FORCES ONLY

Our first numerical analysis of HJ solar cells with exciton-producing absorbers creates a baseline cell with the properties given in Table 5.4. The parameters are those of the P3HT (donor)/PCBM(acceptor) materials system, where known,[15] or values considered typical for organic cells.[12] As can be seen from the device band diagram in Figure 5.33, the doping and contact barrier heights have been chosen so that there is no built-in potential and therefore no built-in electrostatic field in the baseline device at thermodynamic equilibrium. As can also be seen from the figure, the PEDOT HT-EBL usually found adjacent to the p-type P3HT in P3HT-PCBM cell structures[15] is not included in this simple modeling approach. The exclusion of any built-in potential and this simplification allows us to just focus on excitons and HJ effective forces to get started. In the numerical analyses used in this Section 5.3.2.1, the generation layer is taken to have mobilities of $1 \times 10^{-3}\,\mathrm{cm^2/v\text{-}s}$ and band effective densities of states of $1 \times 10^{22}\,\mathrm{cm^{-3}}$. In the baseline

Table 5.4 Baseline Excitonic Cell

Parameter	Material 1 (donor)	Material 2 (acceptor)
Length	100 nm	100 nm
Band gap	$E_G = 1.85\,\text{eV}$	$E_G = 2.10\,\text{eV}$
Electron affinity	$\chi = 3.10\,\text{eV}$	$\chi = 3.70\,\text{eV}$
Absorption properties	Absorption data for P3HT	Absorption data for PCBM
Doping density	$N_A = 3.17 \times 10^{11}\,\text{cm}^{-3}$	$N_D = 3.17 \times 10^{11}\,\text{cm}^{-3}$
Front-contact workfunction, TE Fermi level position, and surface recombination speeds	$\phi_W = 4.33\,\text{eV}$, $E_F - E_V = 0.63\,\text{eV}$ $S_n = 1 \times 10^7\,\text{cm/s}$ $S_p = 1 \times 10^7\,\text{cm/s}$	N.A.
Back-contact workfunction, TE Fermi level position, and surface recombination speeds	N.A.	$\phi_W = 4.33\,\text{eV}$, $E_F - E_V = 1.47\,\text{eV}$ $S_n = 1 \times 10^7\,\text{cm/s}$ $S_p = 1 \times 10^7\,\text{cm/s}$
Electron and hole mobilities	$\mu_n = 1 \times 10^{-4}\,\text{cm}^2/\text{vs}$ $\mu_p = 1 \times 10^{-3}\,\text{cm}^2/\text{vs}$	$\mu_n = 1 \times 10^{-3}\,\text{cm}^2/\text{vs}$ $\mu_p = 1 \times 10^{-4}\,\text{cm}^2/\text{vs}$
Band effective densities of states	$N_C = 1 \times 10^{22}\,\text{cm}^{-3}$ $N_V = 1 \times 10^{22}\,\text{cm}^{-3}$	$N_C = 1 \times 10^{22}\,\text{cm}^{-3}$ $N_V = 1 \times 10^{22}\,\text{cm}^{-3}$
Bulk defect properties	Donor-like gap states at $E_V + 0.93\,\text{eV}$ $N_{TD} = 1 \times 10^{10}\,\text{cm}^{-3}$ $\sigma_n = 1 \times 10^{-9}\,\text{cm}^2$ $\sigma_p = 1 \times 10^{-10}\,\text{cm}^2$	Donor-like gap states at $E_V + 1.05\,\text{eV}$ $N_{TD} = 1 \times 10^{10}\,\text{cm}^{-3}$ $\sigma_n = 1 \times 10^{-9}\,\text{cm}^2$ $\sigma_p = 5 \times 10^{-15}\,\text{cm}^2$
	Acceptor-like gap states at $E_V + 0.93\,\text{eV}$ $N_{TA} = 1 \times 10^{10}\,\text{cm}^{-3}$ $\sigma_n = 1 \times 10^{-10}\,\text{cm}^2$ $\sigma_p = 1 \times 10^{-9}\,\text{cm}^2$	Acceptor-like gap states at $E_V + 1.05\,\text{eV}$ $N_{TA} = 1 \times 10^{10}\,\text{cm}^{-3}$ $\sigma_n = 1 \times 10^{-10}\,\text{cm}^2$ $\sigma_p = 1 \times 10^{-9}\,\text{cm}^2$
HJ interface light reflection	Neglected	
Back light reflection		Neglected

case, there are no defect states in the 1 nm generation layer; i.e., there are no HJ interface states present. At this point we are also assuming there is no loss mechanism of any kind, either unimolecular or bimolecular, present at this interface.

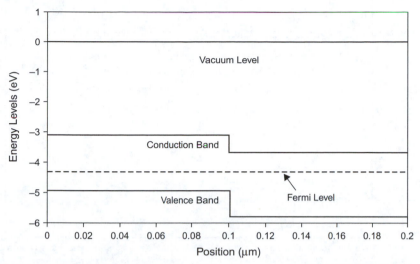

FIGURE 5.33 Band diagram in TE for the baseline excitonic cell of Table 5.4. There is no electrostatic-field barrier present. Effective-force-field barriers are seen to be present for electrons and holes.

Figure 5.34 gives the semi-log light and dark J-V's that come from a numerical analysis of the baseline cell of Figure 5.33. The symmetry breaking caused by the affinity steps is sufficient to give rise to significant photovoltaic action; consequently, the simple baseline structure with exciton-producing absorption works as a solar cell. The simulation results give a dark J-V that is seen from Figure 5.34 to be of the form $J \sim e^{(V/nkT)}$, with a variable diode n-factor larger than unity. The results give a light J-V with FF = 0.46, V_{oc} = 1.33 V, J_{sc} = 8.13 mA/cm^2, and η = 4.96%. As we have seen many times before, the rule of thumb from homojunctions that V_{oc} is limited by the built-in potential is obviously not of general applicability to structures where there is an effective-force-field barrier. Superposition predicts a V_{oc} (gotten by extrapolating through the series- resistance-controlled region) which is higher than that achieved. This is not surprising if one looks at Figure 5.35, which shows the band diagram at open circuit. Charge build-up of the kind we saw in Section 5.3.1.1 due to carriers piling-up at the affinity steps is again present and distorts performance under light. Numerical analysis establishes (data not presented) that the open-circuit voltage is controlled by bulk and contact recombination in the absorber (donor). The band-bending distortions seen in Figure 5.35 are necessary to inject the required electrons needed to cause this recombination. Put succinctly, there is not a significant number of electrons in the donor. There are photogenerated

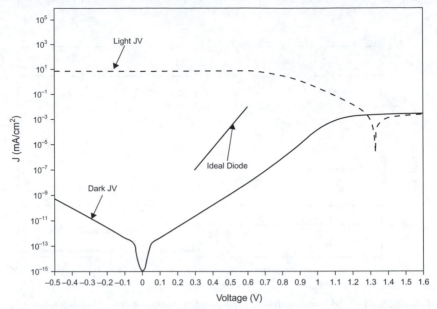

FIGURE 5.34 Computer simulation results giving the light and dark J-V behavior for the cell of Figure 5.33 depicted using semi-log plots. Even allowing for an apparent series resistance and fill factor problem, superposition does not work for this cell. The n-factor is seen to vary with voltage and is >1.

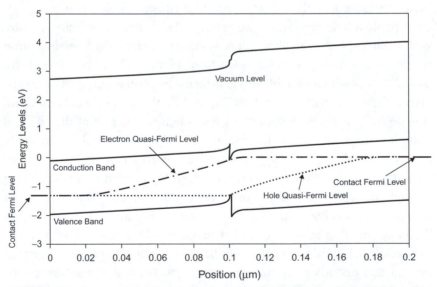

FIGURE 5.35 Computer simulation results for the band diagram under illumination and at open circuit. The charge pile-up at the affinity steps—and the electrostatic field it causes around the HJ interface region—are seen. The strong dependence of the minority-carrier quasi-Fermi levels on position shows how effectively back-injected electrons (back into the donor) and holes (back into the acceptor) are wiped out by the recombination assumed there (Table 5.4).

electrons in the acceptor. To sustain recombination in the donor, the device must bias itself to inject some electrons from the acceptor into the donor. The analogous statement applies to the holes. Of course, this back injection becomes less necessary or even unnecessary if significant interface recombination is present, as we will see. This behavior is a very interesting, unique feature of excitonic cells which employ a HJ for dissociation to produce photogenerated free electrons in the n-type conductor and photogenerated free holes in the p-type conductor.

5.3.2.2 STRUCTURE WITH INTERFACE RECOMBINATION

We now take the baseline device of Figure 5.33 and add a bimolecular loss process at the HJ interface (mechanism 8 of Fig. 5.5). We specifically assume this is a band-to-band mechanism (Appendix B) of the form $\mathscr{R}^R = \gamma np$, where the parameter γ used to characterize the strength of the free-carrier loss path.[12] Here n is the free electron population on the acceptor side and p is the free hole population on the donor (absorber) side of the HJ interface produced by exciton dissociation. In numerical modeling, these are the populations in the generation layer of Figure 5.32. Since this particular bimolecular loss mechanism does not involve localized gap states, its addition does not modify the TE band diagram of Figure 5.33. However, its addition does have a significant effect on cell performance, as seen in Figure 5.36. The impact is strong because of the relatively small "effective gap" (i.e., HOMO (D) – LUMO (A)) at the HJ and because of all the free electrons and holes there due to HJ exciton decomposition. Thanks to numerical analysis, the importance of this recombination mechanism can be varied by simply adjusting γ.[12] The figure shows that, as interface recombination increases, V_{oc} decreases. The same impact of interface recombination is seen (not presented here) if the unimolecular geminate recombination mechanism is imposed at the HJ interface.[12]

Figure 5.37 shows the light and dark J-V characteristics for one of the cases in Figure 5.36 for which interface recombination is the controlling loss mechanism and therefore dictates the open-circuit voltage. In this situation, charge pile-up, similar to that seen in Figure 5.35, does occur (not shown) but is controlled and limited by the interface recombination. Superposition does not fail, if one allows for the apparent fill factor and phenomenological series resistance present, as may be verified from

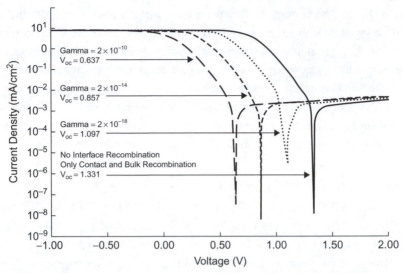

FIGURE 5.36 Computer generated semi-log J-V characteristics for the baseline cell of Table 5.4 and Figure 5.33 under illumination with band-to-band interface recombination added. Results are shown for varying strengths of interface recombination. The J-V data under light from Figure 5.34, for which $\gamma = 0$, are repeated for convenience.

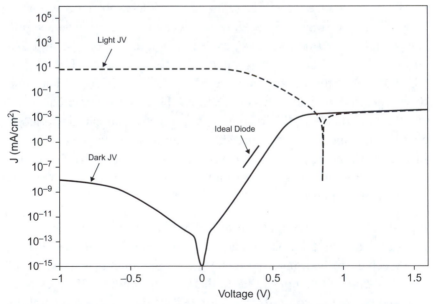

FIGURE 5.37 Computer simulation results for the light and dark J-V behavior for the cell with $\gamma = 2 \times 10^{-14}$ depicted using semi-log plots. Superposition does not work for this cell unless allowance is made for an apparent series-resistance problem.

the simulation results of Figure 5.37. These clearly show a fill factor problem and a series-resistance problem in this particular cell structure.

5.3.2.3 STRUCTURE WITH INTERFACE RECOMBINATION AND A BUILT-IN POTENTIAL

In this section numerical analysis is utilized to explore the structure of Section 5.3.2.2 but with various built-in electrostatic potentials added. This is done by varying the back-contact barrier height (ϕ_{BR}) while keeping the front contact barrier height and all contact recombination speeds

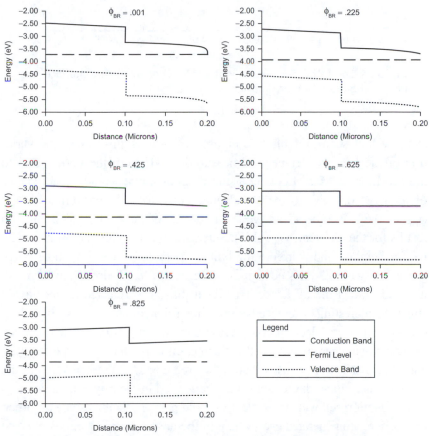

FIGURE 5.38 Band diagrams in TE resulting from the systematic variation of the HJ back-contact barrier height (done by varying the back contact workfunction). Contact recombination speeds were not changed from those of Table 5.4. The front-contact barrier height was kept constant. The value $\phi_{BR} = 0.63$ eV is the baseline cell situation (no built-in potential).

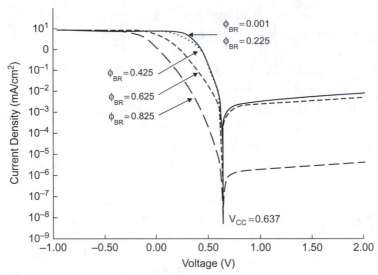

FIGURE 5.39 Log J-V behavior under light as a function of ϕ_{BR}. The interface recombination strength used for the simulations is $\gamma = 2 \times 10^{-14}$.

at the values seen in Table 5.4. Figure 5.38 shows this systematic variation of the back barrier height (measured at TE from the Fermi level to the acceptor LUMO). Looking at the diagrams, one would expect values of $\phi_{BR} > 0.63\,\text{eV}$ to be disadvantageous to cell performance and $\phi_{BR} < 0.63\,\text{eV}$ to be advantageous. Figure 5.39 shows the J-V simulation results for the different ϕ_{BR} values. Surprisingly, changing the back-contact barrier heights (and therefore the built-in potential) over the whole range given in Figure 5.38 has no significant effect on V_{oc}. It does, however, significantly affect the fill factor. This result has been selected to once again stress an interesting difference between heterojunctions with absorbers direcly producing free electron-hole pairs and those with absorbers that produce excitons: in a heterojunction cell with an exciton-producing p-type absorber and HJ-caused exciton dissociation, the free electrons are generated in the n-type material and the free holes are generated in the p-type material. In a heterojunction cell with a free electron-hole pair-producing p-type absorber, the carriers are generated everywhere in the absorber and electrons must be collected to the n-type material. The excitonic cell free carrier generation pattern and collection requirements are very unique. In the cells of Section 5.3.2.2, V_{oc} is controlled by the interface recombination, so the band bending only influences the efficiency of the (majority) holes getting out of the p-material absorber and of the (majority) electrons getting out of the n-material acceptor material; i.e., the effect of the built-in electrostatic potential is to change what we have seen appearing phenomenologically in cell characteristics as a series resistance and fill factor problem.

5.4 ANALYSIS OF HETEROJUNCTION DEVICE PHYSICS: ANALYTICAL APPROACH

We now turn to an analytical analysis approach, rather than numerical analysis, to describe HJ behavior under light and voltage and thereby to describe the origins of the J-V characteristics in terms of equations. The analytical approach allows us to assess the details of all the approximations needed to get an analytical J-V expression. This J-V model will turn out to be of the form of Eq. 5.2. Our use of the analytical analysis approach is limited to abrupt p–n semiconductor-semiconductor HJ structures with a built-in electrostatic field. We will treat selective ohmic contacts with sur-face-recombination-speed boundary conditions. The absorption process is assumed to directly result in free electrons and holes in Section 5.4.1 and to result in free electrons and holes by exciton-producing absorption and HJ dissociation in Section 5.4.2. We point out that, for some variety, the sense of the junction orientation in this section is opposite to that used pre-viously in this chapter, as may be seen from Figure 5.40.

5.4.1 Absorption by free electron–hole excitations

Figure 5.40 gives the band diagram for the abrupt, one-dimensional heterojunction that we will be analyzing in this section. It has a built-in electrostatic barrier as well as effective forces (affinity changes) barri-ers. The device is shown under illumination and, as a consequence, a voltage (V) is being developed. The portion V_1 of this voltage is devel-oped in material 1; the remainder, V_2 (where $V = V_1 + V_2$), is devel-oped in material 2. Our objective is to find an analytical expression for the current density J being produced under illumination at a voltage V for this structure. Figure 5.40 incorporates features that we noticed in many situations in the numerical analysis, such as the "bubbling-up" of the minority-carrier quasi-Fermi levels (exaggerated here) and flat quasi-Fermi levels across the total barrier region. These features are useful in developing an analytical picture of cell operation.

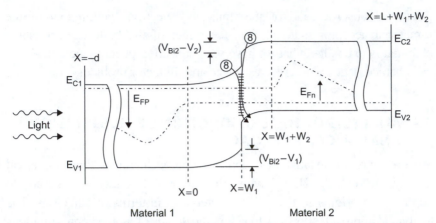

FIGURE 5.40 Schematic of a p–n HJ solar cell configuration with light of intensity $\Phi_0(\lambda)$ impinging at $x = -d$. The cell advantageously has the effective electron and hole forces and the electric field force working together. Interface recombination (path 8) is assumed to be present. The minority carrier quasi-Fermi level variation is greatly exaggerated for clarity.

As we did in the corresponding section of Chapter 4, we use a current density approach for obtaining the J-V characteristic and go through the approximations involved in it systematically. To begin, we pick the x = 0 plane, which lies just outside the electrostatic-field (space-charge) region in material 1. With this choice,

$$J = J_n(0) + J_p(0) \tag{5.15}$$

where these quantities are conventional electron and hole current densities, defined as positive if flowing left to right in Figure 5.40. The left hand boundary plane (x = 0) of the electrostatic built-in barrier region in Figure 5.40 is advantageous because, as we saw in the numerical analysis results of Figure 5.17b, the minority carrier holes are moving principally by diffusion there (note the HJ orientation is the opposite in Figure 5.17). So a good approximation for $J_p(0)$ is

$$J_p(0) = -e\,D_p\,\frac{dp}{dx}\Big|_{x=0} \tag{5.16}$$

It should be noted that Figure 5.11b makes the point that this would not be a good approximation at all for a heterojunction cell in which the built-in electrostatic potential is non-existent or even small since the minority carrier hole drift and diffusion currents are very close in magnitude in that case.

We can get the p(x) needed to evaluate Eq. 5.16 by assuming the electric field is small‡ to the left of x = 0 (the quasi-neutral region assumption discussed in Chapter 4 and Appendix E) and if we additionally assume that recombination in this semiconductor follows a linear recombination lifetime model; i.e., $\mathcal{R}(x) = (p - p_{n0})/\tau_p$ where τ_p is the hole recombination time, taken to be a constant with position. The appropriateness of a lifetime recombination model for formulating unimolecular and bimolecular recombination processes is discussed in Appendices B and C. Under these assumptions, p(x) satisfies

$$\frac{d^2p}{dx^2} - \frac{p - p_{n0}}{L_p^2} + \int_\lambda \frac{\Phi_0(\lambda)}{D_p} \alpha_1(\lambda) e^{-\alpha_1(x+d)} d\lambda = 0 \qquad (5.17)$$

subject to the boundary conditions

$$\frac{dp}{dx}\Big|_{x=-d} = -\frac{S_p}{D_p}[p(-d) - p_{n0}] \qquad (5.18a)$$

$$p(0) = p_{n0}e^{E_{Fp}(0)/kT} \qquad (5.18b)$$

As in the case of homojunctions in Chapter 4, the hole quasi-Fermi level E_{Fp} in Eq. 5.18b is measured positively down from the Fermi-level position in energy at the metal contact to the n-material; i.e., it is measured from the metal-contact workfunction position at x = −d, as seen in Figure 5.40. Equation 5.17 assumes absorption in material 1

---

‡As mentioned in the case of homojunctions, the addition of an advantageously oriented constant electric field in the quasi-neutral region due to doping profiling, which, of course, would also give rise to a hole drift current, can be handled here with ease using minor adjustments. A full analytical discussion of this was first given by Wolf,[16] who showed that incorporation of a constant electric field in the space-charge-neutral region into the analysis is straightforward. Its physical effect is only to aid the diffusion.

follows a Beer-Lambert model and produces free electron–hole pairs according to

$$\int_\lambda G_{ph}(\lambda,x)d\lambda = \int_\lambda G(\lambda,x)d\lambda = \int_\lambda \Phi_0(\lambda)\alpha_1(\lambda)e^{-\alpha_1(\lambda)(x+d)}d\lambda$$

In Eq. 5.17, the spectrum $\Phi_0(\lambda)$ impinges on the structure from the left, as seen in Figure 5.40, and is assumed to suffer no losses at the front surface, no reflection at the HJ interface, and no reflection at the back contact $x = L + W_1 + W_2$ The subscript 1 on the absorption coefficient in this equation reminds us that this absorption coefficient is specific to material 1.

We encountered equations of the form of Eqs. 5.17 and 5.18, along with the same set of assumptions, before when we were constructing an analytical model for homojunction J-V characteristics. Consequently, we can feel comfortable simply taking the solution to equations of this form found in Appendix F and using it in Eq. 5.16 to obtain

$$J_p(0) = e\int_\lambda \Phi_0(\lambda)\left\{\left[\frac{\beta_2^2}{\beta_2^2 - \beta_1^2}\right]\left[\frac{(\beta_3\beta_1/\beta_2 + 1)}{\beta_3\sinh\beta_1 + \cosh\beta_1}\right]\right.$$
$$-\left[\frac{\beta_2^2 e^{-\beta_2}}{\beta_2^2 - \beta_1^2}\right]\left[\left(\frac{\beta_3\cosh\beta_1 + \sinh\beta_1}{\beta_3\sinh\beta_1 + \cosh\beta_1}\right)\left(\frac{\beta_1}{\beta_2}\right) + 1\right]\right\}d\lambda$$
$$-\left\{\frac{eD_pP_{n0}}{L_p}(e^{V/kT} - 1)\right\}\left\{\frac{\beta_3\cosh\beta_1 + \sinh\beta_1}{\beta_3\sinh\beta_1 + \cosh\beta_1}\right\}$$

(5.19)

As noted, Figure 5.40 shows the hole quasi-Fermi level to be flat across the quasi-neutral region and across material 2 (where holes are the majority carrier) and that assumption is used here; i.e., Eq. 5.19 uses $E_{Fp}(0) = V$. This assumption is discussed in the context of homojunctions in Chapter 4 and support for it, except at the short-circuit condition, is also presented for heterojunctions in Section 5.3. The β quantities appearing in Eq. 5.19 are defined in Table 5.5. As is true for homojunctions, these β quantities are useful because they save a lot of

Table 5.5 Material 1 Beta Parameters

β Quantity	Definition	Physical Significance
β_1	d/L_p	Ratio of material 1 quasi-neutral region length to hole diffusion length. Captures the physics of hole collection from absorber 1 by diffusion while undergoing recombination. Need $d \leq L_p$.
$\beta_2(\lambda)$	$d\alpha_1(\lambda)$	Ratio of material 1 quasi-neutral region length to absorption length in material 1 for light of wavelength λ. (This ratio depends on λ.) A measure of how much light of wavelength λ is absorbed in material 1. Want $d + W_1 + W_2 + L \cong 1/\alpha(\lambda)$.
β_3	$L_p S_p / D_p$	Ratio of top-surface hole carrier recombination velocity S_p of material 1 to hole diffusion velocity D_p/L_p in material 1. Captures the physics of contact recombination versus recombination while diffusing. Need $D_p/L_p > S_p$.

writing, but, much more importantly, because they show the interplay between the various material parameters. The expression for $J_p(0)$ is seen to depend on absorption in material 1, as we would expect, and on the voltage V developed across the cell. This piece of the J-V characteristic definitely displays superposition and that makes sense since it comes from a linearized set of equations (Eqs. 5.15–5.18). To summarize, our numerical analysis generally supports all the assumptions behind this piece of the J-V (Eq. 5.19), but a key point is that there must be an electrostatic built-in potential at the HJ which is significant enough to allow hole diffusion to dominate over hole drift everywhere to the left of the plane x = 0 and to suppress carrier pile-up at the HJ interface. This conclusion holds at least for one sun illumination (we utilized a AM1.5G spectrum in the numerical modeling).

As was the case for homojunctions, we handled $J_p(0)$ first because it is easy to get an analytical expression for this term. Just as we saw for homojunctions, evaluating $J_n(0)$ is not so straightforward. Analytical evaluation of $J_n(0)$ requires knowing the electron drift components at x = 0. This point is driven home in the numerical analyses of the previous section, where majority carrier electrons at and in the barrier region are seen to undergo transport dominated by drift (see, for example, Fig. 5.17b and c). To determine the electric field in the quasi-neutral x ≤ 0 region of material 1, numerical analysis has to be employed, as is done self-consistently in Section 5.3. Here we want to see how far we can get with a purely analytical analysis, so, as we did for homojunctions, we will have to sidestep the problem of determining $J_n(0)$. The continuity concept provides the means for this sidestepping and for transferring the problem to evaluating J_n at $x = W_1 + W_2$. Using continuity allows $J_n(0)$ to be expressed as

$$
J_n(0) = e \int_0^{W_1} \int_\lambda G_{ph}(\lambda, x)d\lambda dx + e \int_{W_1}^{W_1+W_2} \int_\lambda G_{ph}(\lambda, x)d\lambda dx
$$

$$
- e \int_0^{W_1} \mathcal{R}(x)dx - J_{IR} - e \int_{W_1}^{W_{11}+W_2} \mathcal{R}(x)dx
$$

$$
+ J_n(W_1 + W_2) \tag{5.20}
$$

Here, the interface recombination J_{IR} has been taken out of $\int_{-d}^{L+W_1+W_2} \mathcal{R}(x)dx$ and has been written explicitly, as is our practice with heterojunctions.

The advantage offered by Eq. 5.20 is that electrons are the minority carrier for the quasi-neutral region $x \geq W_1 + W_2$ of material 2. This fact allows us to employ the same approach in evaluating $J_n(W_1 + W_2)$ that we took in evaluating $J_p(0)$; i.e., transport by diffusion dominates, as substantiated by the numerical analysis of Section 5.3, and therefore we can write

$$
J_n(W_1 + W_2) = e D_n \frac{dn}{dx}\bigg|_{x=W_1+W_2} \tag{5.21}
$$

Once again, the sign in Eq. 5.21 has been adjusted to make J positive in the context of the x-coordinate of Figure 5.40, when the device is operating as a solar cell. Of course, now we also have to address the generation in the material 1 part of the space-charge region $\int_0^W \int_\lambda G_{ph}(\lambda, x)d\lambda dx$ and generation in the material 2 part of the space-charge region $\int_{W_1}^{W_1+W_2} \int_\lambda G_{ph}(\lambda, x)d\lambda dx$ as well as recombination in the material 1 part of the space-charge region $\int_0^{W_1} \mathcal{R}(x)dx$, the HJ interface recombination J_{IR}, and the recombination in the material 2 part of the space-charge region $\int_{W_1}^{W_1+W_2} \mathcal{R}(x)dx$. We will return to these five terms, but first let us finish with obtaining $J_n(W_1 + W_2)$.

Obtaining $J_n(W_1 + W_2)$ from Eq. 5.21 necessitates determining the function $n = n(x)$ for the $x \geq W_1 + W_2$ region of material 2. The required $n(x)$ satisfies

$$\frac{d^2 n}{dx^2} - \frac{n - n_{p0}}{L_n^2} + \int_\lambda \frac{\Phi_0(\lambda)}{D_n} e^{-\alpha_1(d+W_1)} \alpha_2(\lambda) e^{-\alpha_2(x-W_1)} d\lambda = 0 \quad (5.22)$$

subject to the boundary condition

$$n\Big|_{W_1+W_2} = n_{p0} \exp\left[E_{Fn} \frac{W_1 + W_2}{kT}\right] \quad (5.23a)$$

at $x = W_1 + W_2$ and the boundary condition

$$\frac{dn}{dx}\Big|_{L+W_1+W_2} = -\frac{S_n}{D_n}[n(L + W_1 + W_2) - n_{p0}] \quad (5.23b)$$

at $x = W_1 + W_2 + L$. The $E_{Fn}(W_1 + W_2)$ in Eq. 5.23a is the electron quasi-Fermi level position in the gap at $x = W_1 + W_2$. Again we point out that the electron quasi-Fermi level is being measured positively up from the metal-contact workfunction energy position at $x = W_1 + W_2 + L$, as seen in Figure 5.40. The assumptions that go into our being able to use Eq. 5.22 correspond to those used to establish Eq. 5.16. The $n(x)$ that results from solving the Eqs. 5.22–5.23 set for the $x \geq W_1 + W_2$ region of material 2 is found in Appendix G. Using this function, we find from Eq. 5.21 that

$$J_n(W_1 + W_2) = e \int_\lambda \Phi_0(\lambda) \left\{ \left[\frac{\beta_6^2 e^{-\beta_{41}} e^{-\beta_{42}}}{\beta_5^2 - \beta_6^2} \right] \left[\frac{[(\beta_7 \beta_5 / \beta_6) - 1] e^{-\beta_6}}{\beta_7 \sinh \beta_5 + \cosh \beta_5} \right] \right.$$

$$+ \left[\frac{\beta_6^2 e^{-\beta_{41}} e^{-\beta_{42}}}{\beta_5^2 - \beta_6^2} \right] \left[1 - \left(\frac{\beta_5}{\beta_6} \right) \left(\frac{\beta_7 \cosh \beta_5 + \sinh \beta_5}{\beta_7 \sinh \beta_5 + \cosh \beta_5} \right) \right] \right\} d\lambda$$

$$- e \frac{D_n n_{p0}}{L_n} \left[e^{v/kT} - 1 \right] \left[\frac{\beta_7 \cosh \beta_5 + \sinh \beta_5}{\beta_7 \sinh \beta_5 + \cosh \beta_5} \right] \quad (5.24)$$

Figures 5.18a and 5.19a show the electron quasi-Fermi level to be flat across the quasi-neutral region and across material 1 (where holes are the majority carrier) for the maximum power point and open circuit and that behavior is taken to be generally valid and used here; i.e., Eq. 5.24 uses $E_{Fn}(W_1 + W_2) = V$. The dimensionless β quantities in Eq. 5.24 are analogous to those in Eq. 5.19. Their definitions are given here in Table 5.6. The β quantities show the interplay among the various parameters characterizing material 2. Equation 5.24 for $J_n(W_1 + W_2)$ depends on absorption in material 2, as we would expect, and on the voltage V developed across the cell. This piece of the J-V characteristic is the product of a linear set of equations (Eqs. 5.21–5.23) and, as a consequence, it is seen to obey superposition. Our numerical analysis generally supports all the assumptions behind this second piece of the J-V

Table 5.6 Material 2 Beta Parameters

β Quantity	Definition	Physical Significance
$\beta_{41}(\lambda)$	$(d + W_1)\alpha_1(\lambda)$	Ratio of material 1 thickness to absorption length in material 1 for light of wavelength λ. (This ratio depends on λ.) Captures the physics of absorption prior to the light's entry into the p-material. Desired value of this ratio depends on how much the p-portion is being depended on for absorption.
$\beta_{42}(\lambda)$	$W_2\alpha_2(\lambda)$	Ratio of material 2 space-charge-region thickness to absorption length in material 2 for light of wavelength λ. (This ratio depends on λ.) Captures the physics of absorption in the p-material prior to the light's entry into the p-material quasi-neutral region. Desired value of this ratio depends on how much the p-portion quasi-neutral region is being depended on for absorption.
β_5	L/L_n	Ratio of material 2 quasi-neutral region length to electron diffusion length. Captures the physics of electron collection by diffusion, while subject to recombination, from the p-material quasi-neutral region. Need $L \le L_n$.
$\beta_6(\lambda)$	$L\alpha_2(\lambda)$	Ratio of material 2 quasi-neutral region length to absorption length in material 2 for light of wavelength λ. (This ratio depends on λ.) Want $d + W_1 + W_2 + L \cong 1/\alpha(\lambda)$.
β_7	$L_n S_n/D_n$	Ratio of back-surface electron-carrier recombination velocity S_n of material 2 to electron diffusion velocity D_n/L_n in material 2. Captures the physics of electron contact recombination versus recombination while diffusing. Need $D_n/L_n > S_n$.

characteristic but a key point is that there must be an electrostatic built-in potential at the HJ which is significant enough to allow the electron diffusion current density component to dominate everywhere to the right of the plane $x = W_1 + W_2$. Carrier pile-up at the HJ interface must also be unimportant.

Now, working toward our goal of a complete analytical J-V expression for heterojunction solar cells, we substitute Eq. 5.24 into Eq. 5.20 and then substitute the resulting 5.20 expression and Eq. 5.19 into Eq. 5.15. This gives the full J-V characteristic as we have been able to determine it; i.e.,

$$
\begin{aligned}
J = {} & e \int_\lambda \Phi_0(\lambda) \Bigg\{ \Bigg[\frac{\beta_2^2}{\beta_2^2 - \beta_1^2} \Bigg] \Bigg[\frac{(\beta_3 \beta_1 / \beta_2) + 1}{\beta_3 \sinh\beta_1 + \cosh\beta_1} \Bigg] \\
& - \Bigg[\frac{\beta_2^2 e^{-\beta_2}}{\beta_2^2 - \beta_1^2} \Bigg] \Bigg[\left(\frac{\beta_3 \cosh\beta_1 + \sinh\beta_1}{\beta_3 \sinh\beta_1 + \cosh\beta_1} \right) \left(\frac{\beta_1}{\beta_2} \right) + 1 \Bigg] \Bigg\} d\lambda \\
& - \left\{ \frac{e D_p p_{n0}}{L_p} (e^{V/kT} - 1) \right\} \left\{ \frac{\beta_3 \cosh\beta_1 + \sinh\beta_1}{\beta_3 \sinh\beta_1 + \cosh\beta_1} \right\} \\
& + e \int_\lambda \Phi_0(\lambda) \Bigg\{ \Bigg[\frac{\beta_6^2 e^{-\beta_{41}} e^{-\beta_{42}}}{\beta_5^2 - \beta_6^2} \Bigg] \Bigg[\frac{[(\beta_7 \beta_5 / \beta_6) - 1] e^{-\beta_6}}{\beta_7 \sinh\beta_5 + \cosh\beta_5} \Bigg] \\
& + \Bigg[\frac{\beta_6^2 e^{-\beta_{41}} e^{-\beta_{42}}}{\beta_5^2 - \beta_6^2} \Bigg] \Bigg[1 - \left(\frac{\beta_5}{\beta_6} \right) \left(\frac{\beta_7 \cosh\beta_5 + \sinh\beta_5}{\beta_7 \sinh\beta_5 + \cosh\beta_5} \right) \Bigg] \Bigg\} d\lambda \\
& - e \frac{D_n n_{p0}}{L_n} [e^{V/kT} - 1] \Bigg[\frac{\beta_7 \cosh\beta_5 + \sinh\beta_5}{\beta_7 \sinh\beta_5 + \cosh\beta_5} \Bigg] \\
& + e \int_0^{W_1} \int_\lambda G_{ph}(\lambda, x) d\lambda dx + e \int_{W_1}^{W_1 + W_2} \int_\lambda G_{ph}(\lambda, x) d\lambda dx \\
& - e \int_0^{W_1} \mathcal{R}(x) dx - J_{IR} - e \int_{W_1}^{W_{11} + W_2} \mathcal{R}(x) dx
\end{aligned}
\tag{5.25}
$$

This analytical expression is complex. By inspection, we can say with certainty that all of the terms in Eq. 5.25, except for the last five, will not cause the failure of superposition, so long as the assumptions upon which they are based are valid. We cannot make such statements about the last five terms in Eq. 5.25, since we do not have analytical expressions for them. These last five terms cover phenomena in the region $0 < x < W_1 + W_2$ and we have seen from our numerical analysis that interface trapping and recombination effects in this region can cause significant non-linearities.

We now address these five terms and attempt some analytical models. The two space-charge-region generation components among them are easily evaluated analytically by using the Beer-Lambert model for absorption-producing electron-hole pairs. The material 1 term is

$$e \int_0^{W_1} \int_\lambda G_{ph}(\lambda, x) d\lambda dx = \int_\lambda \Phi_0(\lambda) e^{-\alpha_1 d} (1 - e^{-\alpha_1 W_1}) d\lambda$$

or, using the β quantities, it is

$$e \int_0^{W_1} \int_\lambda G_{ph}(\lambda, x) d\lambda dx = \int_\lambda \Phi_0(\lambda) e^{-\beta_2} (1 - e^{-\beta_{41}} e^{\beta_2}) d\lambda \quad (5.26)$$

The material 2 space-charge-region generation contribution is

$$e \int_{W_1}^{W_1 + W_2} \int_\lambda G_{ph}(\lambda, x) d\lambda dx = \int_\lambda \Phi_0(\lambda) e^{-\alpha_1 (d + W_1)} (1 - e^{-\alpha_2 W_2}) d\lambda$$

or, using the β quantities, it is

$$e \int_{W_1}^{W_1 + W_2} \int_\lambda G_{ph}(\lambda, x) d\lambda dx = \int_\lambda \Phi_0(\lambda) e^{-\beta_{41}} (1 - e^{-\beta_{42}}) d\lambda \quad (5.27)$$

Both Eq. 5.26 and Eq. 5.27 are clearly linear in light intensity Φ_0 and they have no voltage dependence. Hence, we can now also say with certainty that they, too, will not cause the failure of superposition.

Obtaining analytical expressions for the two space-charge-region recombination terms $-e\int_0^{W_1} \mathcal{R}(x)dx$ and $-e\int_{W_1}^{W_{11}+W_2} \mathcal{R}(x)dx$, as well as the interface recombination term J_{IR} in Eq. 5.25 is difficult, however. There is a strong electric field in this region and there can be defect states that (1) can support recombination (2) can, trap charge and modify the electric field, or (3) can do both as we saw in Section 5.3. Consequently, these terms should be handled by numerical means (see Section 5.3). To try to address these three recombination terms in analytical analysis, we first follow the procedure used for homojunctions in Chapter 4 and note that the sum of the two space-charge-region recombination terms often has a behavior of the form

$$\int_0^{W_1} \mathcal{R}(x)dx + \int_{W_1}^{W_{11}+W_2} \mathcal{R}(x)dx = \left\{ J_{SCR}\left(e^{V/n_{SCR}kT} - 1\right)\right\} \quad (5.28)$$

where, in reality, there can be voltage and also light dependence in the prefactor J_{SCR} and also in the n-factor n_{SCR}. For the interface recombination term J_{IR}, a similar behavior is also often observed, with J_{IR} of Eq. 5.27 being of the form

$$J_{IR} = \left\{ J_I\left(e^{V/n_I kT} - 1\right)\right\} \quad (5.29)$$

where, in reality, there can be voltage and also light dependence in both the prefactor J_I and the factor n_I. Obviously, J_{SCR}, J_I, n_{SCR}, and n_I can depend on the kinetics of the loss paths, the carrier populations, the electric field in the space-charge region, and illumination, as we saw in our numerical modeling.

Plugging Eqs. 5.26–5.29 into Eq. 5.25 gives the long-pursued general analytical J-V expression for HJ solar cells under illumination:

$$
\begin{aligned}
J = {} & e\int_\lambda \Phi_0(\lambda)\left\{\left[\frac{\beta_2^2}{\beta_2^2 - \beta_1^2}\right]\left[\frac{(\beta_3\beta_1/\beta_2)+1}{\beta_3\sinh\beta_1 + \cosh\beta_1}\right]\right.\\
& \left. - \left[\frac{\beta_2^2 e^{-\beta_2}}{\beta_2^2 - \beta_1^2}\right]\left[\left(\frac{\beta_3\cosh\beta_1 + \sinh\beta_1}{\beta_3\sinh\beta_1 + \cosh\beta_1}\right)\left(\frac{\beta_1}{\beta_2}\right)+1\right]\right\}d\lambda \\
& + e\int_\lambda \Phi_0(\lambda)\left\{\left[\frac{\beta_6^2 e^{-\beta_{41}}e^{-\beta_{42}}}{\beta_5^2 - \beta_6^2}\right]\left[\frac{[(\beta_7\beta_5/\beta_6)-1]e^{-\beta_6}}{\beta_7\sinh\beta_5 + \cosh\beta_5}\right]\right.\\
& \left. + \left[\frac{\beta_6^2 e^{-\beta_{41}}e^{-\beta_{42}}}{\beta_5^2 - \beta_6^2}\right]\left[1 - \left(\frac{\beta_5}{\beta_6}\right)\left(\frac{\beta_7\cosh\beta_5 + \sinh\beta_5}{\beta_7\sinh\beta_5 + \cosh\beta_5}\right)\right]\right\}d\lambda \\
& + e\int_\lambda \Phi_0(\lambda)e^{-\beta_2}(1 - e^{-\beta_{41}}e^{\beta_2})d\lambda + e\int_\lambda \Phi_0(\lambda)e^{-\beta_{41}}(1 - e^{-\beta_{42}})d\lambda \\
& - \left\{\frac{eD_p P_{n0}}{L_p}(e^{V/kT}-1)\right\}\left\{\frac{\beta_3\cosh\beta_1 + \sinh\beta_1}{\beta_3\sinh\beta_1 + \cosh\beta_1}\right\} \\
& - e\frac{D_n n_{p0}}{L_n}\left[e^{V/kT}-1\right]\frac{\beta_7\cosh\beta_5 + \sinh\beta_5}{\beta_7\sinh\beta_5 + \cosh\beta_5} \\
& - \left\{J_{scr}\left(e^{V/n_{SCR}kT}-1\right)\right\} - \left\{J_I\left(e^{V/n_I kT}-1\right)\right\} \tag{5.30}
\end{aligned}
$$

This equation has been arranged with the photocurrent components (the ones containing Φ_0) first, with the opposing dark currents following. It is clearly of the form of Eq. 5.2. Equation 5.30 gives a positive J when the cell is producing power consistent with the positive x-coordinate direction in Figure 5.40. If this J is used in a J-V plot, our convention that J is negative in the power quadrant must be applied. As we discussed in detail in Section 4.4.1, the effects of cell series and shunt resistances can be easily added to this expression, if needed.

5.4.2 Absorption by excitons

We now turn our attention to obtaining an analytical J-V expression for HJ solar cells in which light absorption takes place by the creation of excitons and free electron and hole generation requires exciton diffusion to, and dissociation at, the HJ interface. We take the electron generation rate in the conduction band on the acceptor side of the interface as I, where I is the number of excitons reaching the HJ interface and dissociating per time per area. Correspondingly, I is also the hole generation rate in the donor valence band on the donor side of the HJ interface. The only electron and hole generation occurring in the cell is taking place at the HJ interface. The analytical modeling that is done for this situation generally assumes superposition is valid and modifies Eq. 5.30 to

$$
J = eI - e \frac{D_n n_{p0}}{L_n} \left[e^{V/kT} - 1 \right] \left[\frac{\beta_7 \cosh\beta_5 + \sinh\beta_5}{\beta_7 \sinh\beta_5 + \cosh\beta_5} \right]
$$
$$
- \left\{ J_{SCR} \left(e^{V/n_{SCR}kT} - 1 \right) \right\} - \left\{ J_I \left(e^{V/n_I kT} - 1 \right) \right\} \qquad (5.31)
$$

Here an absorber-window structure has been assumed and the absorber material's being p-type has been taken into account. This equation is, of course, fraught with all the assumptions that led to Eq. 5.30. It is particularly hampered by its lack (1) of addressing charge pile-up at the HJ interface and (2) of explicit consideration of trapping and interface recombination phenomena.

5.5 SOME HETEROJUNCTION CONFIGURATIONS

Although we have focused on one-dimensional single p–n junction HJ configurations to explore the device physics, we would be remiss if we did not point out that there are actually many heterojunctions that are not so straightforward. For example, Figure 5.41a shows an n^+-n isotype (same doping type on both sides of the HJ interface) device of the window-absorber type. An electrostatic-field barrier is present due to the workfunction difference between the two n-type semiconductors. As can also be seen, the electron affinity change at the HJ interface is unfortunately not helpful in this device; i.e., it is in the opposite sense

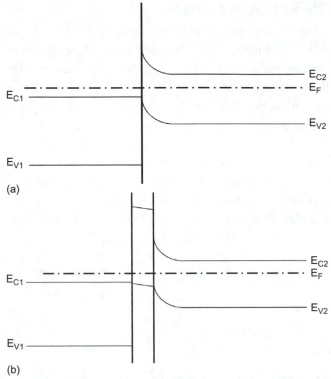

FIGURE 5.41 (a) An isotype $n^+ - n$ device and (b) a corresponding S-I-S configuration.

to the electrostatic-field barrier. Interestingly, photogenerated holes must be filled in at the interface by electrons in the window material conduction band to be collected in this type of heterojunction cell. Figure 5.41b shows a semiconductor-intermediate layer-semiconductor (S-I-S) version of this situation where an HT-EBL material has been inserted between the n^+ window semiconductor and the n absorber. This intermediate layer is assumed not to disadvantageously affect the built-in potential developed in the absorber. This S-I-S modification gives an advantageously oriented effective-field (electron affinity difference) barrier acting in concert with the electrostatic-field barrier. An intermediate layer like this allows materials whose band edges may not be advantageously positioned in energy to be employed in a cell. Other types of S-I-S configurations have also been studied. For example, S-I-S heterojunctions having an I layer with nanoparticles designed to set up plasomon excitations to affect $G_{ph}(\lambda, x)$ have been explored.[17]

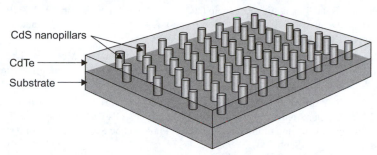

CdS nanopillars

CdTe

Substrate

FIGURE 5.42 A three-dimensional HJ configuration exemplifying the discussion in Figure 3.29 and related text.

Three-dimensional inorganic heterojunctions with nano-scale configurations like that seen schematially in Figure 5.42 have been fabricated to take advantage of the opportunities discussed in conjunction with Figure 3.28.[18] In the case of excitonic cells, successful organic HJ solar cells of the three-dimensional BHJ configuration typified by Figure 5.2 have been achieved[1] Such BHJ structures have even been utilized in tandem cells using multiple heterojunctions.[19] Equally successful organic excitonic HJ solar cells of the PHJ configuration typified by Figure 5.1 have also been successfully fabricated.[20]

REFERENCES

1. B. Lei, Y. Yao, A. Kumar, Y. Yang, V. Ozolins, Quantifying the relation between the morphology and performance of polymer solar cells using Monte Carlo simulations, J. Appl. Phys. 104 (2008) 024504.

2. S. Günes, H. Neugebauer, N.S. Sariciftci, Conjugated polymer-based organic solar cells, Chem. Rev. 107 (2007) 1324.

3. B.C. Thompson, J.M.J. Fréchet, Polymer-fullerene composite solar cells, Angew. Chem. Int. Ed. 47 (2008) 58.

4. W. Ma, C. Yang, X. Gong, K. Lee, A.J. Heeger, Thermally stable, efficient polymer solar cells with nanoscale control of the interpenetrating network morphology, Adv. Funct. Mater. 15 (2005) 1617.

5. D.D. Reynolds, G. Leies, L.L. Antes, R.E. Marburger, Photovoltaic effect in cadmium sulfide, Phys. Rev. 96 (1954) 533.

6. R. Williams, R.H. Bube, Photoemission in the photovoltaic effect in cadmium sulfide crystals, J. Appl. Phys. 31 (6) (1960) 968.

7. R.H. Bube, Hetrojunctions for thin film solar cells, in: L. Murr (Ed.), Solar Materials Science, Academic Press, New York, 1980.

8. E. Carlson, *Research in Semiconductor Films*, WADC Tech. Rep. 56–62, Clevite Corp., 1956.

9. National Solar Technology Roadmap CIGS PV, Management Report NREL/MP-520-41737, June 2007; National Solar Technology Roadmap CdTe PV, Management Report NREL/MP-520-41736 June 2007.

10. S.H. Park, A. Roy, S. Beaupre, S. Cho, N. Coates, J.S. Moon, D. Moses, M. Leclerc, K. Lee, A.J. Heeger, Bulk heterojunction solar cells with internal quantum efficiency approaching 100%, Nature Photonics 3 (May 2009) 297.

11. S. Taira, Y. Yoshimine, T. Baba, M. Taguchi, T. Kinoshita, H. Sakata, E. Maruyama, M. Tanaka, Our Approaches for Achieving Hit Solar Cells with more than 23% Efficiency, Proceeding of the 22nd European Photovoltaic Solar Energy Conference, Milan, Italy, 3–7 Sept. 2007, p. 932

12. J. Cuiffi, T. Benanti, W-J. Nam, S. Fonash, Application of the AMPS Computer Program to Organic Bulk Heterojunction Solar Cells, 34th IEEE Photovoltaic Specialists Conference, Philadelphia, PA, June 7–12, 2009.

13. J. F. Jordan, Thin Film Cu_2S/CdS Solar Cells by Chemical Spraying, Final Report Contract Ex-76-C643579, Dept. of Energy, Washington, DC (1978).

14. R.L. Anderson, Experiments on Ge-GaAs heterojunctions, Solid-State Electron 5 (1962) 341.

15. C. Waldauf, P. Schilinsky, J. Hauch, C.J. Brabec, Material and device concepts for organic photovoltaics: towards competitive efficiencies, Thin Solid Films 503 (2004) 451–452.

16. M. Wolf, Drift fields in photovoltaic solar energy converter cells, Proc. IEEE 51 (1963) 674.

17. S. Forrest, J. Xue, Strategies for solar energy power conversion using thin film organic photovoltaic cells, Conference Record of the Thirty-first IEEE Photovoltaic Specialists Conference, Orlando, FL, January 2005.

18. Z. Fan, H. Razavi, J. Do, A. Moriwaki, O. Ergen, Y.L. Chueh, P.W. Leu, J.C. Ho, T. Takahashi, L.A. Reichertz, S. Neale, K. Yu, M. Wu, J.A. Ager, A. Javey, Three-dimensional nanopillar-array photovoltaics on low-cost and flexible substrates, Nature Mater. 8 (2009) 648–653.

19. J.Y. Kim, K. Lee, N. Coates, D. Moses, T-Q. Nguyen, M. Dante, A.J. Heeger, Efficient tandem polymer solar cells fabricated by all-solution processing, Science 317 (2007) 222.

20. A.L. Ayzner, C.J. Tassone, S.H. Tolbert, B.J. Schwartz, Reappraising the need for bulk heterojunctions in polymer-fullerene photovoltaics: the role of carrier transport in all-solution-processed P3HT/PCBM bilayer solar cells, J. Phys. Chem. C 113 (2009) 20050.

Surface-barrier Solar Cells

6.1 INTRODUCTION

Surface-barrier solar cells all use only one semiconductor, in one doping type. They all have an electrostatic-field barrier, which begins at the semiconductor surface, as the principal source of photovoltaic action. The built-in electrostatic field can extend across the whole semiconductor, analogous to a p–i–n structure, or, at the other extreme, it can be confined to the semiconductor's near-surface region. There are basically two types of surface-barrier cells; they can be distinguished by the materials system used to form the semiconductor electrostatic field barrier: one type is the all-solid-state cell and the other is the electrolyte–solid-state cell. The all-solid-state cell has two functional versions: a metal-semiconductor (M-S) configuration and a metal-intermediate layer-semiconductor (M-I-S) configuration. The electrolyte–solid-state cell uses a liquid-semiconductor configuration. The solid-state cells use the difference between the electrochemical potential of a metal and that of the semiconductor (typically measured for both materials by their workfunctions) to set up the surface electrostatic field in the semicondutor. The electrolyte-based cells use the

DOI: 10.1016/B978-0-12-374774-7.00006-6

difference between the electrochemical potential of an electrolyte (typically measured for electrolytes by the redox potential) and that of the semiconductor to set up the surface electrostatic field. The M-S and M-I-S devices are often referred to as Schottky-barrier (SB) cells and the electrolyte-semiconductor devices are referred to as electrochemical photovoltaic cells (EPC). The version of the metal-semiconductor configuration for which the field reaches across the whole semiconductor is sometimes referred to as an M-I-M cell. Here the I notation is again used but this time to denote a low doped (intrinsic) absorber. The two basic types of surface-barrier solar cells are shown in Figure 6.1. The electrolyte-semiconductor cell can be further broken down into two subtypes, as shown in Figure 6.2: (1) the regenerative cell, which has the oxidation (hole-capture from the semiconductor surface) of the redox couple completed by reduction at the anode without any net chemical change; and (2) the photo-cleaving (or photosynthesis) cell, which has the hole-capture (oxidation) at the semiconductor produce the evolution of one product, while the reduction at the anode produces the evolution of another product,

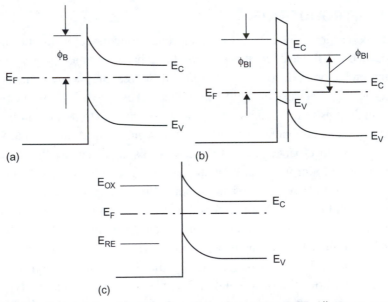

FIGURE 6.1 Some surface-barrier solar cell configurations in TE. All use an n-type semiconductor in these examples. (a) A simple metal-semiconductor configuration, (b) a metal-intermediate layer–semiconductor structure, and (c) an electrolyte-semiconductor structure showing the redox couple energy levels of the solution. The two Schottky barrier heights discussed in the text for metal barrier-former cases are shown.

as seen in the figure. The particular photosynthesis example given in Figure 6.2b shows photolysis, the decomposition of water to produce oxygen and hydrogen which was first demonstrated in 1972.[1]

The lineage of surface-barrier solar cells can be traced back to the electrolyte-solid structures used by Becquerel[2] in 1839 in the first reported studies of photovoltaic-type behavior. Investigation of the solid-state, surface-barrier cell began with the Cu-Cu$_2$O structure, which was shown to be photosensitive by Hallwachs[3] in 1904, and which was developed into a photovoltaic device by 1927.[4] By the 1930s, Cu-Cu$_2$O metal-semiconductor, surface-barrier devices were in production and were being used for applications such as photometry and light control. At this stage of their

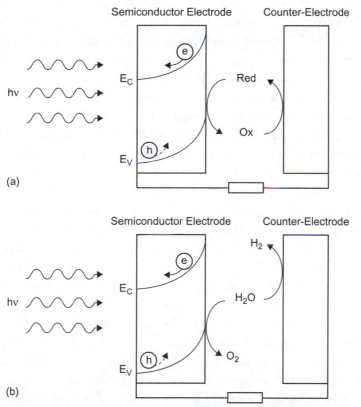

FIGURE 6.2 Examples using an n-type semiconductor: (a) the regenerative cell, where oxidation of the reduced species of the redox couple (hole-capture process by the reduced species at the semiconductor surface) is completed by reduction at the anode without any net chemical change; (b) the photosynthesis cell, where the hole-capture (oxidation) at the semiconductor results in the evolution of one species while the reduction at the anode produces the evolution of another species.

development, solar conversion efficiencies of $\eta \leq 1\%$ had been achieved in solid-state surface-barrier photovoltaic devices.[3]

During the early 1950s, surface-barrier photovoltaic structures were quickly eclipsed by the emerging p–n homojunction solar cell technology, just as Schottky-barrier diodes were quickly surpassed by p–n junction diodes during this time. By 1970–1972, Schottky-barrier solar cells had evolved to only $\eta \leq 6\%$ for single-crystal Si M-S devices and to only $\eta \leq 9\%$ for single-crystal GaAs M-S devices (terrestrial conditions).[5] The electrolyte-semiconductor (EPC) solar cells of the period were unstable devices with efficiencies that had yet to reach 1% under terrestrial sunlight.[6–8]

In 1972, investigators[9] found experimentally that the open-circuit voltage of an Al/(p)Si M-S solar cell could be substantially increased by inserting an insulating SiO_x layer between the Al and the p-type silicon. This development ushered in the metal-insulator–semiconductor solar cell. In the language of the superposition model, this cell used an insulator layer to suppress the dark current and thereby to increase V_{oc}, but the I layer was kept thin enough to not impede collection of the photocurrent. Other groups showed experimentally that the metal-insulator–semiconductor surface-barrier configuration could be used with n-type silicon[10] and with III–V compound semiconductors (n)GaAs,[5,11] and could be developed to totally suppress the majority-carrier component of the dark current.[12–14] By the end of the 1970s, single-crystal metal-insulator–semiconductor silicon cells had attained $\eta = 16\%$ (AM1),[13] while structures fabricated with single-crystal (n)GaAs had achieved $\eta = 17\%$ (AM1).[15] Unfortunately, the metal-insulator–semiconductor cell was generally found to have stability problems due to its use of an ultra-thin, often stoichiometrically deficient, insulator and an ultra-thin ($\sim$10nm) metal electrode. The latter is required to be ultra-thin if illumination comes in the surface-barrier side. Such light entry is advantageous, since the free-carrier-charge-separating, or exciton-dissociating and then free-carrier-charge-separating, electrostatic-field barrier region is then located where the absorption is the most intense. In our use of the designation M-I-S solar cell, we include, as the I-layers, the insulators of the 1970s as well as today's intermediate layers such as HT-EBL materials for n-type semiconductors (see Fig. 6.1b) and ET-HBL materials for p-type semiconductors. With the development of techniques like atomic layer deposition (ALD),[16] self-assembled monolayer (SAM) deposition,[17] and metal–organic

chemical vapor deposition (MOCVD), as well as the development of a manufacturing base for ultra-thin, stable gate dielectrics for today's MOSFETS,[18]
the prospect of successfully fabricating stable M-I-S and analogous EPC
structures with HT-EBL and ET-HBL materials certainly continues to be of
interest. The EPC, M-S and M-I-S types of solar cells, due to their extreme
simplicity and low-temperature processing requirements, serve as excellent vehicles for evaluating new materials and new concepts. For example,
Schottky-barrier metal-nanoparticle (quantum dot) semiconductor structures
have recently been explored by a number of groups.[19,20]

The free electrons and holes produced in electrolyte-semiconductor
surface-barrier solar cells can drive chemical reactions at the semiconductor surface (termed photocorrosion) that are quite problematic. On the
other hand, they can also drive chemical reactions in the electrolyte that
can be beneficial, as seen in Figure 6.2. Work reported in 1976 using a
polychalcogenide electrolyte–(n)CdS device with a conversion efficiency
in the 1%–2% range showed the photocorrison photodecomposition issue
could be made more manageable.[21,22] The particular oxidation-reduction
(redox) couple used was the sulfide-polysulfide (S^{2-}/S_n^{2-}) couple in an
aqueous solution. Its success lay in the fact that this electrolyte redox
couple not only caused the requisite electrostatic-field barrier at the surface of CdS, but also served as an efficient sink for photogenerated holes,
allowing them less opportunity for photocorrosion activity. By choosing redox couples favorably located in energy with respect to the surface
positions of the semiconductor band edges, or by modifying the kinetics
governing electron transfer between the redox couple and the semiconductor, or by doing both, liquid-semiconductor surface-barrier solar cells
developed to the point of achieving reasonably stable terrestrial solar
conversion efficiencies as high as $\eta = 12\%$ (polyselenide electrolyte–
(n)GaAs)[23] by the end of the 1970s. Today, photocorrosion remains a
problem with these cells. However, wider-band semiconductors tend to be
more stable against photocorrosion, since a wider band gap implies stronger bonding. This fact makes the use of EB-HTL (and HB-ETL for p-type
semiconductors) attractive. It also has prompted the use of hybrid tandem
cells with lower-band-gap units behind a wide-band-gap electrolyte-
semiconductor first cell; i. e., these tandem structures are arranged so that
the wider-band-gap cell faces the higher-energy photons and the electrolyte, which is doubly advantageous. Energy conversion efficiencies as
high as 19.6% have been reported for such tandem cell structures.[24]

6.2 OVERVIEW OF SURFACE-BARRIER SOLAR CELL DEVICE PHYSICS

6.2.1 Transport

As is the case for the semiconductor-semiconductor heterojunctions of Chapter 5, surface-barrier solar cells can be designed to use diffusion to bring photogenerated minority carriers to a relatively narrow barrier region, they can be designed to use drift directly to collect photogenerated carriers, or they can be designed for exciton diffusion to the interface and subsequent exciton dissociation. The choice of design depends on the usual assessment of absorption process, absorption length:diffusion length ratio, and absorption length:drift length ratio. In this introductory discussion, we will avoid deciding too much of the specifics of the design of the device of Figure 6.3 and assume it could be an M-S, M-I-S, or EPC cell.

As noted, the absorption mechanism present in the semiconductor of a surface barrier device may directly produce free electrons and holes or exciton decomposition may be required. The latter process could be accomplished in the high electrostatic field at the semiconductor surface,

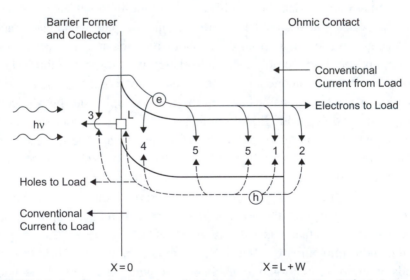

FIGURE 6.3 A surface-barrier solar cell structure under illumination. Paths 1–5 are all mechanisms by which photogenerated carriers are lost. Path L helps to carry the photocurrent, but it involves localized holes and the deleterious possibility of chemical bond breaking.

although, as discussed in Section 2.2.6, it has been argued that such fields are not able to decompose excitons, or it could be accomplished by the interface itself, if the semiconductor is p-type, as may be deduced from a metal-p-type semiconductor surface barrier cell band diagram. Whatever the mechanism, it has been established that exciton decomposition does occur in organic-absorber surface-barrier structures, at least in the case of a p-type absorber.[25] However they are produced, the free carriers in these devices, whether based on organic or inorganic absorbers, face the loss mechanisms we have seen before. The photogenerated holes in absorber in Figure 6.3 can be lost to: Paths 1and 2, recombination at and in the ohmic (back) contact (which can be minimized by the use of a selective ohmic contact); Path 3, recombination at the front surface and in the barrier-forming material; Path 4, semiconductor surface barrier region recombination; and Path 5, recombination in the bulk region. The holes that survive all the loss mechanisms are collected by the barrier-former (a metal, an M-I layer structure, or a redox couple). Current continuity is established at the barrier-former-semiconductor interface simply by oxidation of the reduced species by the holes (in electrolyte cells) or by the filling in of the holes by metal electrons (in M-S and M-I-S structures). Photogenerated holes may also be collected by the barrier-former via Path L in the Figure 6.3. In metal-semiconductor and metal-insulator-semiconductor cells fabricated with an n-type semiconductor, Path L represents localization, at the interface region, of a photogenerated minority-carrier hole and then its subsequent tunneling into the metal. While such a path aids in carrying the photocurrent, it involves localized holes and localized holes are broken bonds. Consequently, this path could lead to chemical instability at the solid-state cell semiconductor surface.[26,27] In liquid-semiconductor cells, there is freedom of motion at the surface. Consequently, a broken bond can lead to decomposition in this structure. For example in semiconductor C of statement 6.1 below, localization of a hole and the breaking of a semiconductor bond at the surface can lead to

$$C + h^+ + solvent \rightarrow C^+ \cdot solvent \tag{6.1}$$

with the resulting, liberated, cation C^+ continuing the conventional current flow to the left in Figure 6.3. This path is obviously the detrimental effect referred to earlier as photocorrosion. It must be suppressed in any successful, stable EPC cell.[28]

If we apply our usual accounting exercise for keeping track of electrons moving across the plane $x = L + W$ in Figure 6.3 (more correctly, moving across a plane just to the right of processes 1 and 2 as discussed in Section 2.3.2.9), we can write

$$
J = -\left[e \int_{0}^{L+W} \int_{\lambda} G_{ph}(\lambda, x)\, d\lambda dx - e \int_{0}^{L+W} \mathscr{R}(x)\, dx - J_{ST}(0) - J_{SB}(L+W) \right]
$$

(6.2)

Here $G_{ph}(\lambda, x)$ is the free-carrier photogeneration function introduced in Section 2.2.6 and its λ integration is over the impinging spectrum. When the photogenerated free electrons and holes are produced directly by absorption, $G_{ph}(\lambda, x)$ is dictated by the absorption pattern, as discussed in previous chapters. When the photogenerated free electrons and holes are produced by exciton dissociation at the barrier-former–semiconductor interface, then $G_{ph}(\lambda, x)$ has an almost delta-function-like behavior in x at the semiconductor surface, as discussed in Chapter 5. The integral $\int_{0}^{L+W} \mathscr{R}(x)\, dx$ accounts for loss mechanisms 4 and 5, whereas $J_{ST}(0)$ accounts for the electrons (and consequently holes) lost at the barrier-former (front, light-entering) surface through loss mechanisms represented by path 3, and $J_{SB}(L + W)$ accounts for the electrons (and consequently holes) lost at the back surface through mechanisms 1 and 2. Because signs are always a concern, a comment on the sign convention in Eq. 6.2 is merited here. The minus sign used in front of all the terms on the right-hand side arises because our procedure is to take the conventional current density in the power quadrant as negative and current flowing in the negative x-direction also as negative.

The loss term $J_{ST}(0)$ in Eq. 6.2 (path 3) needs further discussion. It represents very different processes, depending on the barrier-forming material. If the barrier-former is an electrolyte, $J_{ST}(0)$ is the path that leads to reduction of the oxidized species of the redox couple by surface conduction-band electrons. It still can be modeled by $J_{ST}(0) = eS_n[n(0) - n_0(0)]$, with the surface recombination speed S_n now characterizing the kinetics of the reduction process and having its value adjusted appropriately. Of course, $J_{ST}(0) = eS_n[n(0) - n_0(0)]$ can encompass this reduction process and other parallel loss paths at the front surface. For example, this model can be appropriate too if path 3 is dominated by electron-hole recombination at the interface. If the barrier-former is a metal, we model $J_{ST}(0)$ as the thermionic emission of the conduction-band electrons into the metal. That is, from Section 2.3.2.1,

$$J_{ST} = eS_n [n(0) - n_0(0)] = \frac{A^+ T^2}{N_C}[n(0) - n_0(0)] = A * T^2 e^{-\phi_B/kT}\left[e^{E_{Fn}(0^+)/kT} - 1\right]$$

(6.3)

where $E_{Fn}(0^+)$ is the electron quasi-Fermi-level position just inside the semiconductor, measured from the TE Fermi-level position at the interface. Equation 6.3 shows that, in the thermionic emission case, $S_n = A^+ T^2/eN_C$, which works out to be about $S_n = 10^7$ cm/s at room temperature, as we first observed in Chapter 2. Equation 6.3 makes it explicit that the Schottky barrier height ϕ_B at the metal-absorber interface is an important parameter for assessing thermionic emission if this structure is an M-S cell. As we see from Figure 6.1, there are two barrier heights that can be important in general in M-S and M-I-S cells.

In metal surface barrier-formers, we are interested in intimate metal-semiconductor interfaces but also in cases in which (1) defective surface layers inadvertently occur between the metal and the absorber, as well as in cases in which (2) I-layers are purposefully positioned between the metal and the absorber. We will classify the inadvertent surface layer case as an M-S structure because this type of generally defective surface layer does not serve as a majority carrier blocking EBL (useful in the case of Figure 6.3) or HBL (useful for the case of a p-type semiconductor). What we are calling inadvertent surface (or interface) layers can affect the built-in potential distribution, as we will see. A purposeful I-layer gives rise to an M-I-S structure, because such intermediate layers are chosen to be majority carrier blocking materials and are present for the purpose of significantly affecting transport. In both surface-layer situations, the Schottky barrier height at the metal interface ϕ_B and at the surface layer–semiconductor interface ϕ_{BI}, seen in Figure 6.1b, are useful quantities.

6.2.2 The surface-barrier region

In a surface-barrier cell, energy level and densities of states differences (effective forces) can play a role along with the electrostatic field and band bending in the semiconductor surface in breaking symmetry and in making one direction different from the other. While the electrochemical potential difference between the barrier-former and the semiconductor prior to contact always establishes V_{Bi}, a complication that arises in surface-barrier cells is that V_{Bi} may not always completely manifest itself as band bending in the semiconductor. Some band bending may lie across the surface layer we were just discussing. Figure 6.1b shows an example where some of the V_{Bi} is being developed across a surface layer, as can be seen from the sloped band edges in the layer. As noted, this surface layer seen in Figure 6.1b can be a purposefully introduced intermediate layer, giving an M-I-S device, or it can be present inadvertently—for example, due to chemical reactions or interdiffusion. As we have mentioned, in the latter case the surface layer is usually very defective, giving rise to gap states that can store charge but also render the layer transparent to carrier motion. This may occur due to hopping transport. If the layer is thin enough, it may be completely transparent to carriers, due to tunneling, yet still store charge in the gap states. Charge in the gap states in defective layers develops an electrostatic field across the surface layer, producing band bending in the layer. This band bending, together with the band bending in the semiconductor, must total V_{Bi} in TE. When a surface layer and a semiconductor–surface layer localized state gap distribution are present, all we know from thermodynamics is that the band bending in the semiconductor plus that in the surface layer must equal V_{Bi} at TE. Since the charge in the gap states could even result in band bending in the surface layer that is in opposite sense to that in the semiconductor (the states can be donor or acceptor-like), it is possible that the net result is that the semiconductor band bending component is larger than V_{Bi}. Determining the free carrier populations and the gap-state populations and their charge contributions and working out how V_{Bi} finally translates into band bending in a surface-barrier cell semiconductor at TE necessitates, in general, a numerical approach to solving Poisson's equation (Eq. 2.45 in Section 2.3.4). Solving this equation then gives the width of the field region and the functional dependencies $E_C(x)$, $E_V(x)$, $E_{VL}(x)$, and $\xi(x)$ in TE and allows watching their evolution as a function of voltage V. Computer analysis, of course, does all this for us.

This discussion allows us to return to the concept of Fermi-level pinning, which first came up in Section 3.2.2. This phenomenon can occur whenever the Fermi level is in, or is approaching, an energy range with a very high density of states. In the surface-barrier devices under discussion, we consider the situation for which there is a surface layer (of the inadvertent or intermediate type) present and a high surface-layer gap-state density centered at some energy ϕ_0 at the surface layer-semiconductor interface. In such a case, changing the barrier-former and thereby changing V_{Bi} may not move the Fermi level from ϕ_0 at the physical location of these gap states. All the change in the total band bending, which must occur with barrier-former workfunction change, can take place across the surface layer, because, if there were the slightest shift in energy by the Fermi level at the position of the localized states, it would cause huge changes in the interface state occupancy and therefore huge changes in the charge developed. Thus, the Fermi level is said to be pinned—it does not have to move much in energy at the spatial location of these gap states to adapt to any change required in the electric field.[29] It must be stressed that Fermi-level pinning can render changing the electrochemical potential of the barrier-former useless in terms of trying to modify the barrier inside the semiconductor.

6.3 ANALYSIS OF SURFACE-BARRIER DEVICE PHYSICS: NUMERICAL APPROACH

In this section, we get into the details of device analysis and determine what we can learn from computer simulation studies. This numerical approach is used to solve the equation set of Section 2.4 and thereby to develop the light and dark relationship between J and V several surface-barrier solar cell examples. In this numerical work, we assume the light-absorption process directly produces free electrons and holes and we treat Figure 6.3 as a solid-state surface-barrier M-S or M-I-S cell with, as seen, an n-type absorber. The numerical analysis approach also may be adapted to p-type semiconductors and may be employed for electrochemical cells. In the case of the electrolyte barrier-formers, our numerical analysis applies by assuming that the ionic drift and diffusion taking place in the electrolyte and the reduction process needed at the electrolyte counter-electrode are not rate-limiting. In other words, (1) transport in the semiconductor and (2) the kinetics at the electrolyte-semiconductor surface (represented by S_n and S_p) are taken as the rate-limiting (i.e., controlling) steps. In this picture, S_p characterizes the effectiveness of surface holes in oxidizing the reduced electrolyte species at the electrolyte-semiconductor interface, while S_n characterizes the effectiveness of surface electrons in reducing the oxidized electrolyte species at the interface. The former is the critical process for photocurrent continuity; the latter is a loss process.

Numerical analysis is used here for structures whose TE band diagrams are shown in Figures 6.4 and 6.5. The p–n homojunction shown in Figure 6.4a is included for comparisons. One sun illumination (AM1.5G) is assumed in all cases and it enters through the barrier-former. No allowance has been made for optical losses. For definitiveness, all the devices

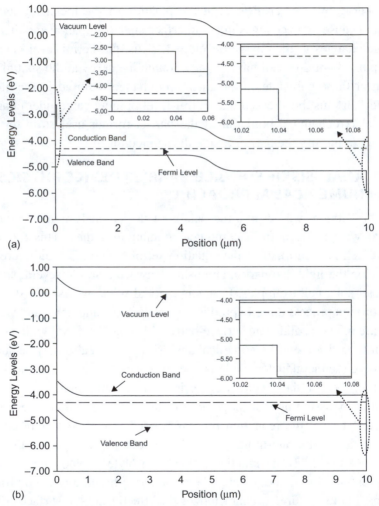

FIGURE 6.4 TE band diagrams of a p–n homojunction and of the corresponding M-S cell based on the same n-type semiconductor. The p–n homojunction has a front HT-EBL and back ET-HBL, as discussed in Section 4.3.3. The surface-barrier cell has the same back ET-HBL and is the lower-barrier M-S cell of Table 6.2.

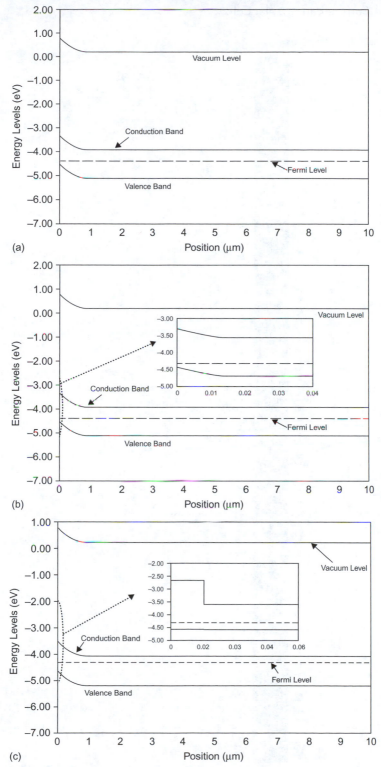

FIGURE 6.5 (a) The higher-barrier M-S, (b) Fermi-level pinning (inset shows the surface-layer band bending), and (c) M-I-S cells of Table 6.2.

Table 6.1 Properties of the n-type Absorber Found in all the Cells of Figures 6.4 and 6.5

n-region length	Band gap	Electron affinity	Absorption properties	Doping density (N_D)	Electron and hole mobilities	Band effective densities of states	Bulk defect properties
10,000 nm	$E_G = 1.12\,\text{eV}$	$\chi = 4.05\,\text{eV}$	Absorption data for Si (see Fig. 3.19)	$N_D = 1.0 \times 10^{15}\,\text{cm}^{-3}$	$\mu_n = 1350\,\text{cm}^2/\text{vs}$ $\mu_p = 450\,\text{cm}^2/\text{vs}$	$N_C = 2.8 \times 10^{19}\,\text{cm}^{-3}$ $N_V = 1.04 \times 10^{19}\,\text{cm}^{-3}$	Donor-like gap states from E_V to mid-gap $N_{TD} = 1 \times 10^{14}\,\text{cm}^{-3}\,\text{eV}^{-1}$ $\sigma_n = 1 \times 10^{-14}\,\text{cm}^2$ $\sigma_p = 1 \times 10^{-15}\,\text{cm}^2$ Acceptor-like gap states from mid-gap to E_C $N_{TA} = 1 \times 10^{14}\,\text{cm}^{-3}\,\text{eV}^{-1}$ $\sigma_n = 1 \times 10^{-15}\,\text{cm}^2$ $\sigma_p = 1 \times 10^{-14}\,\text{cm}^2$

are designed to use diffusion collection to their respective electrostatic barriers, as can be seen from the band diagrams. In the case of the p–n device, diffusion collection is from both the p-material and n-material quasi-neutral regions. In the case of the surface-barrier devices, diffusion collection is from the n-material quasi-neutral region. The p–n homojunction device is analyzed in detail in Section 4.3.3 using the absorption properties of Si, and its band diagram is repeated here for convenience. All the surface-barrier devices in Figures 6.4 and 6.5 also use the absorption properties of Si and have the same n-type region and ET-HTL layer (not shown in a blow-up in Fig. 6.5) as the p–n homojunction. However, the n-region's length has been doubled for the M-S and M-I-S cases so that all devices have the same absorber length and therefore the same J_{sc} potential (see Fig. 3.20). The properties of the n-layer used in these cells are listed in Table 6.1 and are the same as the n-layer properties in Table 4.1. In particular, the n-layer defects providing the loss path for the S-R-H bulk recombination assumed for all these cells are exactly the same.

The p–n homojunction of Figure 6.4a serves as a good basis for comparison of p–n homojunctions and surface-barrier junctions. It and the surface-barrier devices have the same generation $\int_0^{L+W} \int_\lambda G_{ph}(\lambda, x) d\lambda dx$. The p–n device has selective ohmic contacts at the front and back, so its contact losses are negligible. Its performance is simply dictated by generation and bulk recombination. All the devices, p–n junction and surface-barrier, have the same front-surface recombination speed for holes ($S_p = 10^7$ cm/s); i.e., they are all equally good at collecting holes at the front contact (p–n junction) or at the front surface (surface barrier cell). All the surface-barrier cells have the same selective ohmic back contacts as the p–n homojunction cell. Taking all this into consideration, it can be seen that a major difference between the surface-barrier devices and the homojunction device is what happens to the electrons at the left edge of the electrostatic-field barrier (i.e., at the front surface for the surface-barrier devices) under illumination. If Figure 6.3 were a p–n homojunction, the net electron flow at the left end of the barrier region would be coming (as particles) from the p-region into the barrier region, if the cell is operating in the power quadrant. In a surface-barrier cell, there is no p-absorber to the left of the barrier region. Electrons are not coming into the barrier region from a p-absorber in the power quadrant for surface barrier cells. Instead, they are leaving, as path 3 of Figure 6.3 shows. In M-S and M-I-S versions, this electron flow at the left-hand edge (the front surface) is from those electrons that are being injected by thermionic emission into the metal (as particles) or are undergoing interface recombination. In the EPC case, the corresponding electron flow would be due to electrons reducing the oxidized species in solution or undergoing interface recombination.

Table 6.2 gives the parameters characterizing the cells of Figures 6.4 and 6.5 and Table 6.3 gives their performance data. Figure 6.6 has both the linear light and dark J-V characteristics for the cells of Figure 6.4 while Figure 6.7 has these results for the cells of Figure 6.5. These tables and figures show that the M-S cell with the lower Schottky barrier $\phi_B = 0.80\,eV$ has much poorer performance than the p–n homojunction. Table 6.3 and Figure 6.7a show that increasing the Schottky barrier height from $0.80\,eV$ to $1.00\,eV$ has a positive impact on performance. This is expected from Eq. 6.3, since increasing ϕ_B decreases the $J_{ST}(0)$ loss mechanism in Schottky barrier type cells. Ideally increasing ϕ_B is straightforward in an M-S cell, since it follows from Figure 6.1a that $\phi_B = \phi_W - \chi$, where ϕ_W is the metal workfunction and χ is the semiconductor electron affinity. For the two M-S cells, this expression is assumed to work and the barrier height was increased from the lower-barrier M-S case to the higher-barrier M-S case by increasing the metal workfunction. Interestingly, Table 6.2 shows that the higher-barrier-height M-S cell actually has a larger built-in potential than the p–n homojunction cell, but Table 6.3 and Figure 6.7a show it still cannot match the corresponding p–n cell in performance. The cause for this is $J_{ST}(0)$. These simulation results show that p–n cell performance can be matched if an M-I-S cell is used to suppress this front surface loss term. The I layer purposefully introduced between the metal (lower barrier M-S device metal) and the absorber in this demonstration is an electron-blocking layer (10 nm thick) with an electron affinity of $3.15\,eV$ and the same hole affinity as the absorber, as seen in Figure 6.5c. This I layer is taken to have no significant defects. As expected, it suppresses $J_{ST}(0)$ giving the enhanced performance. Table 6.2 shows that the built-in potential values for the lower barrier M-S and the M-I-S cell are the same, which is consistent with our having used the same metal for both. We did to stress the impact of the M-I-S configuration. Tables 6.2 and 6.3 show the M-I-S

Table 6.2 Interface Parameters for Various M-S and M-I-S cells

Parameter	p–n homo-junction	M-S (lower barrier)	M-S (higher barrier)	M-S (pinning)	M-I-S
Barrier height at barrier-former/intermediate layer ϕ_B (see Fig. 6.1)	Not applicable	0.80 eV (No intermediate layer present)	1.00 eV (No intermediate layer present)	1.00 eV	1.70 eV
Barrier height at intermediate layer/absorber interface ϕ_{BI} (See Fig. 6.1)	Not applicable	0.80 eV (No intermediate layer present)	1.00 eV (No intermediate layer present)	0.75 eV	0.80 eV
Total built-in potential	0.62 eV	0.54 eV	0.74 eV	0.74 eV	0.54 eV
Intermediate layer	Not applicable	No intermediate layer	No intermediate layer	15-nm defective layer	10-nm EBL
Intermediate layer defect states	Not applicable	No intermediate layer	No intermediate layer	Donor-like gap states 0.80 eV below E_C in a band 0.01 eV wide $N_{TD} = 1 \times 10^{19} \text{cm}^{-3}$ $\sigma_n = 1 \times 10^{-30} \text{cm}^2$ $\sigma_p = 1 \times 10^{-15} \text{cm}^2$	Defect-free
Built-in potential across Intermediate layer	Not applicable	No Intermediate layer	No Intermediate layer	0.24 eV	Negligible
Built-in potential across the absorber semiconductor	0.62 eV	0.54 eV	0.74 eV	0.50 eV	0.54 eV

device is getting its excellent performance with a smaller built-in potential than the p–n and higher barrier M-S cells. The excellent V_{oc}, which is seen to be larger than the built-in potential, occurs in this M-I-S cell due to charge pile-up at the EBL electron affinity step—a

Table 6.3 Cell Performance Measures

Cell	J_{sc} (mA/cm^2)	V_{oc} (V)	FF	η (%)
p–n	28.4	0.55	0.79	12.4
M-S (lower barrier)	27.7	0.27	0.69	5.1
M-S (higher barrier)	27.9	0.46	0.78	10.0
M-S (pinning)	27.9	0.45	0.78	9.9
M-I-S	27.9	0.56	0.79	12.1

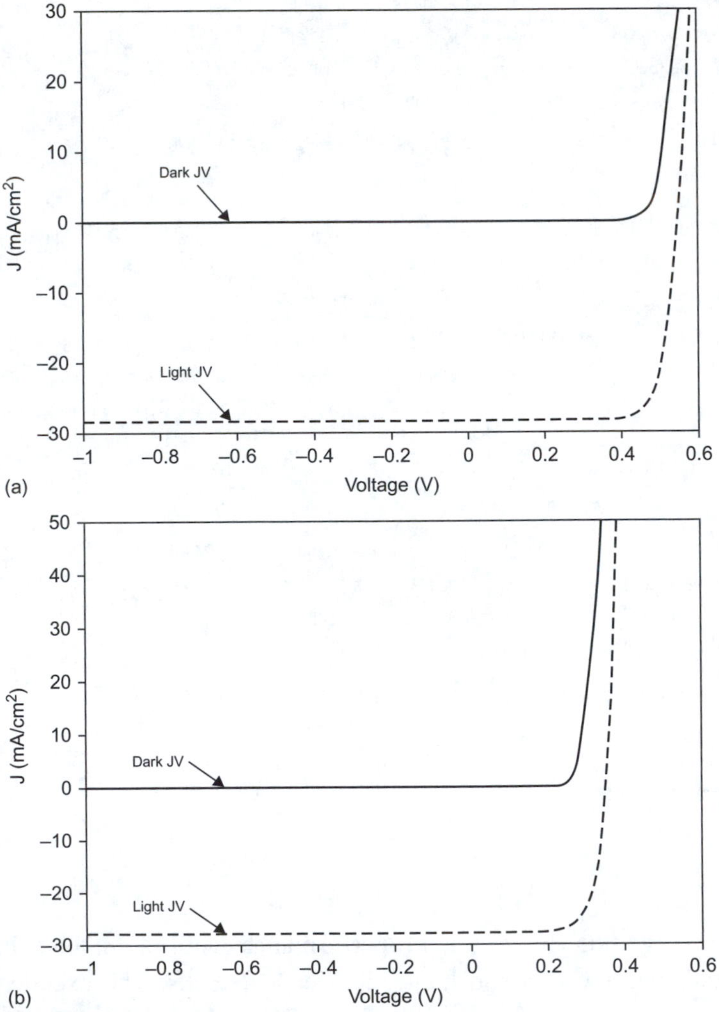

FIGURE 6.6 The linear dark and light J-V characteristics for the cells of Figure 6.4: (a) the p–n homojunction and (b) the lower-barrier M-S cells of Table 6.2.

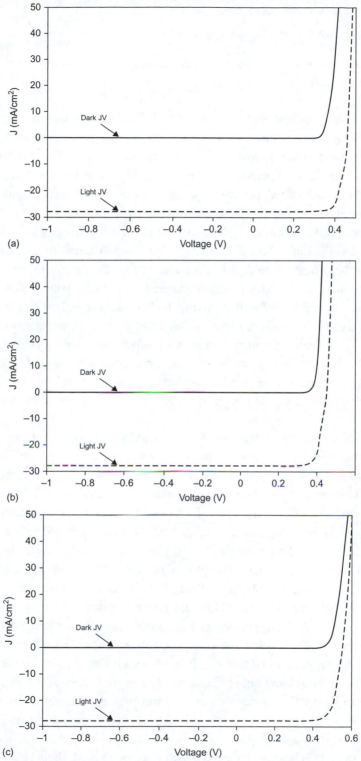

FIGURE 6.7 J-V behavior of (a) the higher-barrier M-S, (b) Fermi-level pinned and (c) the M-I-S cells of Table 6.2.

phenomenon we saw could be possible at effective-field barriers in heterojunctions.

Superposition is seen to work for the M-I-S cell but to fail for both M-S cells in Figures 6.6 and 6.7. We always check the presence or absence of superposition to see if we can learn more about a cell. In this case numerical analysis shows superposition fails for the M-S cells for a reason we discussed was, in general, possible in Chapters 4 and 5 but never actually saw in those devices: failure of the assumption of a flat electron quasi-Fermi level across the electrostatic field and majority carrier region. The quasi-Fermi level actually decreases as the front is approached in the M-S cell due to the efficiency of $J_{ST}(0)$. This means that the electron quasi-Fermi level position at the back contact dictated by the cell's voltage lies above the quasi-Fermi level $E_{Fn}(0^+)$ driving $J_{ST}(0)$ (see Eq. 6.3). The degree to which these to quasi-Fermi levels differ changes with illumination leading to the loss of superposition. This lack of superposition disappears for the M-I-S cell since $J_{ST}(0)$ has essentially been reduced to zero. The charge pile-up mentioned earlier for the M-I-S cell demonstration is not significant enough for this device to trigger that mechanism for superposition loss.

In practice, changing the metal workfunction (or the electrochemical potential of the electrolyte) often does not cause barrier-height changes that follow $\phi_B = \phi_W - \chi$, because of the presence of an inadvertent intermediate layer and gap states, which we discussed earlier. This is explored using the device of Figure 6.5b, which uses the same metal-semiconductor pairing, and therefore the same workfunction difference and total built-in potential, as the device labeled the "higher barrier M-S" device in Table 6.2. However, the device of Figure 6.5b has a 15-nm inadvertent layer containing a large number of defects, which have been placed, for this example, starting at 10 nm below the metal interface. As may be noted from Table 6.2, these gap states have been taken to be donor-like and are located 0.80 eV below the conduction-band edge. From their capture cross-section values in Table 6.2, it can be seen that their occupancy follows the valence-band hole population. The high density of the gap states at 10 nm below the barrier-former position pins the TE Fermi level near 0.80 eV below the conduction-band edge (actually simulation results for this case give $\phi_{BI} = 0.75$ as seen in Table 6.2) at the physical location of the defects. These states are charged and, as a result, their electrostatic field across the surface layer develops a significant potential energy (band bending). In this example, Table 6.2 shows that one third of the built-in potential V_{Bi} is developed across the surface layer. In spite of this reallocation of the built-in potential, the numerical modeling summarized in Table 6.3 shows that the M-S cell with Fermi-level-pinning behavior performs about as well as the higher-barrier M-S device, which does not have the defect layer. This could happen but probably will not in a real situation. In actual devices, such defect layers typically allow electrons to cross by hopping or tunneling, which we did not account for in our modeling. As a result, the barrier height actually controlling the current traversing the intermediate layer and therefore determining $J_{ST}(0)$ is usually the $\phi_{BI} = 0.75\,eV$ and not $\phi_B = 1.00$. In this case, the value of the controlling barrier is dictated by the built-in potential redistribution. Put succinctly, defect layers like that of this example can be present in Schottky barrier devices causing the barrier which is establishing $J_{ST}(0)$ to have little or no correlation with the metal work function.

6.4 ANALYSIS OF SURFACE-BARRIER DEVICE PHYSICS: ANALYTICAL APPROACH

This section describes an analytical analysis of the surface-barrier cell seen in the band diagram of Figure 6.8. As may be noted from the figure, under illumination, light enters from the left and traverses the absorber before coming to the surface barrier rather than entering through the surface barrier as previously assumed. Also, for some variety, the junction orientation is opposite to that used until now in this chapter. Using our standard approach, we pick the x = 0 plane, which lies just outside the electrostatic-field (space-charge) region, to evaluate J from

$$J = J_n(0) + J_p(0) \tag{6.4}$$

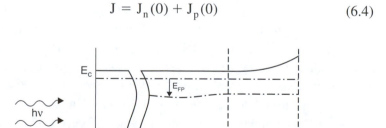

FIGURE 6.8 A surface-barrier solar cell under illumination. The absorber is taken to be n-type and light is impinging from the left.

where these quantities are conventional electron and hole current densities defined to be positive if flowing left to right in Figure 6.8. Since the quasi-neutral region of this n-type material looks just like the n-side of a p–n junction, we neglect drift and say

$$J_p(0) = -e\,D_p \frac{dp}{dx}\Big|_{x=0} \tag{6.5}$$

As we did in Chapters 4 and 5 to obtain the necessary analytical expression for $p(x)$ for the quasi-neutral region of the semiconductor, we assume that recombination in the semiconductor follows a linear recombination lifetime model; i.e., $\mathscr{R}(x) = (p - p_{n0})/\tau_p$, where τ_p is the hole recombination time, a constant with position. Under these assumptions, $p(x)$ satisfies

$$\frac{d^2p}{dx^2} - \frac{p - p_{n0}}{L_p^2} + \int_\lambda \frac{\Phi_0(\lambda)}{D_p}\alpha_1(\lambda)e^{-\alpha_1(x+d)}d\lambda = 0 \tag{6.6}$$

subject to the boundary conditions

$$\frac{dp}{dx}\Big|_{x=-d} = -\frac{S_p}{D_p}[p(-d) - p_{n0}(-d)] \tag{6.7a}$$

$$p(0) = p_{n0}e^{E_{Fp}(0)/kT} \tag{6.7b}$$

As in the case of homojunctions in Chapters 4, the hole quasi-Fermi level E_{Fp} in Eq. 6.7b is being measured positively down from the Fermi-level position in energy at the $x = -d$ contact to the n-material; i.e., it is being measured from the metal-contact workfunction position at $x = -d$. Equation 6.6 assumes that absorption in the semiconductor follows a Beer-Lambert model and produces free electron–hole pairs according to

$$\int_\lambda G_{ph}(\lambda,x)d\lambda = \int_\lambda G(\lambda,x)d\lambda = \int_\lambda \Phi_0(\lambda)\alpha_1(\lambda)e^{-\alpha_1(\lambda)(x+d)}d\lambda$$

which ignores reflection at the barrier-former interface at x = W. The above formulations for generation and the generation term in Eq. 6.6 are easily modified for the situation in which light enters through the surface barrier-former. This is done by replacing x + d in all the generation expressions with the quanity W − x.

Taking the results for the p(x) of Eq. 6.6 obtained in Chapters 4 and 5 and using Eq. 6.5, we can write

$$
\begin{aligned}
J_p(0) = e \int_\lambda \Phi_0(\lambda) &\left\{ \left[\frac{\beta_2^2}{\beta_2^2 - \beta_1^2} \right] \left[\frac{(\beta_3 \beta_1 / \beta_2) + 1}{\beta_3 \sinh \beta_1 + \cosh \beta_1} \right] \right. \\
&\left. - \left[\frac{\beta_2^2 e^{-\beta_2}}{\beta_2^2 - \beta_1^2} \right] \left[\left(\frac{\beta_3 \cosh \beta_1 + \sinh \beta_1}{\beta_3 \sinh \beta_1 + \cosh \beta_1} \right) \left(\frac{\beta_1}{\beta_2} \right) + 1 \right] \right\} d\lambda \\
&- \left\{ \left[\frac{e D_p P_{n0}}{L_p} (e^{V/kT} - 1) \right] \right\} \left[\frac{\beta_3 \cosh \beta_1 + \sinh \beta_1}{\beta_3 \sinh \beta_1 + \cosh \beta_1} \right]
\end{aligned}
$$

(6.8)

which assumes $E_{Fp}(0) = V$. The β quantities appearing in Eq. 6.8 are listed in Table 6.4.

Table 6.4 Beta Parameters for Absorber

β Quantity	Definition	Physical significance
β_1	d/L_p	Ratio of the semiconductor quasi-neutral region length to hole diffusion length. Captures the physics of hole collection from the absorber quasi-neutral region by diffusion. Need $d \leq L_p$.
$\beta_2(\lambda)$	$d\alpha_1(\lambda)$	Ratio of the semiconductor quasi-neutral region length to absorption length for light of wavelength λ. (This ratio depends on λ.) A measure of how much light of wavelength λ is absorbed in the semiconductor. Want $d \cong 1/\alpha(\lambda)$.
β_3	$L_p S_p / D_p$	Ratio of the x = −d surface hole-carrier recombination velocity S_p to the hole diffusion velocity D_p/L_p. Captures the physics of hole diffusion and recombination versus hole contact recombination. Need $D_p/L_p > S_p$.

To complete the expression for J, we need $J_n(0)$. As we know from previous numerical analyses, this majority-carrier current will have a significant drift component at $x = 0$. To side-step the fact that we do not know the electric-field value and thus have no analytical expression for $J_n(0)$, we use the continuity concept to write

$$J_n(0) = e \int_0^W \int_\lambda G_{ph}(\lambda,x)d\lambda dx - e \int_0^W \mathscr{R}(x)dx - eS_n[n(W) - n_0(W)]$$

(6.9)

This statement is valid for M-S, M-I-S, and EPC solar cells, with the last loss term being essentially zero for a properly designed M-I-S cell. This expression becomes

$$J_n(0) = e \int_\lambda \Phi_0(\lambda)e^{-\alpha_1 d}(1 - e^{-\alpha_1 W_1})d\lambda$$
$$- \left\{ J_{SCR} \left(e^{V/n_{SCR}kT} - 1 \right) \right\} - eS_n[n(W) - n_0(W)] \quad (6.10)$$

since

$$\int_0^W \int_\lambda G_{ph}(\lambda,x)d\lambda dx = \int_\lambda \Phi_0(\lambda)e^{-\alpha_1 d}(1 - e^{-\alpha_1 W})d\lambda$$

and

$$e \int_0^W \mathscr{R}(x)dx = \left\{ J_{SCR} \left(e^{V/n_{SCR}kT} - 1 \right) \right\}$$

have been utilized. The latter expression is an attempt at modeling what the $e \int_0^W \mathscr{R}(x)\,dx$ loss term looks like, as we did in Chapters 4 and 5.

Equations 6.8 and 6.10 allow us to construct J from $J_n(0) + J_n(0)$, giving

$$J = e \int_\lambda \Phi_0(\lambda) \left\{ \left[\left[\frac{\beta_2^2}{\beta_2^2 - \beta_1^2} \right] \left[\frac{(\beta_3 \beta_1/\beta_2) + 1}{\beta_3 \sinh\beta_1 + \cosh\beta_1} \right] \right. \right.$$

$$- \left[\frac{\beta_2^2 e^{-\beta_2}}{\beta_2^2 - \beta_1^2} \right] \left[\left(\frac{\beta_3 \cosh\beta_1 + \sinh\beta_1}{\beta_3 \sinh\beta_1 + \cosh\beta_1} \right) \left(\frac{\beta_1}{\beta_2} \right) + 1 \right]$$

$$\left. + \left[e^{-\beta_2} (1 - e^{\alpha_1 W}) \right] \right] d\lambda$$

$$- \left\{ \left[\frac{eD_p p_{n0}}{L_p} (e^{V/kT} - 1) \right] \left\{ \frac{\beta_3 \cosh\beta_1 + \sinh\beta_1}{\beta_3 \sinh\beta_1 + \cosh\beta_1} \right\} \right.$$

$$\left. - \left\{ J_{SCR} \left(e^{V/n_{SCR}kT} - 1 \right) \right\} - eS_n[n(W) - n_0(W)] \right. \tag{6.11a}$$

The last loss term in this expression can be explicitly be written in terms of voltage by using the possibly problematic assumption of a flat electron quasi-Fermi level all the way to x = W. With this addition, Eq. 6.11a is in a convenient form, as is, for an EPC cell. For an M-S cell, this construction can be further focused to

$$J = e \int_\lambda \Phi_0(\lambda) \left\{ \left[\left[\frac{\beta_2^2}{\beta_2^2 - \beta_1^2} \right] \left[\frac{(\beta_3 \beta_1/\beta_2) + 1}{\beta_3 \sinh\beta_1 + \cosh\beta_1} \right] \right. \right.$$

$$- \left[\frac{\beta_2^2 e^{-\beta_2}}{\beta_2^2 - \beta_1^2} \right] \left[\left(\frac{\beta_3 \cosh\beta_1 + \sinh\beta_1}{\beta_3 \sinh\beta_1 + \cosh\beta_1} \right) \left(\frac{\beta_1}{\beta_2} \right) + 1 \right]$$

$$\left. + \left[e^{-\beta_2} (1 - e^{\alpha_1 W}) \right] \right] d\lambda$$

$$- \left\{ \left[\frac{eD_p p_{n0}}{L_p} (e^{V/kT} - 1) \right] \left\{ \frac{\beta_3 \cosh\beta_1 + \sinh\beta_1}{\beta_3 \sinh\beta_1 + \cosh\beta_1} \right\} \right.$$

$$\left. - \left\{ J_{SCR} \left(e^{V/n_{SCR}kT} - 1 \right) \right\} - A^* T^2 e^{-\phi_B/kT} [e^{V/kT} - 1] \right. \tag{6.11b}$$

where it has also been assumed the electron quasi-Fermi level is flat across the device all the way to x = W, an assumption which we saw does not always hold up in Section 6.3. For a properly designed M-I-S cell, our analytical model for the J-V characteristic can be written as

$$
J = e \int_{\lambda} \Phi_0(\lambda) \left\{ \left[\frac{\beta_2^2}{\beta_2^2 - \beta_1^2} \right] \left[\frac{(\beta_3 \beta_1 / \beta_2) + 1}{\beta_3 \sinh\beta_1 + \cosh\beta_1} \right] - \left[\frac{\beta_2^2 e^{-\beta_2}}{\beta_2^2 - \beta_1^2} \right] \right.
$$

$$
\left[\left(\frac{\beta_3 \cosh\beta_1 + \sinh\beta_1}{\beta_3 \sinh\beta_1 + \cosh\beta_1} \right) \left(\frac{\beta_1}{\beta_2} \right) + 1 \right]
$$

$$
\left. + \left[e^{-\beta_2} (1 - e^{\alpha_1 W}) \right] \right\} d\lambda
$$

$$
- \left\{ \frac{eD_p p_{n0}}{L_p} (e^{V/kT} - 1) \right\} \left\{ \frac{\beta_3 \cosh\beta_1 + \sinh\beta_1}{\beta_3 \sinh\beta_1 + \cosh\beta_1} \right\} - \left\{ J_{SCR} \left(e^{V/n_{SCR}kT} - 1 \right) \right\}
$$

$$
\text{(6.11c)}
$$

The terms in Eq. 6.11 involving $\Phi_0(\lambda)$ and the integration over the spectrum are all voltage-independent and the result of photogeneration. They actually represent two contributions: that of hole diffusion collection from the quasi-neutral region and that of additional photogenerated holes produced in the space-charge region. The first contribution is easily spotted by the presence of β_1 (which accounts for recombining while diffusing out of the quasi-neutral region) and β_3 (which accounts for left surface recombination loss while trying to diffuse). The first voltage-dependent loss term $\{[eD_p p_{n0}/L_p](e^{V/kT} - 1)\{[\beta_3 \cosh\beta_1 + \sinh\beta_1]/[\beta_3 \sinh\beta_1 + \cosh\beta_1]\}\}$ is the dark diode current component caused by holes undergoing diffusion and simultaneous recombination in the n-material quasi-neutral region and at its left contact. The second voltage-dependent loss term is our phenomenological model for losses in the electrostatic-field barrier region. It, too, is a component of the dark current. The third voltage dependent loss term, present for EPC and M-S cells, is also part of the dark current and is loss due to the reduction of the electrolyte oxidized species or thermionic emission, respectively. This analysis makes the important point that, although species reduction or thermionic emission may be the dominant loss mechanism in many EPC or M-S cells and control V_{oc}, bulk and space charge region recombination are still taking their toll.

The analytical analysis we just completed allows us to clearly see the role of diffusion in a collection of photogenerated minority carriers in surface-barrier cells with quasi-neutral absorber regions. The analysis gives us criteria (the β quantities) that allow us to assess the importance of region dimensions, bulk recombination, and surface recombination in this collection. Since Eq. 6.11 is in the form of a voltage-independent J_{sc} opposed by a voltage-dependent dark current $J_{DK}(V)$, the analysis also allows us to see the assumptions that lead to a J-V expression that obeys superposition. Equation 6.11 is also useful for accounting for series and shunt resistance effects. This may be undertaken by following the procedure discussed in Chapter 4.

Looking more closely at the M-S cell case, we see that when thermionic emission dominates the dark current, it follows from Eq. 6.11 that superposition says

$$e^{V_{oc}/kT} = \frac{J_{sc}}{A^*T^2}e^{\phi_B/kT}$$

or

$$V_{oc} = \phi_B + kT \ln\left(\frac{J_{sc}}{A^*T^2}\right) \tag{6.12}$$

Our numerical analyses of M-S solar cells used material parameters that resulted in this situation of having thermionic emission dominating over bulk recombination. The analytical analysis just completed allows us to develop a general criterion for the combination of material parameters that allows this to occur. To establish this criterion, we begin by looking at Eq. 6.11b and note that, if thermionic emission is dominating over the hole diffusion–recombination dark diode current (bulk recombination) mechanism, then

$$A * T^2 e^{-\phi_B/kT} > \left\{ \left[\frac{eD_p p_{n0}}{L_p} \right] \left[\frac{\beta_3 \cosh\beta_1 + \sinh\beta_1}{\beta_3 \sinh\beta_1 + \cosh\beta_1} \right] \right\}$$

If a selective ohmic contact to the semiconductor is assumed, which is what we used in the numerical modeling, the inequality becomes

$$A * T^2 e^{-\phi_B/kT} > \left\{ \left[\frac{eD_p p_{n0}}{L_p} \right] \left[\frac{\sinh\beta_1}{\cosh\beta_1} \right] \right\}$$

Consequently, if

$$A * T^2 e^{-\phi_B/kT} > \left\{ \frac{eD_p p_{n0}}{L_p} \right\} \tag{6.13}$$

then thermionic emission will dominate the loss mechanisms and will establish V_{oc} in an M-S solar cell, since $\{[\sinh\beta_1]/[\cosh\beta_1]\} \leq 1$. Statement 6.13 can be rearranged to

$$\left\{ \frac{A * T^2 L_p}{eD_p p_{n0}} \right\} > e^{\phi_B/kT}$$

which allows the ϕ_B upper limit for thermionic emission control to be determined from

$$kT \ln \left\{ \frac{A * T^2 L_p}{eD_p p_{n0}} \right\} > \phi_B \tag{6.14}$$

Using $A* = 120\,A/(cm^2 - K)$, $T = 300\,K$, $L_p = 10^{-3}\,cm$, $D_p = 2.6\,cm^2/s$, and $p_{n0} = 10^5\,cm^{-3}$ allows us to evaluate $\{[A * T^2 L_p]/[eD_p p_{n0}]\}$ for that set of representative material parameters. Doing so gives the result that the Schottky barrier height ϕ_B has to be larger than 1.17 eV for the diffusion-recombination (bulk) mechanism to dominate over thermionic emission and thereby to control surface-barrier-cell performance, for the material parameters used. Equation 6.14 provides, of course, a general tool for assessing the relative importance of these two surface-barrier-cell loss mechanisms for any semiconductor (any set of material parameters) of interest.

6.5 SOME SURFACE-BARRIER CONFIGURATIONS

As noted before, the simplicity of surface-barrier devices and their use of low-temperature processing make them ideal for assessing new materials and configurations. For example, M-S solar cell structures have been used to study the use of p-type PbSe[20] and PbS[30] quantum dot (nanoparticle) films as semiconductor exciton-producing absorbers. Figure 6.9 summarizes some of the results for PbSe quantum dot M-S cells. It gives, as a function of the semiconductor particle size, the band gap as well as the experimentally determined longest wavelength of light able to produce an exciton in the quantum dots. This latter information can be determined

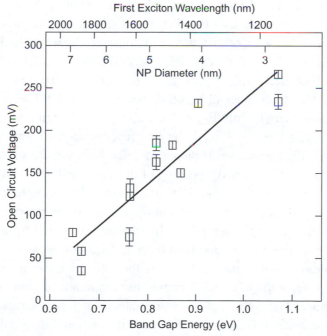

FIGURE 6.9 The experimentally observed V_{oc} and the experimentally observed first (longest-wavelength) exciton on-set as a function of nano-particle (NP) size (band gap). (From Ref. 20, with permission.)

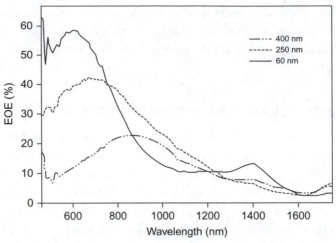

FIGURE 6.10 The experimentally observed EQE for three quantum-dot film thicknesses all with $E_G = 0.72\,\text{eV}$. (From Ref. 20, with permission.)

from the external quantum efficiency (EQE) introduced in Section 4.2.1 and seen in Figure 6.10 for this M-S cell. Figure 6.9 also shows the experimentally observed V_{oc} as a function of particle size (band gap) with the metal barrier-former workfunction held constant. As we have discussed, ideally $\phi_B = \phi_W - \chi$ for an n-type semiconductor. This ideal expression must be modified to $\phi_B = \chi + E_G - \phi_W$ for a p-type semiconductor M-S cell. The dependence of V_{oc} on E_G and its slope of ~ 1 seen in Figure 6.10 suggest that (1) only E_G, and not also χ, is changing with particle size and $\phi_B = \chi + E_G - \phi_W$ is being obeyed with no Fermi-level pinning; or (2) the Fermi level is pinned at defect states at the surface of the quantum dots and the energy position of the states shifts with E_G.[20] Since the illumination used for the EQE data (done with no light or voltage bias) impinged on the surface without the metal barrier in these cells, the short-wavelength EQE in Figure 6.10 is being generated by absorption farthest from the electrostatic-field barrier. As a consequence, the EQE for the short wavelengths decreases as the film thickness increases and the charge-separating barrier becomes further removed from the light-entry surface. The authors of this work transformed these data to internal quantum efficiency (IQE), also discussed in Section 4.2.1, and found it to be less than 100%. It would have to be >100% in some wavelength regime if exciton-caused carrier multiplication (see Sections 2.2.6 and 3.4.2) were important in these particular M-S cells.[20]

In another example of using surface-barrier devices for material and configuration evaluation, EPC devices based on silicon-wire arrays were used to explore the lateral collection approach first discussed in Section 3.4.1.4.[31,32] In this case, silicon wires, oriented perpendicularly to a substrate, were used as the absorber and an engulfing electrolyte was the barrier-former. This configuration permits photogenerated minority carriers to be collected by the barrier-former electrolyte surrounding the wire. The performance level of this structure suggested that control of defects at the semiconductor-electrolyte interface was an issue.[31]

REFERENCES

1. A. Fujishima, K. Honda, Electrochemical photolysis of water at a semiconductor electrode, Nature 238 (1972) 37.

2. E. Becquerel, Memoire sur les effects electriques produit sous l'influence des rayons, C. R. Acad. Sci. 9 (1839) 145.

3. M. Wolf, Historical development of solar cells, Proc. Power Sources Symp., 25th, May 23–25, 1972, p. 120.

4. L.O. Grondahl, The copper-cuprous-oxide rectifier and photoelectric cell, Rev. Mod. Phys. 5 (1933) 141.

5. R.J. Stirn, Y-C.M. Yeh, Proc. 10th IEEE Photovoltaic Spec. Conf. (IEEE, New York, 1974) p. 15; Y-C.M. Yeh, R.J. Stirn, Proc. 11th IEEE Photovoltaic Spec. Conf. (IEEE, New York 1975) p. 391.

6. N.N. Winogradoff, H.K. Kessler, U.S. Patent 3,271,198 (1966).

7. W.W. Anderson, Y.G. Chai, Becquerel effect solar cell, Energy Convers 15 (1976) 85.

8. A.J. Nozik, Photoelectrochemical cells, Philos. Tram. R. Soc. London, Ser. A 295 (1980) 453.

9. E.J. Charlson, A.B. Shak, J.C. Lien, A New Silicon Schottky Photovoltaic Energy Converter, Int. Electron Devices Meet. Washington, D.C. IEEE, New York, 18, 16 (1972).

10. S. Shevenock, S. Fonash, J. Geneczko, Studies of M-I-S Type Solar Cells Fabricated on Silicon, Tech. Dig.-Int. Electron Devices Meet. Washington, D.C. p. 211. IEEE, New York, 1975.

11. R.J. Stirn, Y.-C.M. Yeh, A 15% efficient antireflection-coated metal-oxide-semiconductor solar cell, Appl. Phys. Lett. 27 (1975) 95; and Proc. 11th IEEE Photovoltaic Spec. Conf. (IEEE, New York 1975) p. 437.

12. J. Shewchun, M.A. Green, F.D. King, Minority carrier MIS tunnel diodes and their application to electron- and photo-voltaic energy conversion—II. Experiment, Solid-State Electron. 17 (1974) 563; J. Shewchun, R. Singh, M.A. Green, Theory of metal-insulator-semiconductor solar cells, J. Appl. Phys. 48, 765 (1977).

13. R.E. Thomas, R.B. North, C.E. Norman, Low cost high efficiency MIS/inversion layer solar cells, IEEE Electron Device Lett. 1 (1980) 79.

14. M.A. Green, R.B. Godfrey, M.R. Willison, A.W. Blakers, High Efficiency (Greater than 18%, Active Area, AM1) Silicon MinMIS Solar Cells, Proc. 14th IEEE Photovoltaic Spec. Conf. (IEEE, New York 1980) p. 684.

15. Y.-C.M. Yeh, R.J. Stirn, A Schottky-barrier solar cell on sliced polycrystalline GaAs, Appl. Phys. Lett. 33 (1978) 401; Y.-C. M. Yeh, F.P. Ernest, R.J. Stirn, Progress towards high efficiency thin film GaAs AMOS cells, Proc. 13th IEEE Photovoltaic Spec. Conf. (IEEE, New York 1978) p. 966.

16. H. Kim, H.-B.-R. Lee, W.-J. Maeng, Applications of atomic layer deposition to nano-fabrication and emerging nanodevices, Thin Solid Films 517 (2009) 2563.

17. H.O. Finklea, Self-Assembled Monolayers on Electrodes, Encyclopedia of Analytical Chemistry, vol. 11, John Wiley & Sons Ltd., New York, 2000, pp. 10090–10115.

18. A.C. Jones, H.C. Aspinall, P.R. Chalker, R.J. Potter, T.D. Manning, Y.F. Loo, R. O'Kane, J.M. Gaskell, L.M. Smith, MOCVD and ALD of high-k dielectric oxides using alkoxide precursors, Chem. Vap. Deposition 12 (2006) 83.

19. P.V. Kamat, Quantum dot solar cells. Semiconductor nanocrystals as light harvesters, J. Phys. Chem. C. 112 (2008) 18737.

20. J.M. Luther, M. Law, M.C. Beard, Q. Song, M.O. Reese, R.J. Ellingson, A.J. Nozik, Schottky solar cells based on colloidal nanocrystal films, Nano Lett. 8 (2008) 3488.

21. A.B. Ellis, S.W. Kaiser, M.S. Wrighton, Visible light to electrical energy conversion. Stable cadmium sulfide and cadmium selenide photoelectrodes in aqueous electrolytes, J. Am. Chem. Soc. 98 (1976) 1635.

22. B. Miller, A. Heller, Semiconductor liquid junction solar cells based on anodic sulphide films, Nature 262 (1976) 680.

23. B.A. Parkinson, A. Heller, B. Miller, Enhanced photoelectrochemical solar-energy conversion by gallium arsenide surface modification, Appl. Phys. Lett. 33 (1978) 521.

24. S. Licht, Multiple band gap semiconductor/electrolyte solar energy conversion, J. Phys. Chem. B, 105 (2001) 6281.

25. D. Morell, A.K. Ghosh, T. Feng, E.L. Stogryn, P.E. Purwin, R.F. Shaw, C. Fishman, High-efficiency organic solar cells, Appl. Phys. Lett. 32 (1978) 495.

26. E.H. Nicollian, A.K. Sinha, Effects of interfacial reactions on the electrical characteristics of M–S contacts, in: J.M. Poate, K.N. Tu, J.W. Mayer (Eds.) Thin Films—Interdiffusion and Reactions, Wiley, New York, 1979, p. 481.

27. S.J. Fonash, Metal-insulator-semiconductor solar cells: Theory and experimental results, Thin Solid Films 36 (1976) 387.

28. M. Gratzel, Photoelectrochemical cells, Nature 414 (2001) 338.

29. E.H. Rhoderick, R.H. Williams, Metal-semiconductor contacts, second ed., Oxford Univ. Press, USA, 1988.

30. J.P. Clifford, K.W. Johnston, L. Levina, E.H. Sargent, Schottky barriers to colloidal quantum dot films, Appl. Phys. Lett. 91 (2007) 253117.

31. J.R. Maiolo, B.M. Kayes, M.A. Filler, M.C. Putnam, M.D. Kelzenberg, H.A. Atwater, N.S. Lewis, High aspect ratio silicon wire array photoelectrochemical cells, J. Am. Chem. Soc. 129 (2007) 12346.

32. US patents 6399177, 6919119, and 7341774.

Dye-sensitized Solar Cells

7.1 INTRODUCTION

The dye-sensitized solar cell (DSSC), shown schematically in Figure 7.1, is the newest photovoltaic device configuration. First developed by O'Reagan and Grätzel in 1991,[1] this class of cell has reached efficiencies of over 11%.[2] The basic structure of a DSSC involves a transparent (wide-band-gap) n–type semiconductor configured optimally in a nanoscale network of columns, touching nanoparticles, or coral-like protrusions. The surface area of the network is designed to be huge, and it is covered everywhere with a monolayer of a dye or a coating of quantum dots, which functions as the dye. We will henceforth refer to either as the dye sensitizer. The dye sensitizer is the absorber. An electrolyte is then used to permeate the resulting coated network structure to set up a conduit between the dye and the anode. The dye absorbs light, producing excitons, which dissociate at the dye-semiconductor interface, resulting in photogenerated electrons for the semiconductor and oxidized dye molecules that must be reduced—and thereby regenerated—by the electrolyte.

DOI: 10.1016/B978-0-12-374774-7.00007-8

The transparent semiconductor network provides the path to the cathode[†] for the photogenerated electrons. The liquid electrolyte is the pathway from the anode[†] for the reducing species which provide electrons for the oxidized (hole-bearing) dye molecules. The diagram of Figure 7.1 shows the constant production of the electrolyte reduced species at the anode and the constant supplying of electrons via the cathode to the external circuit to do work. The semiconductor forming the network coated with dye molecules or quantum dots needs to be transparent to allow light to reach the absorber materials, and the network must have a huge surface area to provide the dye amount required for absorption. The electrolyte must be able to permeate the whole network to give electrical continuity; consequently, liquid electrolytes have proven very effective.

A ruthenium dye was initially employed in the DSSC, but by now several organic dyes[3–5] and inorganic quantum dot "dyes"[6,7] have been explored. Initially, TiO_2 (anatase) was used as the transparent, n-type semiconductor network, but other TCO-type semiconductors have been

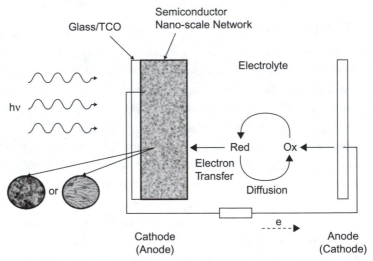

FIGURE 7.1 Schematic of a DSSC. Blow-ups show particle (left) and nano-filament (right) networks. The anode–cathode labeling is consistent with the labeling used in this text. The terminology in parentheses is that often used in electrochemistry.

[†]There is a terminology issue here. In solid-state photovoltaics, the electrode that is negative in the power quadrant is called the cathode. In electrochemical and DSSC devices, the term cathode is often applied to the electrode where reduction of the electrolyte species takes place.

utilized, including SnO_2 and ZnO.[8,9] Alternatives to the liquid electrolyte have included gels[10] and solid-phase hole conductors.[11,12] The latter approach removes the holes from the oxidized dye molecules by straightforward solid-state hole transport.

7.2 OVERVIEW OF DYE-SENSITIZED SOLAR CELL DEVICE PHYSICS

7.2.1 Transport

In the DSSC device, the dye material, whether organic molecules or inorganic quantum dots, absorbs light by exciton production. As we know from previous encounters with similar situations (see, for example, Fig. 3.29 and related discussion), the dye LUMO level must be sufficiently above the semiconductor conduction-band edge, as shown in Figure 7.2, to allow the exciton on the dye to dissociate by producing a photogenerated electron in the network semiconductor and a photogenerated hole localized on the dye site. The DSSC

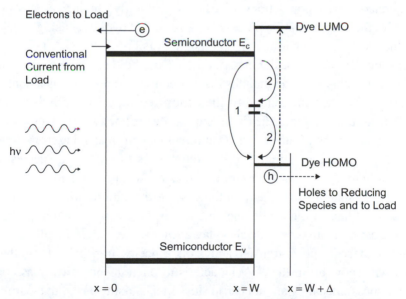

FIGURE 7.2 A DSSC cell under illumination, with light entering from the left. Shown are the photocarrier interface loss mechanisms 1 and 2. Here W is the distance traveled in the network (e.g., TiO_2) by electrons as they move to the cathode. In this diagram, the quantity Δ represents the dye monolayer or quantum-dot coating thickness. There is no electrostatic barrier; only effective-force-field barriers.

has no built-in electrostatic-field barrier; it has only an effective-field barrier, as is also seen in Figure 7.2. Just as in the case of the exciton heterojunction cells of Section 5.3.2, exciton dissociation at the interface causes photogenerated carriers in DSSC devices to appear on their "collected" side of the barrier; i.e., electrons and holes each are generated on their respective "downstream" sides of the effective-force-field barrier. The semiconductor provides the network that the photogenerated electrons then must traverse to the collecting cathode. There are, of course, carrier-loss mechanisms in the cell and, as usual, they equal the photogeneration at open circuit. Neglecting injection of electrons from the semiconductor back into oxidized dye molecules, the carrier-loss mechanisms are seen to be at the semiconductor-dye interface.[13] They are shown as carrier-loss Paths 1 and 2 in Figure 7.2. Both result in recombination of a photogenerated electron and a photo-generated hole. As seen in the figure, Path 1 is a direct process, whereas Path 2 involves interface states at the semiconductor surface. The numerical modeling in Section 7.3 shows that either Path 1 or Path 2 controls the open-circuit voltage, because these paths control the photo-generated carrier populations. The holes on the dye molecule sites that survive the recombination processes are removed by the reducing species in the electrolyte (or in the all solid-state version of this cell, by a hole transport medium). Ideally, the kinetics of the hole removal process are fast. If not, this step can manifest itself phenomenologically as a contact problem. For current continuity, the cation resulting from removing the hole from the dye molecule (in the liquid-electrolyte case) must move to the anode seen in Figure 7.1, and capture an electron arriving from doing work in the external circuit. It then returns, as the reducing species, to the dye monolayer. The transport of the reducing species, and of the cation, necessitate diffusion and, in the case of the cation, drift also can be involved. Once at the anode, the kinetics of the electron capture process by the cation from the anode also play a role. If there are difficulties with any of these electron-transfer and redox-species transport steps, they, too, will show up in the J-V characteristic phenomenologically as contact issues, since they are not carrier-loss mechanisms. From this outline, it can be seen that the DSSC device is very similar to organic planar and bulk heterojunction cells discussed in Chapter 5. The energy level and densities of states steps at the semiconductor-dye interface cause exciton dissociation and the photogenerated electrons and holes are created on opposites sides of the interface—i.e., they are "born" separated, as is the case in the organic PHJ and BHJ cells. The energy step in the allowed electron levels seen in Figure 7.2 breaks symmetry and makes one direction of motion much more favorable than the other for the photogenerated electrons, thereby setting up photovoltaic action. As is the case for excitonic PHJ and BHJ devices, the one-dimensional band diagram of Figure 7.2 suffices to represent the cell processes. It provides a "straightened-out" version of what the photogenerated electrons and holes experience.

The discussion of transport can be formalized into a mathematical statement using the continuity concept from Section 2.3.3 in its integral form, as we have done previously for the other cell classes. Choosing electrons as the particles that we will "bean-count" and expressing the results in terms of the total conventional current density J gives

$$J = eI - e\int_0^W \mathcal{R}(x)\,dx - J_{ST}(0) - J_{IR}(W) \tag{7.1}$$

where J is being carried by electrons just to the left of $x = 0$. The quantity I, first used in Section 5.4.1, is the number of excitons dissociating per time per area and therefore the number of photogenerated electrons being created per time per area just to the left of $x = W$. The quantity $J_{ST}(0)$ is the electron loss at the contact at $x = 0$ due to interface recombination, $\int_0^W \mathcal{R}(x)\,dx$ is the integrated bulk carrier recombination loss in the semiconductor, and $J_{IR}(W)$ is interface carrier loss, expressed as a current density, due to Paths 1 and 2 in Figure 7.2. The quantity $J_{IR}(W)$ depends, in general, on the rate constants describing Paths 1 and 2, on the photogenerated electron population in the semiconductor at the interface, and on the photogenerated hole population present on the dye molecules (or quantum dots). The latter quantity depends on I, the traffic through Paths 1 and 2, the kinetics of the dye-molecule reduction, the previously described transport processes in the electrolyte, and the kinetics of the reduction process at the anode. The term $J_{IR}(W)$ will depend on the cell voltage V. The back injection of electrons into oxidized dye molecules at $x = W$ is being neglected and is absent in Eq. 7.1. The bulk recombination loss $\int_0^W \mathcal{R}(x)\,dx$ and the front-contact loss $J_{ST}(0)$ can be neglected, since the semiconductor form-ing the network has an insignificant hole population due to its n-type doping and lack of any significant photogeneration since it is a wide band gap, transparent material.

7.2.2 The dye-sensitized solar cell barrier region

As established in Section 3.2, there are two principal sources of pho-tovoltaic action: (1) built-in electrostatic-field barriers and (2) built-in effective-force-field barriers. Either one breaks symmetry, makes one direction different from the other, and gives rise to photovoltaic action. As we recall, effective-force-field barriers arise when the available energy levels change in their energy position, density of states, or both, with spatial position.[14] The simple p–n homojunctions of Chapter 4 relied solely on electrostatic built-in barriers for photovoltaic action. The DSSC is a device that relies solely on an effective-force-field bar-rier for photovoltaic action. In that regard it is analogous to the solid-state heterojunction devices in Chapter 5 that do not have electrostatic-field barrier regions. The device of Figure 5.33 is exactly equivalent to the DSSC device in that it, too, has no electrostatic-field barrier, and, more impor-tantly, it uses the effective-field barriers arising from steps in the allowed states to dissociate excitons; i.e., the cell of Figure 5.33 has an exciton-producing absorption process and exciton dissociation at the interface due to energy-level steps. The resulting photogenerated electrons and photo-generated holes are created on their respective downstream sides of the effective-force barrier, thereby obviating the need for collection. Using the terminology employed for the cell of Figure 5.33 in the context of DSSC devices, we can say that the dye (or quantum dot) coating is the donor and the transparent semiconductor is the acceptor material in a DSSC.

The devices of Figure 5.33 and Figure 7.2 are what have been termed excitonic solar cells, devices whose absorption process produces excitons that then dissociate at an interface.[15] In excitonic solar cells, this disso-ciation causes the photogeneration of electrons and holes to take place on their respective downstream sides of the effective-force-field barrier at the interface. The latter, as we have mentioned, is a very distinctive feature of these cells.[16,17]

7.3 ANALYSIS OF DSSC DEVICE PHYSICS: NUMERICAL APPROACH

In this section, numerical solutions to the equations of Section 2.4 are used to determine DSSC behavior under light and voltage and thereby to generate the J-V characteristics. Numerical analysis is especially suited to DSSC cells because it is able to fully handle effective-force barriers and their impact on device physics. As we have seen, it also has the capacity to "open up a cell" and allow us to peer into the inner workings. To ensure that exciton dissociation produces electrons in the conduction band of the network semiconductor and holes at the HOMO level of the dye molecules (or in the valence band of the quantum dots) in Figure 7.2, a factious 2-nm generation layer will be utilized in our DSSC numerical modeling, as was done for the excitonic solar cell devices in Section 5.3.2.[16,17] This layer is seen in the TE band diagram presented in Figure 7.3. The modeling used does not directly handle redox-couple

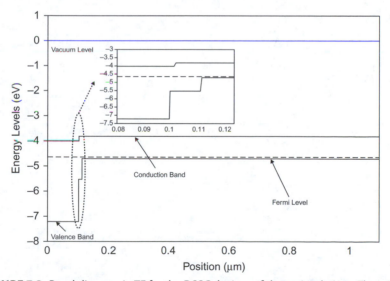

FIGURE 7.3 Band diagram in TE for the DSSC devices of these simulations. There is no built-in electrostatic-field barrier present. The inset shows the effective-force-field barriers that dissociate the excitons in greater detail. The factious generation layer is the 2-nm region between x = 100 nm and x = 102 nm. The dye or quantum-dot coating is the next 10 nm, whereas the layer representing the hole-transport medium starts at x = 112 nm.

transport, the dye-molecule reduction at the dye-electrolyte interface, or cation reduction at the anode. These are treated phenomenologically with the series resistance of a hole-transporting semiconductor and the series process of hole transport at the back contact. These elements also are seen in Figure 7.3. The hole-transporting function is analogous to the hole-transporting role played by the donor in an organic HJ device.

Table 7.1 gives the basic material parameters assumed in the numerical simulations for the transparent network semiconductor, for the dye (or quantum dot) layer, and for the hole-transporting medium representing the electrolyte and its transport, oxidation, and reduction processes. Alternatively this layer represents precisely the hole transport material found in all solid-state versions of this cell type (see Section 7.4).

Table 7.1 Material Parameters Used in DSSC Numerical Simulations

Parameter	Network semiconductor	Dye or quantum-dot layer	Hole-transport medium
Length	100 nm	10 nm	1000 nm
Band gap	$E_G = 3.20\,eV$	$|HOMO - LUMO| = 1.7\,eV$	$E_G = 0.88\,eV$
Electron affinity	$\chi = 4.00\,eV$	$\chi = 3.80\,eV$	$\chi = 3.80\,eV$
Absorption	No absorption	Adjusted to produce el $= 18.9\,mA/cm^2$	No absorption
Doping density	$N_D = 1.0 \times 10^{12}\,cm^{-3}$	None	$N_A = 1.0 \times 10^{18}\,cm^{-3}$
Front-contact workfunction and surface recombination speeds	$\phi_W = 4.65\,eV$ $S_n = 1 \times 10^7\,cm/s$ $S_p = 0.0\,cm/s$	N.A.	N.A.
Back-contact workfunction and surface recombination speeds	N.A.	N.A.	$\phi_W = 4.65\,eV$ $S_n = 0.0\,cm/s$ $S_p = 1 \times 10^7\,cm/s$
Electron and hole mobilities	$\mu_n = 1350\,cm^2/vs$ $\mu_p = 450\,cm^2/vs$	$\mu_n = 1350 \times 10^{-3}\,cm^2/vs$ $\mu_p = 450 \times 10^{-4}\,cm^2/vs$	$\mu_n = 1350\,cm^2/vs$ $\mu_p = 450 \times 10^{-4}\,cm^2/vs$
Band effective densities of states	$N_C = 2.8 \times 10^{19}\,cm^{-3}$ $N_V = 1.0 \times 10^{19}\,cm^{-3}$	$N_C = 2.8 \times 10^{19}\,cm^{-3}$ $N_V = 1.0 \times 10^{19}\,cm^{-3}$	$N_C = 2.8 \times 10^{19}\,cm^{-3}$ $N_V = 1.0 \times 10^{19}\,cm^{-3}$
Defect properties	No defects	No defects	No defects

As can be seen from the table, the network semiconductor is taken to be highly resistive in the dark. The table also shows that simulations assume there is no carrier recombination in any of the material layers. Interface Paths 1 and 2 are taken to be the only carrier-loss mechanisms present. The Path 1 interface carrier-loss mechanism of Figure 7.2 is assumed to dominate in the numerical simulations. Path 1 is modeled with $\mathscr{R}^R = \gamma(n_s p_D - n_{s0} p_{D0})$ which is the same type of expression that was used in a corresponding situation in Section 5.3. Here, n_s and p_D are the semiconductor electron, and dye molecule hole, photo-generated populations on their respective sides of the interface and the subscript 0 denotes TE. The computer generated dark and light J-V plots are presented in Figure 7.4 for the material parameters of Table 7.1 and $\gamma = 10^{-11}\,\mathrm{cm^3/s}$. This device gives $J_{sc} = 11.9\,\mathrm{mA/cm^2}$, $V_{oc} = 1.27\,\mathrm{V}$, FF $= 0.84$, and $\eta = 12.6\%$. Figure 7.5 gives the corresponding dark and light J-V output for $\gamma = 10^{-9}\,\mathrm{cm^3/s}$. This device gives $J_{sc} = 11.9\,\mathrm{mA/cm^2}$, $V_{oc} = 1.15\,\mathrm{V}$, FF $= 0.76$, and $\eta = 10.4\%$. Figure 7.6 gives the dark and light J-V results for $\gamma = 10^{-7}\,\mathrm{cm^3/s}$. This device's performance parameters are $J_{sc} = 11.9\,\mathrm{mA/cm^2}$, $V_{oc} = 1.03\,\mathrm{V}$, FF $= 0.58$, and $\eta = 7.14\%$. Obviously, the fill factor and open-circuit voltage decrease with increasing strength of the interface carrier-loss path.

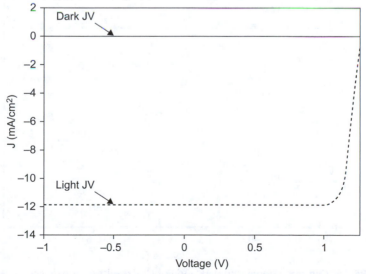

FIGURE 7.4 Computer generated dark and light J-V behavior for the device of Figure 7.3 with $\gamma = 10^{-11}\,\mathrm{cm^3/s}$. The dark J-V is essentially that of a large resistor due to the low dark conductivity used in this modeling for the network semiconductor. As is our practice, current is negative in the power quadrant for these plots.

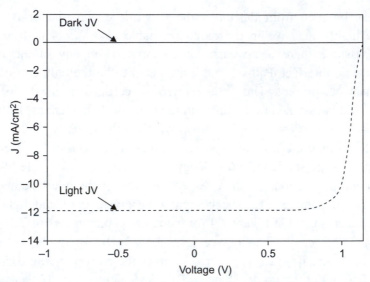

FIGURE 7.5 Computer generated dark and light J-V results for the device of Figure 7.3 with $\gamma = 10^{-9}$ cm^3/s. The dark J-V is essentially that of a large resistor due to the low dark conductivity used in this modeling for the network semiconductor.

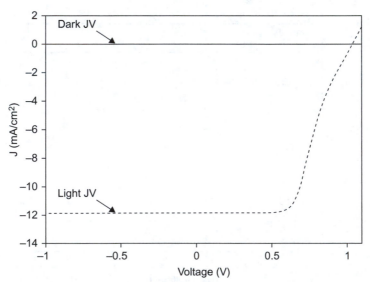

FIGURE 7.6 Computer generated dark and light J-V results for the device of Figure 7.3 with $\gamma = 10^{-7}$ cm^3/s. The dark J-V is essentially that of a large resistor due to the low dark conductivity used in this modeling for the network semiconductor.

The band diagram in Figure 7.3 applies to all three γ values. The electron total current density and drift and diffusion components at short circuit shown in Figure 7.7 were determined for the $\gamma = 10^{-11}$ cm^3/s case, but the behavior it shows is typical of all three cases. Interestingly, Figure 7.7 makes it clear that the electrons are moving away from the interface (as particles) due to diffusion in this transparent network semiconductor; however, this electron-particle motion to the cathode must overcome a significant, oppositely oriented drift component. The corresponding hole currents are not plotted, since they are so small. Figure 7.8 shows the electrostatic field across these devices at TE (zero for all three) and for the $\gamma = 10^{-11}$ cm^3/s device at open circuit. While differing somewhat with the specific γ value, this overall behavior of the electric field that exists at open circuit is typical of all three devices. As can be noted, the open-circuit condition requires the development of the strong electric field shown at the interface region in order to have carrier loss exactly equal generation. This field must develop due to a build-up of electrons at the transparent semiconductor–dye interface and a corresponding build-up of positive charge at the dye-electrolyte interface. In our modeling, the latter is represented by the interface at $x = 112$ nm in Figure 7.3. As is true for all solar cells, and, as was stressed in Section 2.4, the integral $V = -\int_{\text{structure}} [\xi(x) - \xi_0(x)]dx$ evaluated from the left to the right

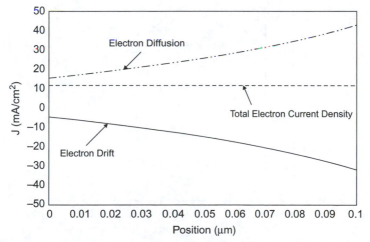

FIGURE 7.7 The electron total current density and its drift and diffusion components in the network semiconductor at short circuit. Shown for the case with $\gamma = 10^{-11}$ cm^3/s. Positive current density is in the direction of the positive x-axis of Figure 7.3.

contact in Figure 7.3 under open-circuit conditions is V_{oc}. This integral must equal the shift up of the Fermi level in the left contact with respect to the Fermi-level position in the right contact at open circuit for the configuration of Figure 7.3. For a DSSC device, this reduces to evaluating

$$V_{oc} = - \int_{structure} [\xi(x)]dx \tag{7.2}$$

at open circuit since the electrostatic field in TE is $\xi_0(x) \equiv 0$; i.e., there is no built-in electrostatic field in these cells. The minus sign is required in Eq. 7.2 due to the selection of the positive x-direction in Figures 7.2 and 7.3 and the fact that the left contact is the cathode.

This analysis shows that a DSSC device operates as expected for a structure in which the "charge separation engine" breaking symmetry is effective forces. Overall, its behavior is very similar to that seen for excitonic HJ cells.

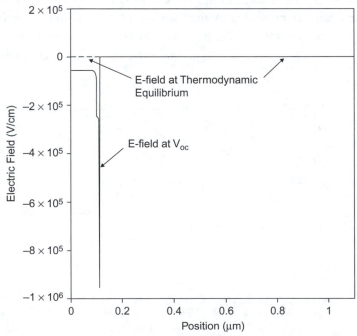

FIGURE 7.8 The electrostatic field across these devices at TE (zero for all three) and for the $\gamma = 10^{-11}$ cm^3/s device at open circuit. The positive field direction is in that of the positive x-axis of Figure 7.3.

7.4 SOME DSSC CONFIGURATIONS

Since the appearance of the dye-sensitized solar cell based on a liquid electrolyte as the hole-transporting medium, there has been activity aimed at replacing the liquid electrolyte with a solid-state hole-transporting medium.[11,12,18–19] The impetus for this effort is the increased practicality of an all-solid-state device and the avoidance of chemical irreversibility originating from ionic discharging and the formation of active species.[19] The dye-sensitized solid-state solar cell (DSSSC) configuration has the band picture seen in Figure 7.3, the diagram we used in our numerical modeling; i.e., it has the network semiconductor coated with a dye or quantum-dot layer, and a solid-state hole-conducting semiconductor. The details of the electron affinity and band gap of the hole-conducting semiconductor will obviously vary with the solid-state hole-conducting

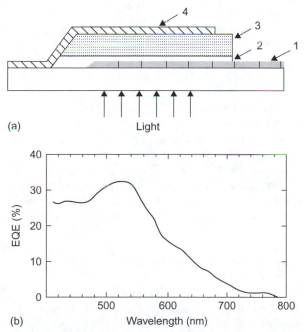

FIGURE 7.9 (a) DSSSC configuration: Material 1 is a transparent contact, material 2 is a transparent planar hole-blocking coating on the transparent contact (for preventing shorting), material 3 is the dye- or quantum-dot-coated nano-structured network with its incorporated hole-transport medium, and material 4 is the back electrode. (b) EQE (expressed in percent) for a DSSSC device with the organic hole conductor OMeTAD as the hole-transport medium. (From Ref. 18, with permission.)

material utilized. An example of the physical configuration of a DSSSC is seen in Figure 7.9a. Some authors have referred to this version of the DSSC device as a heterojunction. We refrain from that classification and include DSSSC devices here, since the definition of a heterojunction solar cell used in this text requires that at least one of the junction-forming semiconductors be both an absorber and transport layer. The dye in a DSSC or DSSSC does not play a role in transport. The EQE seen in the example of Figure 7.9b closely follows the absorption spectrum of the dye used, which confirms that the dye is the absorber.

REFERENCES

1. B. O'Reagan, M. Grätzel, A low-cost, high-efficiency solar cell based on dye-sensitized colloidal TiO_2 films, Nature 353 (1991) 737.

2. Y. Chiba, A. Islam, Y. Watanabe, R. Komiya, N. Koide, L.Y. Han, Dye-sensitized solar cells with conversion efficiency of 11.1%, Jpn. J. Appl. Phys. 45 (2006) L638. Part 2.

3. S.L. Li, K. Jiang, K.F. Shao, L.M. Yang, Novel organic dyes for efficient dye-sensitized solar cells, Chem. Commun. (2006) 2792.

4. S.C.R. Yanagida, Recent research progress of dye-sensitized solar cells in Japan, Chim. 9 (2006) 597.

5. L. Schmidt-Mende, U. Bach, R. Humphry-Baker, T. Horiuchi, H. Miura, S. Ito, S. Uchida, M. Grätzel, Organic dye for highly efficient solid-state dye-sensitized solar cells, Adv. Mater. 17 (2005) 813.

6. A. Zaban, O.I. Micic, B.A. Gregg, A.J. Nozik, Photosensitization of nanoporous TiO_2 electrodes with InP quantum dots, Langmuir 14 (1998) 3153.

7. Q. Shen, D. Arae, T. Toyoda, Photosensitization of nanostructured TiO_2 with CdSe quantum dots: effects of microstructure and electron transport in TiO_2 substrates, J. Photochem. Photobiol. A 164 (2004) 75.

8. G. Kumara, K. Tennakone, I.R.M. Kottegoda, P.K.M. Bandaranayake, A. Konno, M. Okuya, S. Kaneko, K. Murakami, Efficient dye-sensitized photoelectrochemical cells made from nanocrystalline tin(IV) oxide–zinc oxide composite films, Semicond. Sci. Technol. 18 (2003) 312.

9. K. Keis, E. Magnusson, H. Lindstrom, S.E. Lindquist, A. Hagfeldt, A 5% efficient photoelectrochemical solar cell based on nanostructured ZnO electrodes, Sol. Energy Mater. Sol. Cells 73 (2002) 51.

10. W. Kubo, S. Kambe, S. Nakade, T. Kitamura, K. Hanabusa, Y. Wada, S. Yanagida, Photocurrent-determining processes in quasi-solid-state dye-sensitized solar cells using ionic gel electrolytes, J. Phys. Chem. B 107 (2003) 4374.

11. B. O'Regan, D.T. Schwartz, Large enhancement in photocurrent efficiency caused by UV illumination of the dye-sensitized heterojunction TiO_2/RULL'NCS/CuSCN: initiation and potential mechanisms, Chem. 10 (1998) 1501.

12. A. Konno, G.R.A. Kumara, R. Hata, K. Tennakone, Effect of imidazolium salts on the performance of solid-state dye-sensitized photovoltaic cell using copper iodide as a hole collector, Electro-chemistry 70 (2002) 432.

13. L.M. Peter, Characterization and modeling of dye-sensitized solar cells, J. Phys. Chem. C 111 (2007) 6601.

14. S.J. Fonash, S. Ashok, An additional source of photovoltage in photoconductive materials, Appl. Phys. Lett. 35 (1979) 535; S.J. Fonash, Photovoltaic Devices, CRC Critical Reviews Solid State Mater 9, 107 (1980).

15. B.A. Gregg, Excitonic solar cells, J. Phys. Chem. B 107 (2003) 4688.

16. J. Cuiffi, T. Benanti, W.J. Nam, S. Fonash, Open circuit voltage behavior of bulk hererojunction solar cells, Appl. Phys. Lett. 9 (2009).

17. U. Bach, D. Lupo, P. Comte, J.E. Moser, F. Weissortel, J. Salbeck, H. Spreitzer, M. Grätzel, Solid-state dye-sensitized mesoporous TiO_2 solar cells with high photon-to-electron conversion efficiencies, Nature 395 (1998) 583.

18. G.R.A. Kumara, A. Konno, K. Shiratsuchi, J. Tsukahara, K. Tennakone, Dye-sensitized solid-state solar cells: use of crystal growth inhibitors for deposition of the hole collector, Chem. Mater. 14 (2002) 954.

19. K. Tennakone, G.K.R. Senadeera, D.B.R.A. De Silva, I.R.M. Kottegoda, Highly stable dye-sensitized solid-state solar cell with the semiconductor $4CuBr\ 3S(C_4H_9)_2$ as the hole collector, Appl. Phys. Lett. 77 (2000) 2367.

The Absorption Coefficient

A few of the terms used in photovoltaics can have more than one definition. One such situation, and a very important one, is the definition of "absorption coefficient"; one definition uses the natural log, while another uses log to the base 10.

If illumination of wavelength λ and intensity $I(\lambda)$ (photons per cm^2 per second) impinges on a material of thickness d, some of the light is reflected $R(\lambda)$ and some $T(\lambda)$ emerges from the other side of the material at d. The fluxes I, R, and T are related by the simple expression

$$I(\lambda) = T(\lambda) + R(\lambda) + A(\lambda) \qquad (A.1)$$

where $A(\lambda)$ is the absorption taking place for this wavelength. When the Beer-Lambert law is applicable,[1] then the relationship between T and I-R can be expressed as

$$T(\lambda) = [I(\lambda) - R(\lambda)][\exp(-\alpha(\lambda)d)] \qquad (A.2)$$

Where $\alpha(\lambda)$ is the material's absorption coefficient at the wavelength λ. This means that

$$A(\lambda) = [I - R][1 - \exp(-\alpha(\lambda)d] \qquad (A.3)$$

DOI: 10.1016/B978-0-12-374774-7.00009-1

Going back to Eq. A.2, we see that it can be rearranged to read

$$\frac{T(\lambda)}{I(\lambda) - R(\lambda)} = [\exp(-\alpha(\lambda)d)] \tag{A.4}$$

This is a convenient form since minus the natural logarithm ($-\ln$) of Eq. A.4 is defined as the absorbance A_{abs} for the wavelength λ; i.e.,

$$A_{abs}(\lambda) = -\ln\left[\frac{T(\lambda)}{I(\lambda) - R(\lambda)}\right] \tag{A.5}$$

As can be seen from Eqs. A.4 and A.5, this last expression is very useful, since the absorption coefficient $\alpha(\lambda)$ at wavelength λ can be extracted from it by noting

$$\alpha(\lambda) = \frac{A_{abs}(\lambda)}{d} \tag{A.6}$$

The definition of absorption coefficient given by Eq. A.6 is used throughout this text.

The complication that arises with the absorption coefficient is that a different definition for absorbance can be found in the literature. This other definition is given by

$$A_{abs}(\lambda) = -\log_{10}\left[\frac{T(\lambda)}{I(\lambda) - R(\lambda)}\right] \tag{A.7}$$

Equation A.7 is then used to determine $\alpha(\lambda)$ through Eq. A.6. The impact of this is that the absorption coefficient deduced from T, R, and I using Eq. A.7 must be multiplied by ln 10 to convert that data to the absorption coefficient deduced from Eq. A.5 and used in Eq. A.3.

Obviously, when looking at absorption coefficient data, one must be careful in determining which $\alpha(\lambda)$ is being presented.

REFERENCE

1. J.D.J. Ingle, S.R. Crouch, Spectrochemical Analysis, Prentice Hall, New Jersey, 1988.

Radiative Recombination

In this text, "radiative recombination" refers to the net band-to-band recombination traffic resulting from the $path_1$ and $path_2$ traffic seen in Figure B.1. This type of recombination involves photons and perhaps phonons for energy emission ($path_1$) and assimilation ($path_2$) but there is no involvement of energy-gap states. $Path_1$ (electrons or holes recombining per time per volume) requires the presence of an electron concentration n in the conduction band and the presence of a hole concentration p in the valence band and is proportional to their product; i.e.,

$$path_1 = \kappa_R \, np \tag{B.1}$$

This np product–dependence of radiative recombination is the hallmark of what is termed a bimolecular process.

$Path_2$ of Figure B.1 is the opposite of $path_1$ and results in the promoting of electrons from the valence band to the conduction band. It is expected to depend only on temperature; i.e.,

$$path_2 = g_{th}^R(T) \tag{B.2}$$

where $g_{th}^R(T)$ gives the number of electrons in the conduction band and holes in the valence band generated per time per volume. Utilizing

DOI: 10.1016/B978-0-12-374774-7.00010-8

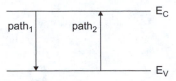

FIGURE B.1 Band-to-band traffic.

Eqs. B.1 and B.2 allows the general expression for the net radiative recombination traffic $\mathcal{R}^R$ to be written as

$$\mathcal{R}^R = \kappa_R np - g_{th}^R(T) \tag{B.3}$$

In thermodynamic equilibrium, $\mathcal{R}^R$ is zero, according to the principle of detailed balance, which allows us to deduce that

$$g_{th}^R(T) = \kappa_R n_0 p_0 \tag{B.4}$$

Using this fact in Eq. B.3 finally gives

$$\mathcal{R}^R = \left[\frac{g_{th}^R}{n_i^2}\right](pn - n_i^2) \tag{B.5}$$

where $n_i^2 = n_0 p_0$ is the square of intrinsic number density and the subscript zero refers to thermodynamic equilibrium (TE) values. In all of these expressions $g_{th}^R(T)$ is assumed to be a known quantity. The dimensions of $\mathcal{R}^R$ are the net number (of free holes or equivalently of free electrons) annihilated per volume per time. When a solar cell is developing power, it is out of thermodynamic equilibrium, so Eq. B.5 applies, but often it simplifies to

$$\mathcal{R}^R = \left[\frac{g_{th}^R}{n_i^2}\right](pn) \tag{B.6}$$

since $np > n_i^2$. The electron quasi-Fermi level E_{Fn} and the hole quasi-Fermi level E_{Fp} can be introduced into Eq. B.6 through Eqs. 2.17 and 2.26; i.e., through

$$n = N_c \exp\left[-\frac{(E_c - E_{Fn})}{kT}\right] \tag{B.7}$$

$$p = N_V \exp\left[-\frac{(E_{Fp} - E_V)}{kT}\right] \quad (B.8)$$

With these expressions, Eq. B.6 becomes

$$\mathscr{R}^R = g_{th}^R \exp\left[\frac{(E_{Fn} - E_{Fp})}{kT}\right] \quad (B.9)$$

Equation B.9 makes it clear that splitting the quasi-Fermi levels apart increases radiative recombination. This helps to give a useful, physical "feel" to quasi-Fermi levels.

Shockley-Read-Hall (Gap-state–assisted) Recombination

This appendix derives the Shockley-Read-Hall (S-R-H) gap-state–assisted recombination-generation mathematical model.[1,2] The derivation begins with Figure C.1, which depicts gap states at some energy E_T. These states are being used by carriers to provide a recombination-generation path between the conduction band and the valence band. We write the conduction-band electron traffic (electrons per volume per time) following path_1 to these states of number N_T per volume as $\text{path}_1 = n\tilde{p}_T v \sigma_n$, where v is the electron thermal velocity, σ_n is the capture cross-section of these gap states for electrons (its magnitude depends on whether the N_T states per volume at energy E_T are charged or not), $\tilde{p}_T$ is the number per volume of these states that are empty (can accept

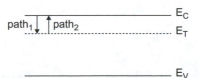

FIGURE C.1 Gap state–conduction band traffic. Shown are the electron capture process path_1 from the conduction band and the electron emission process path_2 to the conduction band from the states at energy level E_T.

DOI: 10.1016/B978-0-12-374774-7.00011-X

an electron), and n is the usual conduction band population per volume. We write the traffic (electrons per volume per time) following path_2 from these states to the conduction band as $\text{path}_2 = \kappa_n \tilde{n}_T$ where κ_n characterizes the emission of electrons per time from the states N_T to the conduction band and $\tilde{n}_T$ is the number of these states per volume containing an electron. We expect that κ_n will only depend on temperature.

The principle of detailed balance says that, in thermodynamic equilibrium, $\text{path}_1 = \text{path}_2$. Therefore, we find that

$$\kappa_n \tilde{n}_{T0} = n_0 \tilde{p}_{T0} v\sigma_n$$

where the 0 subscript indicates thermodynamic equilibrium values. Doing some rearranging gives:

$$\kappa_n = \frac{n_0 \tilde{p}_{T0}}{\tilde{n}_{T0}} v\sigma_n = v\sigma_n n_1$$

where the definition $n_1 \equiv n_0 \tilde{p}_{T0} / \tilde{n}_{T0}$ has been used. Using the expressions from Fermi-Dirac statistics (see Appendix D) for $\tilde{p}_{T0}$ and $\tilde{n}_{T0}$ as well as the Boltzmann approximation (assumes $E_C - E_F \geq kT$ where k is Boltzmann's constant) for n_0, results in our being able to write n_1 as

$$n_1 = N_C e^{-(E_C - E_F)/kT} \left[\frac{1 + e^{(E_T - E_F)kT}}{1 + e^{-(E_T - E_F)/kT}} \right]$$

or, finally, as

$$n_1 = N_C e^{-(E_C - E_T)/kT}$$

This algebra allows us to write an expression for the net electron traffic $\mathscr{R}_C$ per unit time per volume out of the conduction band into these states at energy level E_T, which occurs when the material is out of thermodynamic equilibrium. This expression is

$$\mathscr{R}_C = \text{path}_1 - \text{path}_2 = v\sigma_n \left(n\tilde{p}_T - n_1 \tilde{n}_T \right) \qquad (C.1)$$

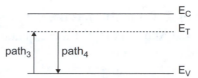

FIGURE C.2 Gap state–valence band traffic. Shown are the hole-capture process path$_3$ from the valence band and the hole-emission process path$_4$ to the valence band from the states at energy level E$_T$.

Our development of expressions for the traffic between the gap states at energy level E$_T$ and the valence band follows that leading to Eq. C.1, except we frame everything in the context of holes. The path$_3$ seen in Figure C.2 supports the traffic of holes going from the valence band to these gap states. This path can be modeled as path$_3$ = $p\tilde{n}_T v\sigma_p$ where v is the hole thermal velocity (assumed to equal electron thermal—any thermal velocity differences between carriers can be taken up in the products $v\sigma_p$ or $v\sigma_n$ by adjusting the capture cross-sections). This expression also uses σ_p, which is the capture cross-section of these localized states for holes (whose magnitude depends on the charge configuration of these gap states), $\tilde{n}_T$, which is the number per volume of these states occupied by an electron (notation first introduced above), and p, which is the hole population of the valence band per volume. Following what was done for the electron traffic from these states to the conduction band, the hole traffic per time per volume from the states at energy level E$_T$ to the valence band is given by path$_4$ = $\kappa_p \tilde{p}_T$, where κ_p characterizes the emission of holes per time from the states N$_T$ to the valence band and, as noted above, $\tilde{p}_T$ is the number of these states per volume containing a hole (notation also first introduced above). Just as with κ_n, we expect that κ_p will only depend on temperature.

Once again, we invoke the principle of detailed balance, which says that, in thermodynamic equilibrium, path$_3$ = path$_4$, and which leads us to find:

$$\kappa_p \tilde{p}_{T0} = p_0 \tilde{n}_{T0} v\sigma_p$$

where, as before, the 0 subscript indicates thermodynamic equilibrium values. This may be rearranged to give

$$\kappa_p = \frac{p_0 \tilde{n}_{T0}}{\tilde{p}_{T0}} v\sigma_p = v\sigma_p p_1$$

where the definition $P_1 \equiv p_0 \tilde{n}_{T0}/\tilde{p}_{T0}$ has been utilized. Using the expressions from statistics (see Appendix D) for these quantities results in

$$P_1 = N_v e^{-(E_T - E_v)/kT}$$

Putting these expressions together allows the net hole traffic $\mathscr{R}_V$ per volume per time out of the valence band into these gap states in question to be written as

$$\mathscr{R}_V = \text{path}_3 - \text{path}_4 = v\sigma_p \left(p\tilde{n}_T - P_1\tilde{p}_T \right) \qquad (C.2)$$

We can eliminate $\tilde{n}_T$ and $\tilde{p}_T$ from Eqs. C.1 and C.2 by first noting that we have two equations involving these two unknowns. The first is

$$\tilde{n}_T + \tilde{p}_T = N_T \qquad (C.3a)$$

This statement is always true. The second comes from $\mathscr{R}_C = \mathscr{R}_V$; i.e., equating Eqs. C.1 and C.2, which is true only in steady state. However, that is good enough, because we are interested in steady-state solar cell operation. Its use gives

$$v\sigma_n \left(n\tilde{p}_T - n_1\tilde{n}_T \right) = v\sigma_p \left(p\tilde{n}_T - P_1\tilde{p}_T \right) \qquad (C.3b)$$

Solving C.3a and C.3b for $\tilde{n}_T$ and $\tilde{p}_T$ results in the following:

$$\tilde{n}_T = \left[\frac{N_T \left(\sigma_p P_1 + \sigma_n n \right)}{\sigma_p \left(p + P_1 \right) + \sigma_n \left(n + n_1 \right)} \right] \qquad (C.4a)$$

and

$$\tilde{p}_T = \frac{N_T \left(\sigma_n n_1 + \sigma_p p \right)}{\sigma_p \left(p + P_1 \right) + \sigma_n \left(n + n_1 \right)} \qquad (C.4b)$$

Using these steady-state expressions for $\tilde{n}_T$ and $\tilde{p}_T$ in the expression for $\mathscr{R}_C$ or $\mathscr{R}_V$ gives

$$\mathscr{R}^L = \frac{v\sigma_n \sigma_p N_T \left(np - n_i^2 \right)}{\sigma_p \left(p + P_1 \right) + \sigma_n \left(n + n_1 \right)} \qquad (C.5)$$

Where $\mathscr{R}^L = \mathscr{R}_C = \mathscr{R}_V$. Equation C.5 is the general steady-state expression for gap-state–assisted recombination between the conduction and valence bands. Since it is steady state, the mechanism requires that every conduction-band electron annihilation causes a corresponding valence-band hole annihilation.

If the electron quasi-Fermi level E_{Fn} and the hole quasi-Fermi level E_{Fp} are introduced into Eq. C.5 through Eqs. 2.17 and 2.26 (i.e., through the

$$n = N_c \exp\left[-\frac{(E_c - E_{Fn})}{kT}\right] \tag{C.6}$$

$$p = N_V \exp\left[-\frac{(E_{Fp} - E_V)}{kT}\right] \tag{C.7}$$

expressions), and if $np > n_i^2$, Eq. C.5 becomes

$$\mathscr{R}^L = \frac{v\sigma_n\sigma_p N_T}{\sigma_p(p + p_1) + \sigma_n(n + n_1)} \exp\left[\frac{(E_{Fn} - E_{Fp})}{kT}\right] \tag{C.8}$$

Expression C.8 has p and n (and therefore E_{Fn} and E_{Fp}) in its denominator but it certainly suggests that splitting the quasi-Fermi levels apart increases band-to-band recombination.

It is very important to realize that the expressions we have developed allow us to determine the relationship between $\tilde{p}_T$, which counts the gap states per volume at energy E_T that are missing an electron, and p_T. The latter is used in the expression ep_T for the contribution of the gap states at E_T to the positive charge in the space charge (charge density) term $e\left[p - n + \sum p_T - \sum n_T + N_D^+ - N_A^-\right]$ of Eq. 2.48e. Similarly, these expressions allow us to determine the difference between $\tilde{n}_T$, which counts the single electron states per volume at energy E_T that have acquired an electron, and n_T. This quantity n_T appears in the expression $-en_T$ for the contribution of the gap states at E_T to the negative charge in the space charge term $e\left[p - n + \sum p_T - \sum n_T + N_D^+ - N_A^-\right]$. These relationships can be established by noting that the quantity $\tilde{p}_T$

contributes to p_T only if the states shown in Figure C.1 (and in Fig. C.2) at energy E_T are donor-like. In this case

$$p_T = \tilde{p}_T \tag{C.9a}$$

and

$$n_T = 0 \tag{C.9b}$$

Correspondingly, $\tilde{n}_T$ contributes to n_T only if the states shown in Figure C.1 (and in Fig. C.2) at energy E_T are acceptor-like. In this case

$$n_T = \tilde{n}_T \tag{C.10a}$$

and

$$p_T = 0 \tag{C.10b}$$

Obviously, Eqs. C.9 and C.10 provide the machinery for calculating the space charge density contributions for these states at E_T, once it is determined if they are acceptor-like or donor-like.

The equations we have just developed also help us to see the difference between a recombination center and a trap. As an example, we take the localized states at E_T to have a large capture cross-section for holes but, to make our point by using the extreme, a zero capture cross-section for electrons. Under these conditions, Eq. C.5 shows that such states do not act as a conduit for recombination; i.e., $\mathscr{R}^L = 0$. However, under these conditions Eqs. C.4a and C.4b show the occupancy of these states at E_T is certainly not zero; i.e., they have occupancies given by

$$\tilde{n}_T = \left[\frac{N_T(p_1)}{(p + p_1)} \right]$$

and

$$\tilde{p}_T = \frac{N_T(p)}{(p + p_1)}$$

If $p > p_1$, these expressions reduce to $\tilde{n}_T = 0$ and $\tilde{p}_T = N_T$.

It is important to note that Eqs. C.5 and C.8 are written for N_T localized states at energy level E_T in the band gap. The total gap state assisted recombination $\mathscr{R}^L$ must be obtained by summing over all the gap states. Keeping this in mind, Eq. C.5, for example, can be generalized to

$$\mathscr{R}^L = \sum_i \frac{v\sigma_n^i\sigma_p^i N_T^i\left(np - n_i^2\right)}{\sigma_p^i\left(p + p_1^i\right) + \sigma_n^i\left(n + n_1^i\right)} \qquad (C.11a)$$

This statement can also be expressed as an integral over the band gap:

$$\mathscr{R}^L = \int_E \frac{v\sigma_n\sigma_p N_T\left(np - n_i^2\right)}{\sigma_p(p + p_1) + \sigma_n(n + n_1)}\, dE \qquad (C.11b)$$

In this last version, the units of N_T are now states per energy per volume.

It is interesting to determine which states among a distribution of localized gap states are actually responsible for most of the recombination. This is very straightforwardly done numerically by examining the contributions by different states to Eq. C.11a or C.11b, as appropriate. It can be done analytically if we make the assumption that the states over which the summation or integration takes place in Eqs. C.11 have a capture cross-section ratio σ_n/σ_p that does not vary with E_T.[3] The analysis, first done in reference 3, shows that above a so-called electron demarcation[†] level E_{Tn}, the states are essentially empty because the electrons are emitted so easily back into the conduction band. Correspondingly there is a hole demarcation[†] level E_{Tp} below which the states are essentially full because the holes are emitted so easily back into the valence band. The states carrying the Shockley-Read-Hall defect-assisted recombination traffic are therefore those that lie between these demarcation levels. This is the case because these states contain both the holes and electrons

---

[†]The concept of demarcation levels was introduced by Taylor and Simmons[3] who referred to them as "quasi-Fermi levels for traps."

needed for recombination to take place. In fact, the analysis gives the electron probability of occupancy for these states between the demarcation levels to be $Rn/(Rn + p)$ where $R \equiv \sigma_n/\sigma_p$. The demarcation levels are given by[3]

$$E_{Tn} = E_F + kT \ln\left[\frac{\sigma_p p + \sigma_n n}{\sigma_n n_0}\right] \tag{C.12}$$

and

$$E_{Tp} = E_F - kT \ln\left[\frac{\sigma_p p + \sigma_n n}{\sigma_p p_0}\right] \tag{C.13}$$

Here E_F is the location of the Fermi level at TE in the energy gap.

REFERENCES

1.   W. Shockley, W.T. Read, Phys. Rev. 87 (1952) 835.
2.   R.N. Hall, Phys. Rev. 87 (1952) 387.
3.   G.W. Taylor, J.G. Simmons, J. of Non-Crystalline Solids 8-10 (1972) 940.

Conduction- and Valence-band Transport

We now turn our attention to the drift-diffusion formalism for modeling transport in semiconductors. In general, charge transport takes place in both the conduction-band electron and in the valence band in a semiconductor. The word semiconductor is used here and everywhere in this text to include organic and inorganic materials. The terms conduction band and valence band are used to include the corresponding molecular orbitals of an organic material.

We begin by noting that n, the number of electrons per volume in the conduction band, and p, the number of holes per volume in the valence band, are, in general, given by[1-3]

$$n = \int_{E_c}^{\infty} \frac{g_e^c(E)dE}{[1 + \exp(E - E_{Fn})/kT_n]} \tag{D.1}$$

and

$$p = \int_{-\infty}^{E_v} \frac{g_e^v(E)dE}{[1 + \exp(E_{Fp} - E)/kT_p]} \tag{D.2}$$

DOI: 10.1016/B978-0-12-374774-7.00012-1

Here $g_e^c(E)$ is the single electron density of states per volume $g_e(E)$ for the conduction band and $g_e^v(E)$ is the single electron density of states per volume $g_e(E)$ for the valence band (The density of states concept $g_e(E)$ is introduced in Section 2.2.3.1). The quantity E_{Fn} is the electron quasi-Fermi level and E_{Fp} is the hole quasi-Fermi level. The quantity T_n is the electron effective temperature and T_p is the hole effective temperature. The use of a quasi-Fermi level and an effective temperature to describe the carriers in a band assumes that the carriers in that band have equilibrated among themselves so that these concepts have meaning.[1] In general, $g_e^c(E)$, $g_e^v(E)$, E_{Fn}, E_{Fp}, T_n, and T_p can all be functions of position x in these expressions.

These quasi-Fermi levels and effective temperatures were invented so that the thermodynamic equilibrium (TE) expressions for n and p given by Fermi-Dirac statistics[1–3]

$$n = \int_{E_c}^{\infty} \frac{g_e^c(E)dE}{[1 + \exp(E - E_F)/kT]} \tag{D.3}$$

$$p = \int_{-\infty}^{E_v} \frac{g_e^v(E)dE}{[1 + \exp(E_F - E)/kT]} \tag{D.4}$$

can still be utilized, with the appropriate substitutions for the Fermi level and temperature when the material system is not in TE. The quasi-Fermi levels collapse to the Fermi level E_F and the effective temperatures to the temperature T when TE is achieved. While $g_e^c(E)$ and $g_e^v(E)$ can still be functions of position, E_F and T cannot vary with position in TE in Eqs. D.3 and D.4.

In the case of inorganic crystalline semiconductors, the density of states near their respective band edges can be shown rigorously to be parabolic in energy[3]; i.e.,

$$g_e^c(E) = A_C(E - E_C)^{1/2} \tag{D.5}$$

and

$$g_e^v(E) = A_V(E_V - E)^{1/2} \tag{D.6}$$

Here the quantities A_C and A_V are material properties and therefore will vary with position, if the material composition varies with position. These expressions are very useful in evaluating Eqs. D.1–D.4 since most of the carriers will be near their respective band edges. In the case of organic semiconductors the functions $g_e^c(E)$ and $g_e^v(E)$ are expected to have Gaussian shapes.[4]

Using Eqs. D.5 and D.6 in Eqs. D.1 and D.2 gives

$$n = \int_{E_C}^{\infty} \frac{A_C(E - E_C)^{1/2}dE}{[1 + \exp(E - E_{Fn})/kT_n]} \tag{D.7}$$

and

$$p = \int_{-\infty}^{E_V} \frac{A_V(E_V - E)^{1/2}dE}{[1 + \exp(E_{Fp} - E)/kT_p]} \tag{D.8}$$

These expressions can now be integrated analytically if it is possible to use Boltzmann approximations in place of the Fermi function terms. To be specific, the Boltzmann approximation $\exp - (E - E_{Fn})/kT_n$ can be used for $[1 + \exp(E - E_{Fn})/kT_n]^{-1}$ if $(E - E_F) > kT_n$ is true for all the energies E in the conduction band and the Boltzmann approximation $\exp - (E_{Fp} - E)/kT_p$ can be used for $[1 + \exp(E_{Fp} - E)/kT_p]^{-1}$, if $(E_F - E) > kT_p$ is true for all the energies in the valence band. Assuming these conditions can be met, use of these Boltzmann approximations allows Eqs. D.7 and D.8 to integrate analytically to

$$n = N_C \exp - (E_C - E_{Fn})/kT_n \tag{D.9}$$

and

$$p = N_V exp - (E_{Fp} - E_V)/kT_p \qquad (D.10)$$

where N_C and N_V are temperature-dependent material properties called the conduction- and valence-band effective densities of states, respectively. Since they arise out of A_C and A_V, they will vary with position, if the material composition varies with position. In TE, Eqs. D.9 and D.10 collapse to

$$n = N_C exp - (E_C - E_F)/kT \qquad (D.11)$$

and

$$p = N_V exp - (E_F - E_V)/kT \qquad (D.12)$$

Equations D.9–D.12 can also be written in terms of $V_n \equiv E_C - E_{Fn}$ (or $V_n \equiv E_C - E_F$ in TE) and $V_p \equiv E_{Fp} - E_V$ (or $V_p \equiv E_F - E_V$ in TE). The quantities V_n and E_{Fn} are illustrated for the conduction band in Figure D.1.

The electrons per volume n in the conduction band and holes per volume p in the valence band can carry a conventional electric current. In thermodynamic equilibrium both the electron conventional current density J_n and the hole conventional current density J_p are identically zero. When the materials system is driven out of TE by illumination, voltage, temperature gradients, or some combination of these, neither J_n nor J_p needs be zero.

When a materials system is driven out of TE, then J_n is given by[5–8]

$$J_n = e\mu_n n\, dE_{Fn}/dx - en\mu_n S_n dT_n/dx \qquad (D.13)$$

Here μ_n is the electron mobility and S_n is the Seebeck coefficient for electrons, a negative quantity. The Seebeck coefficient is also called the thermoelectric power. We emphasize that Eq. D.13 is very general and is valid

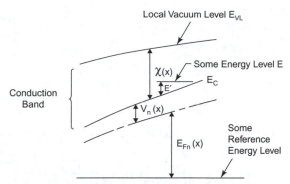

FIGURE D.1 A very general semiconductor conduction band with an electron affinity that is a function of position. Such a situation could arise in an alloy semiconductor whose composition varies with position. The top of the conduction band is termed the local vacuum level. A constant, reference energy is shown.

for current transport by electrons in the conduction band in the presence of electrostatic fields, variable material properties, and a gradient in the effective electron temperature. Equation D.13 demonstrates that, even in this most general situation, the electron current density is simply being driven by a gradient in the electron quasi-Fermi level (electrochemical potential for electrons) and by a gradient in the electron effective temperature.

There are a number of alternative ways of writing Eq. D.13. One particularly useful form replaces dE_{Fn}/dx with an electrostatic field term, an effective force field term, and a concentration gradient term. This is accomplished by noting from Figure D.1 that

$$E_{VL}(x) = \chi(x) + V_n(x) + E_{Fn}(x) \tag{D.14}$$

Substituting this into the expression for J_n given by Eq. D.13, we get

$$J_n = e\mu_n n\left(\xi - \frac{d\chi}{dx} - \frac{dV_n}{dx}\right) - en\mu_n S_n \frac{dT_n}{dx}$$

which, with the aid of Eq. D.9, becomes[†]

[†]By using Eq. D.9 to obtain Eq. D.15, we have limited ourselves to a specific density of states model and to a nondegenerate conduction band; i.e., to the case where $(E - E_F) > kT_n$ is obeyed for the conduction band. More general discussions can be found in references 5–9.

$$J_n = e\mu_n n \left(\xi - \frac{d\chi}{dx} - kT_n \frac{d\ln N_c}{dx} \right) + ekT_n\mu_n \frac{dn}{dx}$$

$$- \left(\frac{eV_n\mu_n n}{T_n} + e\mu_n n S_n \right) \frac{dT_n}{dx} \qquad (D.15)$$

The conduction-band current density J_n can now be expressed as[8,9]

$$J_n = e\mu_n n \left(\xi - \frac{d\chi}{dx} - kT_n \frac{d\ln N_c}{dx} \right)$$

$$+ ekT_n\mu_n \frac{dn}{dx} + e D_n^T \frac{dT_n}{dx} \qquad (D.16)$$

or alternatively as

$$J_n = e\mu_n n \xi + e\mu_n n \xi'_n + e D_n \frac{dn}{dx} + e D_n^T \frac{dT_n}{dx} \qquad (D.17)$$

Here ξ is the electrostatic field and ξ'_n is the effective force field acting on an electron. In these last two expressions for J_n, the electron diffusion coefficient D_n and the electron thermal diffusion coefficient (or electron Soret coefficient) D_n^T have been introduced where

$$D_n = kT_n\mu_n \qquad (D.18)$$

and

$$D_n^T = -\mu_n n(V_n + S_n T_n)/T_n \qquad (D.19)$$

As expressed in Eqs. D.15 and D.16, the electron current density J_n can now be viewed as being driven by a total force[9] $F_e = -e[\xi - d\chi/dx - kT_e d \ln N_c/dx]$ acting on electrons (giving the total drift term), by an electron concentration gradient (diffusion term), and by a temperature gradient (thermal diffusion term). The total force F_e on an electron is seen to be the electrostatic force $-e\xi$ except for those materials that have properties (affinity, density of states) that vary with position.[10,11] Equations D.13 and D.16 or D.17 are equivalent expressions for the current density J_n; however, one sometimes proves to be more convenient than the other when analyzing solar cell structures. Equation D.17 stresses the point that there are two types of drift for electrons that are possible: drift in an electrostatic field and drift in an effective field arising from spatial variations in the electron affinity, density of states, or both.

Turning now to J_p, we note that expressions analogous to Eq. D.13 and Eq. D.16 or D.17 exist for this component of the total current density carried by holes in the valence band. Starting with the electrochemical potential formulation, the formulation analogous to Eq. D.13 is[5–11]

$$J_p = e\mu_p p \frac{dE_{FP}}{dx} - ep\mu_p S_p \frac{dT_p}{dx} \tag{D.20}$$

which gives J_p in terms of the hole quasi-Fermi level E_{Fp} and the hole effective temperature T_p. The hole mobility and the hole Seebeck coefficient appear in this equation. The hole Seebeck coefficient is a positive quantity. For the formulation in terms of the electric field ξ, gradients in the hole affinity $(\chi + E_G)$ and band effective density of states, carrier concentration gradient, and effective hole temperature gradient, the expression is

$$J_p = e\mu_p p \left(\xi - \frac{d(\chi + E_G)}{dx} + kT_p \frac{d\ln N_v}{dx} \right) - ekT_p\mu_p \frac{dp}{dx}$$
$$+ \left(\frac{eV_p\mu_p p}{T_p} - e\mu_p p S_p \right) \frac{dT_p}{dx} \tag{D.21}$$

This expression is obtained by making use of Eq. D.10 and the valence band expression corresponding to Eq. D.14 in Eq. D.20. Recasting Eq. D.21 in terms of the electrostatic field ξ, the effective hole force field ξ'_p, the concentration gradient dp/dx, and the temperature gradient (dT_p/dx) gives[‡5–11]

---

‡Equation D.20 is of general validity but Eq. D.21, uses a specific model for the valence band density of states and it applies only to a nondegenerate valence band, since $p = N_V \exp(-V_p/kT_p)$ has been used. More general formulations of Eq. D.21 may be found in the references.

$$J_p = e\mu_p p\xi + e\mu_p p\xi'_p - eD_p \frac{dp}{dx} - eD_p^T \frac{dT_p}{dx} \qquad (D.22)$$

The hole effective force field ξ'_p used in this equation has the definition

$$\xi'_p = -\frac{d(\chi + E_G)}{dx} + kT_p \frac{d\ln N_v}{dx} \qquad (D.23)$$

We have also used in Eq. D.22 the definition of the hole diffusion coefficient D_p and hole thermal diffusion coefficient (or hole Soret coefficient) D_p^T where

$$D_p = kT_p\mu_p$$

and

$$D_p^T = \mu_p p(S_p T_p - V_p)/T_p$$

Equation D.22 shows that J_p can now be viewed as being driven by a total force [9]

$$F_h = e\left(\xi - \frac{d(\chi + E_G)}{dx} + kT_p \frac{d\ln N_v}{dx}\right)$$

acting on holes (total drift term), by a hole concentration gradient (diffusion term), and by a temperature gradient (thermal diffusion term). The total force F_h is simply the electrostatic force $e\xi$ except for those materials that have properties (affinity, density of states) that vary with position.[9–11] Equation D.22 stresses the point that there are two types of hole drift possible: drift in an electrostatic field and drift in an effective field arising from spatial variations in the hole affinity, density of states, or both.

REFERENCES

1. S. Wang, Fundamentals of Semiconductor Theory and Device Physics, Prentice Hall, Englewood Cliffs, NJ, 1989.

2. S. Fonash, Solar Cell Device Physics, Academic Press, NY, 1981.

3. S. Sze, K.K. Ng, Physics of Semiconductor Devices, third ed., John Wiley & Sons, Hoboken, NJ, 2007.

4. R. Hoffmann, Solids and Surfaces: A Chemist's View of Bonding in Extended Structures, Wiley-VCH, NY, 1988.

5. C.T. Sah, F.A. Lindholm, Solid-State Electron. 16 (1973) 1447.

6. A.H. Marshak, K.M. van Vleit, Solid-State Electron. 21 (1978) 417; K.M. van Vliet, A.H. Marshak, Solid-State Electron. 23 (1980) 49.

7. See, for example, B.R. Nag, Theory of Electrical Transport in Semiconductors, Pergamon, Elmsford, NY, 1972; A. van der Ziel, Solid State Physical Electronics, Prentice-Hall, Englewood Cliffs, NJ, 1976; A.C. Smith, J.F. Janak, R.B. Adler, Electronic Conduction in Solids, McGraw-Hill, NY, 1967; J.S. Blakemore, Semiconductor Statistics, Pergamon, Oxford, 1962.

8. J. Bardeen, in: E.V. Condon (Ed.), Handbook of Physics, McGraw-Hill, NY, 1967.

9. See, for example, H. Kromer, RCA Rev. 18 (1957) 332; J. Tauc, Rev. Mod. Phys. 29 (1957) 308; P.R. Emtage, J. Appl. Phys. 33 (1962) 1950; L.J. Van Ryuven, F.E. Williams, Am. J. Phys. 35 (1967) 705; Y. Marfaing, J. Chevallier, IEEE Trans. Electron. Devices 18 (1971) 465.

10. S.J. Fonash, CRC Crit. Rev. Solid State Mater. Sci. 9 (1980) 107.

11. S.J. Fonash, S. Ashok, Appl. Phys. Lett. 35 (1979) 535.

The Quasi-neutral-region Assumption and Lifetime Semiconductors

In our analytical treatment of the mathematical systems describing solar cell behavior, we are forced to make the assumption that certain regions of the cell structure can be considered to be quasi-neutral even in the presence of current flow. We do this to obtain a situation amenable to analysis. When quasi-neutrality is invoked, it means that one is assuming the the right hand side of Eq. 2.48e is essentially zero; i.e., one is assuming that the space charge (charge density)

$$e\left[p - n + \sum p_T - \sum n_T + N_D^+ - N_A^-\right] \approx 0$$

As we have seen, this assumption, when valid, can greatly facilitate the mathematical analysis of solar cells.

The assumption that there are quasi-neutral regions outside of the barrier region, even in the presence of current flow, is predicated upon minority carrier lifetimes $\tau_{n,p}$ (see Section 2.2.5.1) being much larger than the dielectric relaxation time τ_D. The dielectric relaxation time is defined by[1,2]

$$\tau_D = \varepsilon/\sigma = \varepsilon\rho \tag{E.1}$$

DOI: 10.1016/B978-0-12-374774-7.00013-3

where ε is a material's permittivity, σ is its conductivity, and ρ is its resistivity. If $\tau_{n,p} > \tau_D$, then mobile carriers can exist long enough to enable them to neutralize charge; hence, quasi-neutral regions are possible in this case even in the presence of current flow. Figure E.1 shows the ranges $\tau_{n,p} > \tau_D$ (lifetime semiconductor) and $\tau_{n,p} < \tau_D$ (relaxation semiconductor) for a hypothetical material whose permittivity ε is such that $\tau_D = \rho \times 10^{-12}$s ($\rho$ in ohm cm) and whose carrier lifetime $\tau_{n,p} \approx 10^{-8}$s. In the relaxation semiconductor regime,[3] quasi-neutrality is not a justifiable a priori assumption. In the extreme case of the space-charge-limited regime seen in Figure E.1, electric fields, arising from the space charge created by the carriers themselves, control currents. Carrier lifetimes less than 10^{-8}s exist in solids like amorphous and organic materials.

All of this is moot when we use computer modeling to solve the whole set of equations describing solar cell device physics. The mathematics computes and accounts for the space charge and its implications automatically. In other words, in the computer modeling used in this text, space-charge-limited, relaxation, and lifetime semiconductor behavior are all automatically handled.

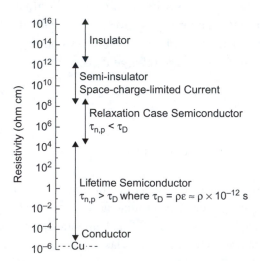

FIGURE E.1 Classification of materials by electrical resistivities. Here ρ is the resistivity in the units of ohm-cm. For this figure, the material permittivity ε has been taken to be 10^{-12} F/cm and the carrier lifetime $\tau_{n,p}$ has been taken to be 10^{-8}s. The ranges noted on the figure shift according to permittivity and lifetime values.

REFERENCES

1. R.H. Bube, Electronic Properties of Crystalline Solids, John Wiley & Sons, Ltd., New York, 1974.

2. S. Sze, K.K. Ng, Physics of Semiconductor Devices, third ed., John Wiley & Sons, Ltd., Hoboken NJ, 2007.

3. W. van Roosbroeck, H.C. Casey, Jr., Phys. Rev. B: Solid State 5 (1972) 2154.

Determining p(x) and n(x) for the Space-charge-neutral Regions of a Homojunction

In Section 4.4.1 of Chapter 4, the hole density p as a function of x is needed for the top quasi-neutral region and the electron density n as a function of x is needed for the bottom quasi-neutral region. These regions are shown in Figure F.1. As established in Section 4.1.1, p(x) satisfies

$$\frac{d^2 p}{dx^2} - \frac{p - p_{n0}}{L_p^2} + \int_\lambda \frac{\Phi_0(\lambda)}{D_p} \alpha(\lambda) e^{-\alpha(x+d)} d\lambda = 0 \qquad \text{(F.1)}$$

subject to the boundary conditions

$$\frac{dp}{dx}\bigg|_{x=-d} = \frac{S_p}{D_p}[p(-d) - p_{n0}] \qquad \text{(F.2a)}$$

DOI: 10.1016/B978-0-12-374774-7.00014-5

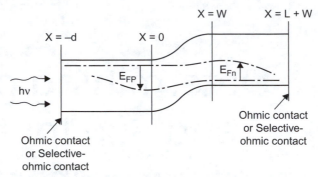

FIGURE F.1 An n–p homojunction cell under illumination. The quasi-Fermi levels are measured as depicted. Their variations with position are shown exaggerated. The actual variation is dictated by $J_n = en\mu_n\, dE_{Fn}/dx$ and $J_p = -ep\mu_p\, dE_{Fp}/dx$. The minus sign in the second expression is necessary since E_{Fp} is being measured as shown in the figure.

$$p(0) = p_{n0}e^{E_{Fp}(0)/kT} \tag{F.2b}$$

The solution to Eq. F.1 is

$$p = Ae^{-x/L_p} + Be^{x/L_p} + p_{n0} + \int \frac{L_p^2}{D_p(1 - \alpha^2 L_p^2)}\Phi_0(\lambda)\alpha(\lambda)e^{-\alpha(x+d)}d\lambda \tag{F.3}$$

for the assumed space-charge-neutral region $-d \le x \le 0$. Equation F.3 may be verified by substituting it back into Eq. F.1. Applying the boundary conditions given above to Eq. F.3 shows that

$$A = \left| \frac{-1}{\dfrac{2}{L_p}\cosh d/L_p + \dfrac{2S_p}{L_p}\sinh d/L_p} \right.$$

$$\left[\int \frac{\Phi_0 L_p^2\alpha^2}{D_p(1-\alpha^2 L_p^2)}d\lambda + \int \frac{S_p\Phi_0 L_p^2\alpha}{D_p^2(1-\alpha^2 L_p^2)}d\lambda - \left\{ \frac{e^{-d/L_p}}{L_p} - \frac{S_p e^{-d/L_p}}{D_p} \right\} \right.$$

$$\left. \left. \left\{ p_{n0}(e^{E_{Fp}(0)/kT} - 1) - \int \frac{\Phi_0 L_p^2\alpha^2}{D_p(1-\alpha^2 L_p^2)}e^{-d/L_p}d\lambda \right\} \right] \right] \tag{F.4}$$

and

$$B = \left| \dfrac{1}{\dfrac{2}{L_p}\cosh d/L_p + \dfrac{2S_p}{L_p}\sinh d/L_p} \right|$$

$$\left[\int \dfrac{\Phi_0 L_p^2 \alpha^2}{D_p(1 - \alpha^2 L_p^2)}d\lambda + \int \dfrac{S_p\Phi_0 L_p^2 \alpha}{D_p^2(1 - \alpha^2 L_p^2)}d\lambda + \left\{ \dfrac{e^{d/L_p}}{L_p} + \dfrac{S_p e^{d/L_p}}{D_p} \right\} \right.$$

$$\left. \left\{ p_{n0}(e^{E_{Fp}(0)/kT} - 1) - \int \dfrac{\Phi_0 L_p^2 \alpha^2}{D_p(1 - \alpha^2 L_p^2)}e^{-d/L_p}d\lambda \right\} \right] \qquad (F.5)$$

To find $n = n(x)$ for the region $W \le x \le W + L$, which is assumed to be space-charge-neutral also, we need to find the solution to

$$\dfrac{d^2n}{dx^2} - \dfrac{n - n_{p0}}{L_n^2} + \int_\lambda \dfrac{\Phi_0(\lambda)}{D_n}\alpha(\lambda)e^{-\alpha(x+d)}d\lambda = 0 \qquad (F.6)$$

subject to the boundary conditions

$$n(W) = n_{p0}e^{E_{Fn}(W)/kT} \qquad (F.7a)$$

and

$$\left. \dfrac{dn}{dx} \right|_{x=W+L} = -\dfrac{S_n}{D_n}[n(W + L) - n_{p0}] \qquad (F.7b)$$

The solution to Eq. F.6 must be of the form

$$n = Ce^{-x/L_n} + De^{x/L_n} + n_{p0} + \int \dfrac{L_n^2}{D_n(1 - \alpha^2 L_n^2)}\Phi_0(\lambda)\alpha(\lambda)e^{-\alpha(x+d)}d\lambda \qquad (F.8)$$

as may be verified by substituting this expression back into Eq. F.6. Following the same procedure outlined above for $p(x)$ allows C and D to be determined from the boundary conditions.

Determining n(x) for the Space-charge-neutral Region of a Heterojunction p-type Bottom Material

We consider here a space-charge-neutral region $(W_1 + W_2) \leq x \leq (W_1 + W_2 + L)$ in the bottom layer of the heterojunction seen in Figure 5.40. To have such a layer, we have assumed that material 2 is a lifetime semiconductor (see Appendix E). We also now assume that electrons remain the minority carrier under illumination and that we can use a linear lifetime model for recombination. Under these conditions, the governing equation for $n(x)$ is:

$$\frac{d^2n}{dx^2} - \frac{n - n_{p0}}{L_n^2} + \frac{1}{D_n} \int_\lambda \Phi_0(\lambda) e^{-\alpha_1(\lambda)(W_1+d)} \alpha_2(\lambda) e^{-\alpha_2(\lambda)(x-W_1)} d\lambda = 0$$

$$(G.1)$$

DOI: 10.1016/B978-0-12-374774-7.00015-7

Our goal here is to find the solution to Eq. G.1 subject to the boundary conditions

$$n(W_1 + W_2) = n_{p0} e^{E_{Fn}(W_1+W_2)/kT} \qquad \text{(G.2)}$$

and

$$\frac{dn}{dx}\bigg|_{L+W_1+W_2} = -\frac{S_n}{D_n}[n(L + W_1 + W_2) - n_{p0}] \qquad \text{(G.3)}$$

We directly solve the system of Eqs. G.1–G.3 relying on our experience gained in Appendix F. From that appendix we know that the solution to Eq. G.1, a second-order linear differential equation, can be written as

$$n(x) = Ae^{-x/L_n} + Be^{x/L_n} + n_{p0} + \Theta e^{-\alpha_2 x} \qquad \text{(G.4)}$$

where Θ is defined by

$$\Theta \equiv \int_\lambda \frac{\alpha_2(\lambda)L_n^2 \Phi_0(\lambda)\, e^{-\alpha_1(\lambda)(d+W_1)} e^{\alpha(\lambda)_2 W_1}}{D_n(1 - \alpha_2^2(\lambda)L_n^2)} d\lambda \qquad \text{(G.5)}$$

as may be verified by using Eq. G.5 in Equation G.4 and by substituting Eq. G.4 back into Eq. G.1. After boundary conditions are imposed on Eq. G.4, the A and B of Eq. G.4 are found to be

$$
\begin{aligned}
A = {} & n_{p0}\left[e^{E_{Fn}(W_1+W_2)/kT} - 1\right]\left[\frac{e^{(W_1+W_2)/L_n}e^{\beta_5}(\beta_7 + 1)}{2(\beta_7 \sinh\beta_5 + \cosh\beta_5)}\right] \\
& - \int_\lambda \frac{\alpha_2(\lambda)L_n^2 \Phi_0(\lambda)\, e^{-\alpha_1(\lambda)(d+W_1)} e^{\alpha(\lambda)_2 W_1}}{D_n(1 - \alpha_2^2(\lambda)L_n^2)} \\
& \left[(e^{(W_1+W_2)/L_n})(e^{-\alpha_2(W_1+W_2)})\right]\frac{\left[(\beta_7 + 1)e^{\beta_5} + \left(\dfrac{\beta_6}{\beta_5} - \beta_7\right)e^{-\beta_6}\right]}{2(\beta_7 \sinh\beta_5 + \cosh\beta_5)} d\lambda
\end{aligned}
$$

$$\text{(G.6)}$$

and

$$
\begin{aligned}
B = {} & n_{p0}\left[e^{E_{Fn}(W_1+W_2)/kT} - 1\right]\left[\frac{e^{-(W_1+W_2)/L_n}e^{-\beta_5}(1-\beta_7)}{2(\beta_7\sinh\beta_5 + \cosh\beta_5)}\right] \\
& - \int_\lambda \frac{\alpha_2(\lambda)L_n^2\Phi_0(\lambda)e^{-\alpha_1(\lambda)(d+W_1)}e^{\alpha(\lambda)_2 W_1}}{D_n(1-\alpha_2^2(\lambda)L_n^2)} \\
& \left[(e^{-(W_1+W_2)/L_n})(e^{-\alpha_2(W_1+W_2)})\right]\frac{\left[(\beta_7-1)e^{-\beta_5} + \left(\dfrac{\beta_6}{\beta_5}-\beta_7\right)e^{-\beta_6}\right]}{2(\beta_7\sinh\beta_5+\cosh\beta_5)}\,d\lambda
\end{aligned}
$$

$$(G.7)$$

The dimensionless β parameters defined in Table G.1 have been introduced in Eqs. G.6 and G.7.

Table G.1

β Quantity	Definition	Physical significance
β_5	L/L_n	Ratio of material 2 quasi-neutral region length to electron diffusion length.
$\beta_6(\lambda)$	$L\alpha_2(\lambda)$	Ratio of material 2 quasi-neutral region length to absorption length in material 2 for light of wavelength λ. (This ratio depends on λ.)
β_7	L_nS_n/D_n	Ratio of back-surface electron carrier recombination velocity S_n of material 2 to electron diffusion velocity D_n/L_n in material 2. Captures the physics of electron diffusion/recombination versus electron contact recombination. Need $D_n/L_n > S_n$.

Index

Note: Page numbers with 'f' and 't' refer to figures and tables, respectively.